Textbook of Industrial, Mechanical and Manufacturing Science

Textbook of Industrial, Mechanical and Manufacturing Science

Edited by **Jeff Hansen**

WILLFORD PRESS

New York

Published by Willford Press,
118-35 Queens Blvd., Suite 400,
Forest Hills, NY 11375, USA
www.willfordpress.com

Textbook of Industrial, Mechanical and Manufacturing Science
Edited by Jeff Hansen

© 2016 Willford Press

International Standard Book Number: 978-1-68285-063-3 (Hardback)

The publisher's policy is to use permanent paper from mills that operate a sustainable forestry policy. Furthermore, the publisher ensures that the text paper and cover boards used have met acceptable environmental accreditation standards.

Trademark Notice: Registered trademark of products or corporate names are used only for explanation and identification without intent to infringe.

Printed in the United States of America.

Contents

Preface

This book aims to highlight the current researches and provides a platform to further the scope of innovations in this area. This book is a product of the combined efforts of many researchers and scientists from different parts of the world. The objective of this book is to provide the readers with the latest information in the field.

This book provides a multidisciplinary perspective in the fields of industrial, mechanical and manufacturing sciences and their emerging applications. The chapters in this book aim to familiarize the readers with topics such as solid mechanics, fluid mechanics, dynamics and control, micro-and nanosystems, etc. to help develop a better understanding of these fields. The case studies and researches provided by eminent experts herein are bound to elucidate the latest advancements in the optimization and synthesis of various designs.

I would like to express my sincere thanks to the authors for their dedicated efforts in the completion of this book. I acknowledge the efforts of the publisher for providing constant support. Lastly, I would like to thank my family for their support in all academic endeavors.

Editor

Flexure delicacies

S. Henein

Centre Suisse d'Electronique et de Microtechnique (CSEM SA), Neuchâtel, Switzerland

Correspondence to: S. Henein (simon.henein@csem.ch)

Abstract. Flexures are nowadays enjoying a new boom in numerous high-precision and extreme-environment applications. The paper presents some delicate issues concerning stiffness compensation, large reduction ratios, as well as rectilinear and circular movements in compliant mechanisms. Novel concrete technical solutions to these well-known issues are described, giving a glimpse into the vast and still largely unexploited potential of flexure mechanisms manufactured by wire-electrical-discharge machining.

1 Introduction

Although the basic principles of flexible bearings have been known for several decades (Koster, 1996), the design methods that could be found in literature have long remained fragmented. Only recently the interest for flexures and their applications has grown, leading to a more systematic approach of the respective design methodologies such as Smith (2000), Henein (2001) and Howell (2001). This development has been driven by the increasing need for motion accuracies in the nanometer range in extreme environments like vacuum, cryogenic or high temperatures, under radiation or in outer space. Moreover, the increasing prevalence of manufacturing technologies like wire-electrical-discharge machining (wire-EDM) and more recently of deep silicon etching has made flexure-mechanisms a key element of the broad technological system our society relies upon for living.

Typical limitations of flexure mechanisms are the presence of an elastic restoring force that opposes the intended motion, the large outer dimensions of the mechanisms relative to the admissible strokes, and the complex kinematic movements of the most basic flexures (e.g. parallel spring stage and cross spring pivot) which do not produce purely rectilinear or circular trajectories. The three sections below illustrate some simple, yet subtle, solutions addressing these three issues: first a zero-stiffness translation bearing is presented, then a virtual lever with a very large reduction ratio (up to 1:1000) and finally a rectilinear translation mechanism and a circular flexure pivot.

2 Stiffness compensation and bi-stability

Ideal bearings would be movable by infinitesimal forces or torques. However, this is not the case for basic flexures, which present a non-zero spring constant that resist motion intended away from their neutral position. Nevertheless, if a preloaded spring is used in an arrangement where the spring loses elastic energy while the flexure is moved away from its neutral position, then the overall stiffness of the structure can be reduced (Koster, 1996; Henein, 2001). Special designs allow to practically reduce the natural stiffness of flexures by several orders of magnitude (Fig. 1), approaching a zero-stiffness ideal behaviour. If additional preload is applied, then negative stiffness can be achieved, i.e. the flexures present a bi-stable behaviour. Figure 1 shows an example of such a mechanism composed of a fixed frame (a), and a main parallel spring stage (b) defining the trajectory of the mobile block (c). A secondary parallel spring stage (d) is attached to (c), and both stages are loaded in compression by a preload spring (e). A fifth blade (or rod) (f) is used to define the position of the secondary stage. One can give the following qualitative explanation to the behaviour of this mechanism: when (c) is moved away from its neutral position, the four blades (b and d) store elastic energy. Meanwhile the elongation of the preload spring decreases, due to the shortening effect of both parallel spring stages, therefore the preload spring loses elastic energy. At the level of the mechanism, it is the global balance between the rates of elastic energy variations of the parallel spring blades versus the preload spring which dictates the elastic behaviour of the mechanism.

For simplification, the elastic element (f) might be combined with (e) as a single preloaded blade. This leads to a planar mechanism that can be manufactured monolithically

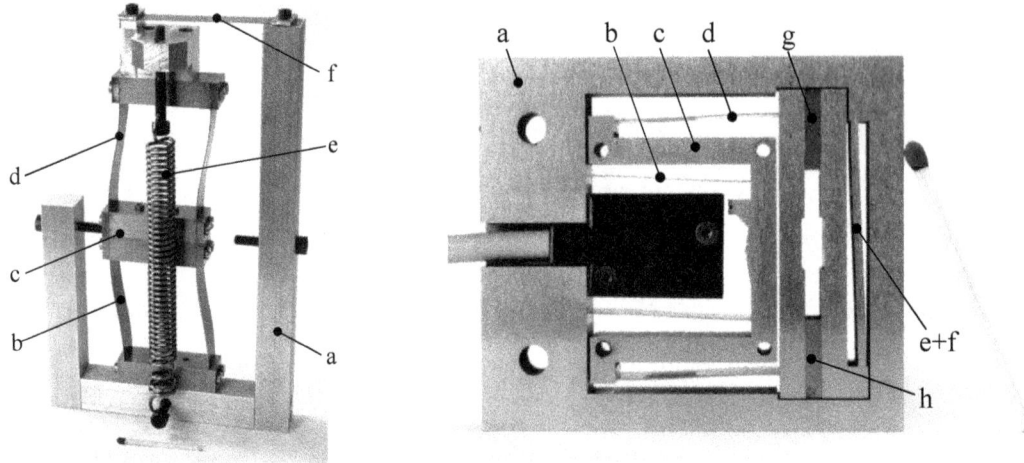

Figure 1. Parallel spring stage equipped with a stiffness compensation mechanism. Left: mock-up model showing the architecture. Right: same arrangement machined monolithically by wire-electrodischarge machining (wire-EDM).

(Fig. 1). In this arrangement, the middle-point of the blade (e + f) is placed on the axis of symmetry of the mechanism. Therefore, the insertion of the preloading shims (g) and (h) result in an equal compression loading of all four guiding blades (b and d). By choosing the right shim thickness, the preload might be tuned to obtain a reduced stiffness, a near zero-stiffness or a negative stiffness. Equations (1) and (2) give the formulas for the calculation of the theoretical preload force which leads to zero-stiffness (Henein, 2001).

In a parallel spring stage, the linear stiffness k of one blade of length L, width b, thickness h and Young's Modulus E, preloaded in compression by a force N is:

$$k = \frac{N}{\frac{2}{S}\tan\frac{SL}{2} - L}; \text{ with } S = \sqrt{\frac{N}{EI}}; \text{ and } I = \frac{bh^3}{12} \qquad (1)$$

Therefore a preload N_0 exists, which leads to a zero stiffness of the blade:

$$N_0 = \frac{\pi^2 EI}{L^2} \qquad (2)$$

Figure 2 shows the theoretical force versus displacement characteristic of the monolithic mechanism shown in Fig. 1 with and without preloading the blade (e + f). The dimensions (L, b, h) of the four guiding blades (b and d) are $26 \times 10 \times 0.25$ mm. The dimensions of the preload blade (e + f) are $32 \times 10 \times 1.1$ mm. The material used is a spring steel with a Young's modulus of 196 GPa. The nominal preload force in the neutral position is 5 % above the load N_0 (Eq. 2), which leads to a slightly negative stiffness around the central position (i.e. with compensation, the slope of the curve is negative around the origin). The stiffness reduction effect is clearly visible on the graph.

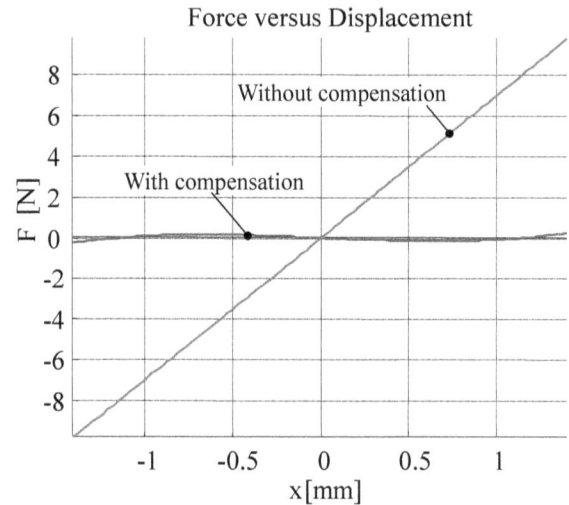

Figure 2. Force applied on the mobile block (c) (Fig. 1, right) as a function of its displacement with and without stiffness compensation.

3 Achieving large reduction ratios

Striving after nanometric motion accuracies often leads to the use of mechanical transmissions with very large reduction ratios, which scale down the motion of commercial actuators. It is well-known in the state-of-the-art that flexures can be used as reduction mechanisms, but the classical solutions generally have non-linear characteristics (i.e. the reduction factor is not constant over the motion range) (Eden, 1956). In comparison, a non-conventional design such as the Nanoconverter (Fig. 3) (Henein, 2006) presents the following key advantages: it exhibits a constant reduction factor; it can achieve very high reduction factors (typically up to 1:1000); it can be designed to be tunable using a simple tuning screw

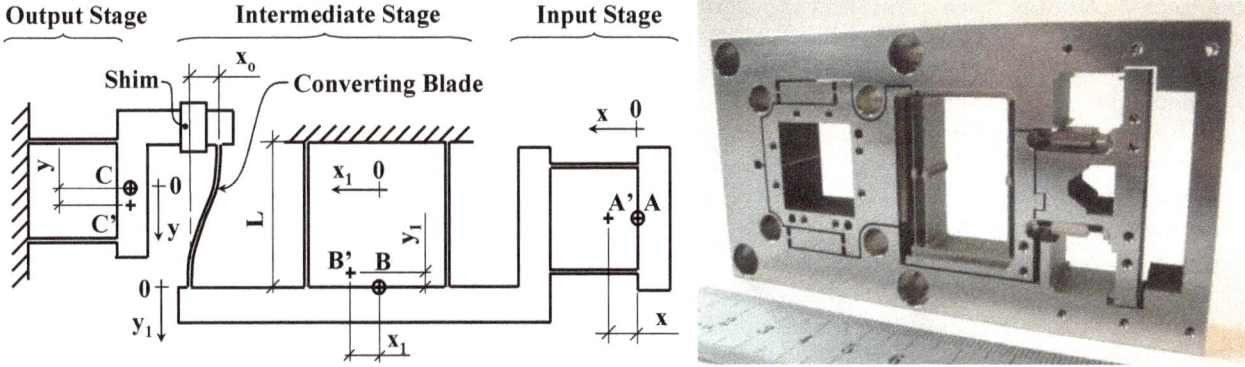

Figure 3. Working principle (using an offset shim) and photo of a monolithic Nanoconverter.

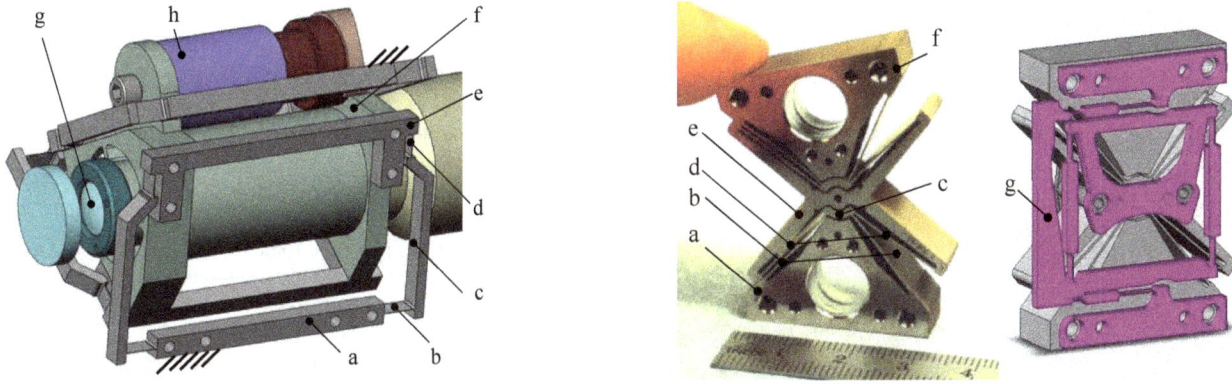

Figure 4. Left: rectilinear flexure mechanisms of the Mars Close-up Imager. Center: picture of the monolithic Butterfly flexure pivot (angular stroke up to ±15°; centre shift below 2 microns). Right: Butterfly Pivot equipped with a slaving plate.

or shim to select the reduction factor over a wide range (typically 1:20 to 1:1000); its simple planar structure can be manufactured monolithically (no need for assembly) using a wide variety of techniques (e.g. wire-EDM, laser cutting, silicon etching, LIGA).

The working principle is the following: a commercial linear actuator with micrometric motion accuracy rectilinearly drives point A (Fig. 3) of the Input Stage to A'. This motion is transmitted to the Intermediate Stage that is guided by a classical parallel-spring-stage (with blade length L): point B moves to B'. Due to the shortening of the blade projection, the motion of this stage is a well-known parabolic translation: $y_1 = -3x_1^2/(5L)$, where $x_1 \cong x$. A third blade of the same length L (called "Converting Blade"), that has an offset deformation x_0, links the Intermediate Stage to the Output Stage. The Output Stage is guided vertically by a classical parallel-spring-stage. The motion x_1 causes the Converting Blade to shorten, following the same parabolic law as the two blades of the Intermediate Stage, but with an offset x_0. The resulting motion y (Eq. 3) of the Output Stage (motion from C to C') is equal to the differential shortening of the blades (subtraction of two parabolas with an offset):

$$y = \frac{3(x+x_0)^2}{5L} - \frac{3x^2}{5L} = \frac{6x_0}{5L}x + \frac{3x_0^2}{5L};$$ (3)

$$i = \frac{x}{y} = \frac{5L}{6x_0}$$ (4)

Therefore, if the origin of the y-axis is adequately chosen, the displacement y of the Output Stage is simply proportional to the displacement x of the actuator, with reduction ratio i (Eq. 4) that is constant over the whole displacement range and is inversely proportional to the offset x_0. Choosing an offset x_0 that is small compared to the blade length L leads to very large demagnification ratios. This is mechanically very easy to achieve by using a shim or by monolithic manufacturing as in Fig. 3.

4 Rectilinear and circular flexures

Producing purely rectilinear movements can be achieved with the well-known compound-parallel-spring-stage (Smith, 2000), but some other less common structures exist like the 5 or 6 folded leaf springs (Schellenkens et al., 1998). For example, the latter concept has been adapted for the

Close-up Imager (CLUPI Instrument) (Fig. 4) of the European Mars Mission EXOMARS developed by CSEM for the Space Exploration Institute (SPACE-X). It is composed of three identical plates connecting the fixed outer frame to the mobile body (f) holding the optical lens (g). Each plate is a planar monolithic symmetric compliant structure composed of a fixed bar (a), a first flexure hinge (b), a second rigid bar (c), a second flexure hinge (d) and an output bar (e). Each plate blocks one rotation and one translation (f). The three thin plates can accommodate for the over-constraint of this kinematic arrangement without causing high stresses. Hence, this arrangement offers long stroke rectilinear motion.

Nearly pure circular rotations can be achieved by structures such as the Butterfly Pivot (Henein et al., 2003; Pei and Yu, 2011). This monolithic planar compliant mechanism (Fig. 4) was originally designed for space pointing applications. It is composed of a fixed base block (a), two blades (b) forming a first Remote-Center-Compliance (RCC) pivot, an intermediate block (c), a second RCC pivot (d) that compensates for a large part of the parasitic shifts produced by (b), a second intermediate block (e), then a mirror arrangement leading to the output block (f) doubles the total angular stroke. An optional internal coupling mechanism (g) can be integrated in order to suppress the undesired internal DOF of the central cross (e) raising its Eigen-mode from 320 to 2000 Hz. This additional kinematic chain plays the same role as the slaving lever commonly used for the same reason with the compound parallel spring stage (Smith, 2000), i.e. it forces the intermediate body (e) to move by exactly half the stroke of the output (f).

5 Conclusions

During the last decades, the old and simple idea of using the elastic deformation of solid bodies to produce well defined movements has been used in more and more application fields. As the simplest original components (e.g. cross spring-pivots and leaf spring stages) are still widely used, many novel designs are flourishing which exploit the possibilities offered by wire-EDM to produce mechanical structures with original characteristics like zero-stiffness, extreme reduction ratios or ideal kinematic behaviours. This illustrates how this new manufacturing technology has led not only to an improvement of existing concepts but also to a revolution in the design of flexure-mechanisms. In this light, one should be very attentive to the coming revolution in this field, which will probably come from the use of deep silicon etching technologies in the design of tomorrow's flexure-mechanisms.

Acknowledgements. The author wishes to thank J.-L. Josset from the Space Exploration Institute, Ph. Schwab, L. Giriens, W. Glettig and P. Spanoudakis from CSEM, U. Frommherz from the Paul Scherrer Institut, as well as S. Bottinelli from MECAR-TEX SA for their contributions to the work presented in this short communication.

Edited by: J. A. Gallego Sánchez

References

Eden, E. M.: Application of spring strips to instrument design, Notes of Applied Science No. 15, 1956.

Henein, S.: Conception des guidages flexibles, Presses Polytechniques et Universitaires Romandes, ISBN 2-88074-481-4, 2001.

Henein, S.: Device for converting a first motion into a second motion responsive to said first motion under a demagnification scale, Patent EP06021785, Holder: Paul Scherrer Institut, 2006.

Henein, S., Spanoudakis, P., Droz, S., Myklebust, L., and Onillon, E.: Flexure pivot for aerospace mechanisms, Proc. 10th European Space Mechanisms and Tribology Symposium, San Sebastian, Spain, 24–26 September 2003, ESA SP-524, 285–288, 2003.

Howell, L.: Compliant mechanisms, John Wiley & Sons Inc, 2001.

Koster, M. P.: Constructieprincipes voor het nauwkeurig bewegen en positioneren, 5th edn., HBuitgevers, Baarn, NL, 1996.

Pei, X. and Yu, J.: ADLIF: a new large-displacement beam-based flexure joint, Mech. Sci., 2, 183–188, doi:10.5194/ms-2-183-2011, 2011.

Schellenkens, P., Rosielle, N., Vermeulen, M., Wetzels, S., and Pril, W.: Design for precision: current status and trends, Annals of the CIRP, Vol. 47/2/1998, 1998.

Smith, S. T.: Flexures, Gordon and Breach Science Publishers, ISBN 90-5699-261-9, 2000.

Building block method: a bottom-up modular synthesis methodology for distributed compliant mechanisms

G. Krishnan[1]**, C. Kim**[2]**, and S. Kota**[3]

[1]Mechanical Engineering, University of Michigan, Ann Arbor, MI 48105, USA
[2]Mechanical Engineering, Bucknell University, Lewisburg, PA 17837, USA
[3]Department of Mechanical Engineering, University of Michigan, Ann Arbor, MI 48105, USA

Correspondence to: G. Krishnan (gikrishn@umich.edu)

Abstract. Synthesizing topologies of compliant mechanisms are based on rigid-link kinematic designs or completely automated optimization techniques. These designs yield mechanisms that match the kinematic specifications as a whole, but seldom yield user insight on how each constituent member contributes towards the overall mechanism performance. This paper reviews recent developments in building block based design of compliant mechanisms. A key aspect of such a methodology is formulating a representation of compliance at a (i) single unique point of interest in terms of geometric quantities such as ellipses and vectors, and (ii) relative compliance between distinct input(s) and output(s) in terms of load flow. This geometric representation provides a direct mapping between the mechanism geometry and their behavior, and is used to characterize simple deformable members that form a library of building blocks. The design space spanned by the building block library guides the decomposition of a given problem specification into tractable sub-problems that can be each solved from an entry in the library. The effectiveness of this geometric representation aids user insight in design, and enables discovery of trends and guidelines to obtain practical conceptual designs.

1 Introduction

Kinematics of conventional rigid-link mechanisms and their systematic synthesis has been studied for almost a couple of centuries (Erdman et al., 2001). This has resulted in a database of a number of tried and tested mechanisms, each adept in a specific task (Sclater and Chironis, 2007). A designer can either choose an existing conceptual design from the database, or systematically combine a number of designs to meet a more complex specification. Once the conceptual design is chosen, it can be refined for the application at hand by first analyzing the various geometrical configurations that rigid links and kinematic pairs assume, and then analyzing forces in links to determine the amount of material required to maintain strength and rigidity. For most applications multiple solutions can be generated, and secondary criteria such as aesthetics and ergonomics can be used to determine the best solution. The simplicity of this process is as a result of the decoupling between kinematic and structural aspects of design.

In the past couple of decades researchers are interested in monolithic mechanisms devoid of joints and links that derive their mobility on elastic deformation alone. They have significant advantages that are well documented in literature such as elimination of friction, backlash and reduced manufacturing costs by avoiding assembly (Howell, 2001). Furthermore, their versatility is increased by their scalability in various length scales. However, elastic deformation may lead to material failure at certain regions having high stresses thus limiting range of motion and the load carrying ability of these mechanisms. The main design challenge for compliant mechanisms is the intricate coupling between kinematics and elasticity. In other words, the topology of a mechanism and the size of its individual elements together determine how the mechanism moves and the internal forces within its members, thus precluding synthesis akin to rigid-link mechanisms.

As a first attempt at design, Howell and Midha used conventional rigid-link topologies and the associated design methodologies to first design a rigid-link mechanism that meets the kinematic specifications. Appropriate torsional springs were placed at the joints of this mechanism to meet the stiffness specification. The rigid links were then replaced with beams of equivalent length to match the required kinematics. The cross-section of the beams were then determined based on the stiffness of the torsional springs and the stress considerations. This technique where an equivalent rigid link and a torsional spring is used to analyze and design monolithic elastic mechanisms is known as pseudo-rigid body model. The designs that resulted from this methodology, though practical yield distinct areas where flexibility is lumped. These are also the areas where stresses are concentrated and thus limit lead bearing and large range of motions.

Topology optimization was developed by Ananthasuresh (1994) that aimed at a better utilization of the design space without restricting to conventional topologies. This procedure initially lists all possible interconnections of beams or a planar finite element mesh of the design domain, whose widths or thickness are the design variables for optimization. Elements with the values of design variables lower than a specific cut-off will not be considered to contribute towards the final topology. The optimization algorithm determines an optimal topology that maximizes an objective function. If the objective function defines the kinematic behavior of the mechanism alone, without incorporating strength considerations, the optimization yields mechanisms that are composed of rigid members and thin flexures (Saxena and Ananthasuresh, 2003). Yin and Ananthasuresh (2003) and Canfield et al. (2007) have demonstrated the use of additional constraints that prevent large relative rotations between two finite elements that constitute the topology in the design domain. Nevertheless, there is considerable complexity in incorporating stress considerations in topology optimization. Furthermore, this process yields little insight on the functional contribution of different elements in the topology. Furthermore, the optimum connectivity may not be present in the design parametrization, which leads to suboptimal or convoluted designs.

Thus, there is a need to understand the contributions of each member that constitutes the design. This understanding can yield insightful conceptual synthesis of compliant mechanisms, where simple deformable members are systematically combined to obtain the overall topology. Such a building block method was proposed by Kim (2005); Kim et al. (2008).

2 Overview of building block methods

Complex systems such as an automobile, aircraft and electronic gadgets are always broken down into simpler subsystems that can be easily designed. For example, an automobile consists of combustion, transmission, electrical subsystems all working together. These subsystems are designed and fabricated separately and integrated during assembly. A similar

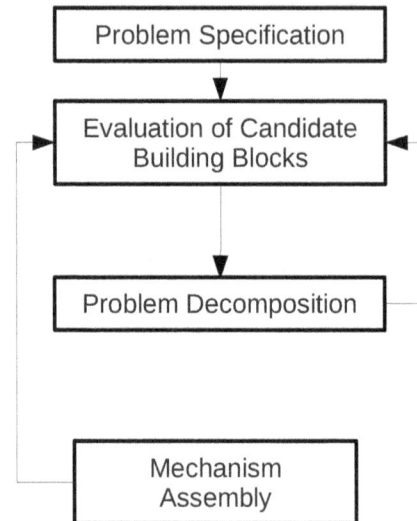

Figure 1. Schematic of the building block method for design.

building block approach was proposed for mechanism design by Kota and Chiou (1992) for conventional rigid link mechanism synthesis. The building block method is captured in the flowchart shown in Fig. 1 (Kim, 2005). Once kinematic specifications were determined, they were compared with entries in the database of existing designs. If no entry is found, the problem specification is broken down to tractable subproblems, whose solutions are found in the database. The final design is an assembly of the individual subproblems.

Building block methods are common in conceptual designs where there is a functional independence between the constituent building blocks. This means that when two building blocks are combined, one does not change the inherent behavior of the other. For example, a slider joint remains a slider irrespective of the number of links attached to it. However, in domain specific design problems such as compliant mechanism synthesis, the deformation behavior of a building block is in general determined by rest of the topology. In other words the same building block may have two different deformation behaviors based on its loading condition. Consider a simple cantilever beam that is ubiquitous in most compliant mechanism topologies. Its deformation behavior can range from fixed-free to fixed-guided based on loads acting on the free end as seen in Fig. 2. In most problems this deformation behavior cannot be determined before hand. Is it then possible to use the elements of the building block method as shown in Fig. 1 for compliant mechanism synthesis?

The answer for the above question lies in the representation of compliance that favors building block method. Such a representation must provide

1. Expression of compliance quantities from first principles: the representation can aid visual insight if geometric quantities are used to express various aspects of

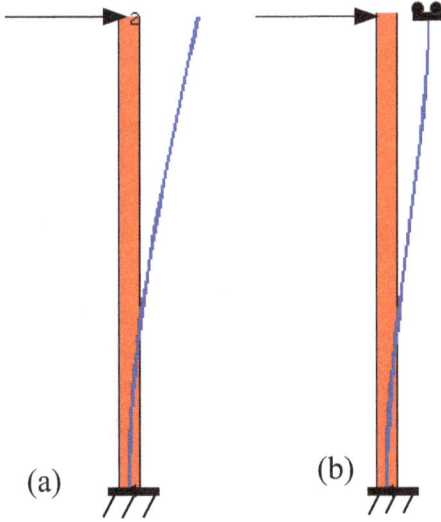

Figure 2. A beam with an end load has different deformation profiles when its end is (**a**) free, or (**b**) guided (constrained rotation and axial motion). Whether it behaves as (**a**) or (**b**), or between the two in a compliant mechanism depends on the subsequent members attached to it.

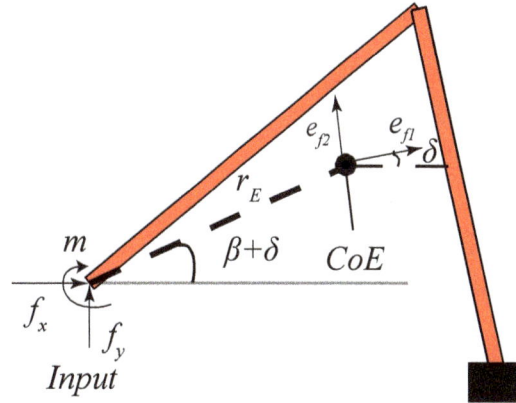

Figure 3. Eigen-twist and Eigen-wrench parameters for a particular building block geometry.

compliance. Furthermore the representation must be intrinsic to the topology and independent of reference frames used to evaluate it.

2. Parametric characterization: simple deformable building blocks are characterized with changing geometric parameters to span the design space. Use of geometric quantities enable visualization of the tractable design space.

3. Systematic decomposition of problem into tractable subproblems: the compliance representation must provide user insight in problem decomposition.

4. Enabling seamless assembly: the physics must enable seamless integration of the subproblems into the final solution.

There have been efforts in the past decade to propose various representations that permit a building block approach. At a single point of interest the compliance matrix has been characterized as a three dimensional ellipsoid (Kim et al., 2008; Kim, 2005). This representation enabled designing mechanisms with a given stiffness behavior by parallel combination of curved beams and dyads (series combination of beams). Design for single-input single output mechanisms have been similarly proposed based on combining a number of single point mechanisms between an input and output. Though these techniques yield conceptual designs that meet kinematic specifications, the representation lacks mathematical rigor and thus yields limited insight in the process of combining building blocks. This paper reviews recent advances in proposing a physically insightful, mathematically

robust compliance representation at a single point of interest (Krishnan et al., 2011) and relative compliance between two points (Krishnan et al., 2010).

3 Single port compliance: eigen-twist and eigen-wrench characterization

Single port compliance involves characterizing the force displacement relationship at a single point of interest. This is of importance in the design of constraints (Awtar et al., 2007), suspensions for microsystems, and elastic vision based sensors (Cappelleri, 2008). This force displacement relationship is given by the compliance matrix (or its inverse stiffness matrix). The dimensions of the compliance matrix in planar two dimensional case is 3×3. The three degrees of freedom are two translations in the plane of the geometry and a rotation about an axis perpendicular to the plane. The terms in the compliance matrix thus consist of both translational and rotational terms having different dimensions. To avoid dimensional inconsistencies, it is desired that translations and rotations are dealt with separately. This is accomplished by shifting the point of interest from the input to a new point where decoupling translational and rotation terms of the compliance is possible (Lipkin and Patterson, 1992). This point, known as the Center of Elasticity (CoE) and is always unique for a planar geometry Kim (2008) and is shown for a compliant dyad in Fig. 3. This point is similar to the remote center of compliance (RCC) in robotics, and the well established concepts of center of stiffness or center of compliance (Ciblak and Lipkin, 2003) as defined for a planar geometry. If a rigid connection is established between the CoE and the input, then any force applied at this point yields pure translation, and any moment applied yields a pure rotation. Thus the translational compliance at this point can be represented by an ellipse whose semi-major and semi-minor axes (a_{f_1} and a_{f_2}) denote the primary and secondary compliance directions. In these directions (e_{f_1} and e_{f_2}), any force applied

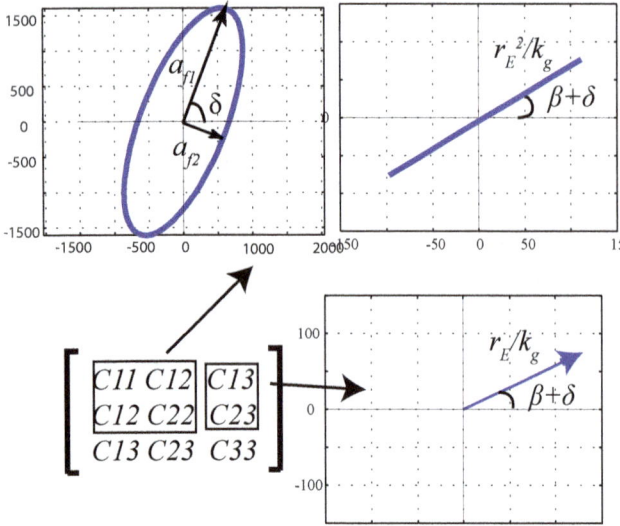

Figure 4. Compliance ellipse and Compliance coupling vector (c_v).

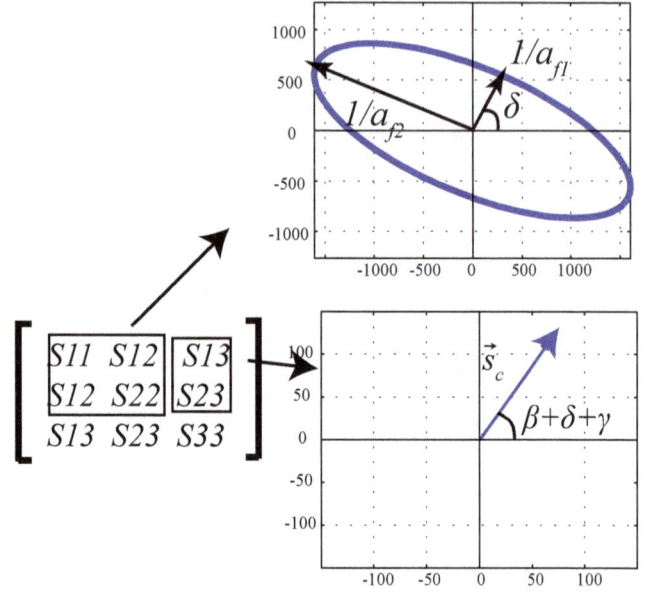

Figure 5. Stiffness ellipse and Stiffness coupling vector (s_c).

yields translation along the same direction. The rotational stiffness at this point can be represented by a scalar value k_g. The orientation of the CoE is given by an angle β and the angle δ refers to the orientation of the geometry in the two dimensional plane. These quantities are illustrated in Fig. 3. They are called eigen-twist and eigen-wrench characterization as the parameters can be obtained by eigen value analysis of the compliance matrix with selective normalization of its twists and wrenches (Kim, 2008).

Though it is convenient to characterize compliance at the center of elasticity as seen above, it is required to relate this to the compliance at the input point, as this was our original location of interest. The compliance at the input can be represented by the very same terms that characterize it at the CoE with some additional terms as seen in Fig. 4. The translation ellipse at the CoE (a_{f_1} and a_{f_2}) is supplemented with a degenerate ellipse oriented perpendicular to the line jointing the input and the CoE. Furthermore the coupling between translational rotational compliance at the input is represented by a coupling vector (c_v) whose magnitude denotes the amount of translation obtained due to a unit moment, or the amount of rotation obtained due to a unit force. However, if any force is applied along the direction of the coupling vector no rotation is observed. This vector is thus named as the coupling vector. Similar to the compliance matrix, the terms of the stiffness matrix can be represented as a stiffness ellipse and a stiffness coupling vector as seen in Fig. 5. Further insight into this characterization can be obtained in Krishnan et al. (2011).

This representation of compliance easily sets stage for a systematic building block based synthesis method. The first stage for the building block method after determining the problem specification is evaluation of candidate building blocks, or developing a library of building blocks. The most versatile building block for compliant mechanism syn-

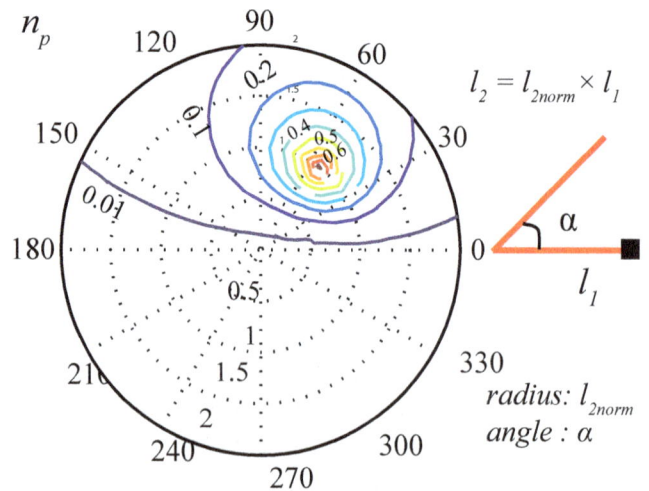

Figure 6. Parametric characterization of a compliant dyad for its eigen-twist and eigen-wrench parameters. The figure shows one such plot adapted from Krishnan et al. (2011).

thesis is shown to be a series combination of two beams, as a number topologies are shown to be composed of them (Kim, 2005; Krishnan et al., 2010). The eigen-twist and the eigen-wrench parameters can be evaluated by varying the angle between the two beams that make up the dyad and their relative lengths. Figure 6 plots $n_p = a_{f_2}/a_{f_1}$ for varying length ratios (radius of the polar plot) and dyad angles α. This gives an indication of the design space spanned by the dyad for n_p. Similarly other parameters (r_E, β, a_{f_1}) are plotted in Krishnan et al. (2011).

Consider an example where equal biaxial (X and Y) stiffness is required at a point without any coupling translational and rotational terms. Such a specification is required for a vision-based force sensor (Cappelleri et al., 2010), where external applied force can be evaluated by measuring the deformation of a point. Such a problem specification requires a circular compliance ellipse with zero coupling vector magnitude shown in Fig. 7a. Comparing from the database of compliant dyads no design matches these specifications (Krishnan et al., 2011). Figures 7b–c and 8 illustrate achieving these specifications using series and parallel combination of dyads. In series combination, the coupling vectors of individual building blocks add. Thus the zero coupling vector specification can be achieved by aligning equal and opposite building block specifications. Since the degenerate shift ellipse depends on the coupling vector orientations alone, its magnitude can be evaluated and subtracted from the required ellipse to obtain a net ellipse (Fig. 7b). Two dyads are then chosen from the building block library to meet the ellipse specifications. The next step involves assembly of dyads between themselves, and between one of the dyads and the input using rigid connecters as they donot change the compliance characteristics at the CoE of a building block. Thus all the steps illustrated in Fig. 1 are accomplished with geometrically intutive quantities.

One of the limitations of series combination is that the CoE always lies within the footprint of the mechanism (for proof of this, please refer Krishnan et al., 2011). This does not provide an easy access of the input for interacting with the objects in the vision based force sensor application. To overcome this, parallel combination of building blocks are recommended. During parallel combination, the stiffness ellipses and striffness coupling vectors of the building blocks add. Two sub-mechanisms whose stiffness ellipses are circular and whose stiffness coupling vectors are aligned equal and opposite to each other are connected together as shown in Fig. 8. A practical realization of this involves parallel combination of symmetric halves with some accommodation for a rigid probe as shown in Fig. 8d. The resulting input has equal biaxial compliance and decoupled translational and rotational compliance.

Thus, it is shown how compliance representation in the form of eigen-twist and eigen-wrench parameters enables a systematic and insightful building block method to synthesize single-port compliant mechanisms.

4   Multi-port compliance: load flow analogy

In the previous section we dealt with compliance where deformation was directly actuated by an applied force at the required point. There are a number of transmission problems such as grasping objects and amplifying, where force applied is spatially separated from the required output deformation. The challenge for the building block method is

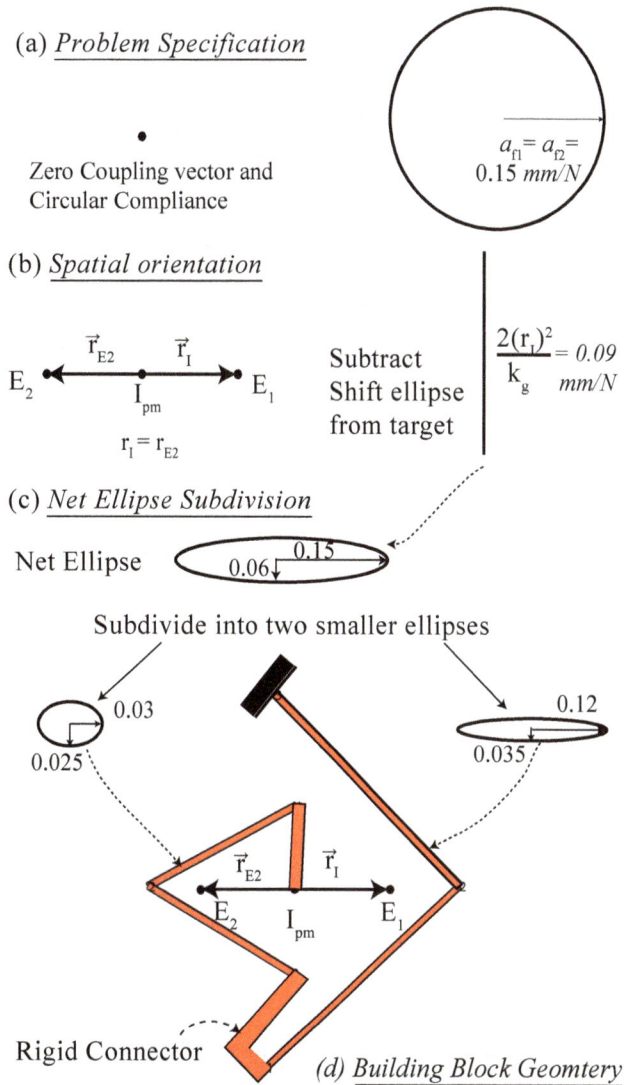

Figure 7. Guidelines with an example. (**a**) Problem Specification in terms of Compliance Ellipse and Coupling vector (**b**) Choose E_1, E_2, I_{pm} and evaluate shift ellipse (**c**) Net ellipse evaluation and subdivision into smaller building block ellipses (**d**) Design geometry of the two building blocks and their orientation.

formulating a representation for relative compliance between any two points in a continuum that can facilitate a building block method. In other words, it must enable representation of problem specification that can easily be decomposed into tractable subproblems. In this section, we present a load flow based analogy of relative compliance representation.

Between two points in a continuum, the relation between the applied forces and deformation is obtained from the extended form of the compliance matrix as shown below.

$$\begin{bmatrix} \mathbf{u}_{in} \\ \mathbf{u}_{out} \end{bmatrix} = \begin{bmatrix} \mathbf{C}_{in} & \mathbf{C}_{in\text{-}out} \\ \mathbf{C}_{in\text{-}out}^T & \mathbf{C}_{out} \end{bmatrix} \begin{bmatrix} \mathbf{f}_{in} \\ \mathbf{f}_{out} \end{bmatrix} \quad (1)$$

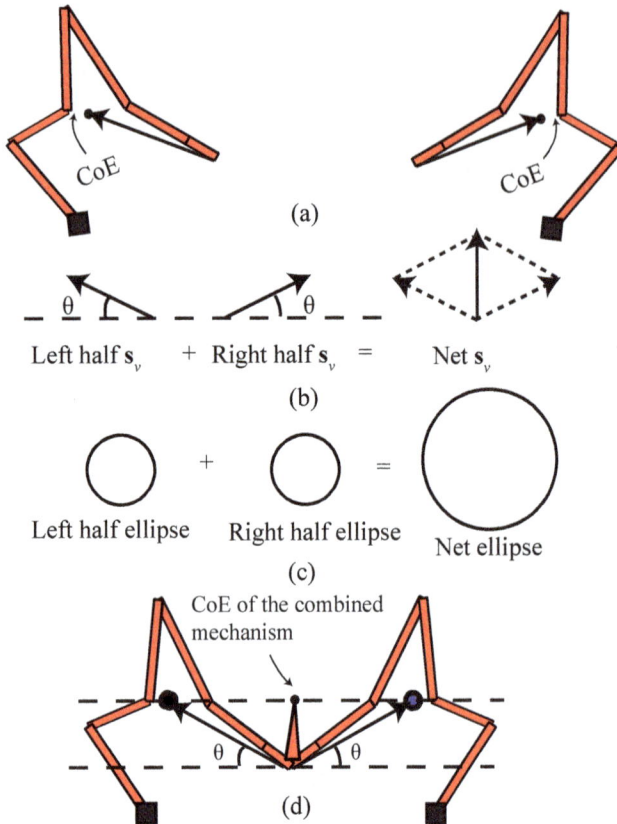

Figure 8. Parallel Combination (**a**) Two symmetric halves (**b**) Addition of Stiffness Coupling Vectors (**c**) Addition of Stiffness ellipses (**d**) Final mechanism with a rigid probe.

where $\mathbf{u}_{\text{in}}$ and $\mathbf{f}_{\text{in}}$ are the displacements and applied load respectively at the input, and $\mathbf{u}_{\text{out}}$ and $\mathbf{f}_{\text{out}}$ are the displacements and applied load respectively at the output. Load transfer between the two ports is defined as an equivalent applied load at the output that produces the same output displacement as a unit input load. Load transfer can be thus defined as the output load that would cause the same output deformation as an applied input load as shown in Fig. 9. This is similar to the notion of transferred forces defined in Harasaki and Arora (2001). The relation between this transferred load and the applied load is given by the LT matrix given by Eq. (2).

$$\tilde{\mathbf{f}}_{\text{tr}} = \mathbf{C}_{\text{out}}^{-1}\mathbf{C}_{\text{in-out}}^{T}\mathbf{f}_{\text{in}}$$
$$\mathbf{T}_{\text{L}} = \mathbf{C}_{\text{out}}^{-1}\mathbf{C}_{\text{in-out}}^{T} \qquad (2)$$

where $\mathbf{T}_{\text{L}}$ is the Load Transfer (LT) matrix that relates the transferred load and the input load $\mathbf{f}_{\text{in}}$. The detailed derivation of the Eq. (2) is shown in Krishnan et al. (2010). Furthermore, it must be noted that the transferred force in Eq. (2) is not the same as an applied output load in Eq. (1). The implication of defining transferred load is that a two-port problem between the input and the output is converted into a single port problem where the transferred force acting at the output produces the required output displacement.

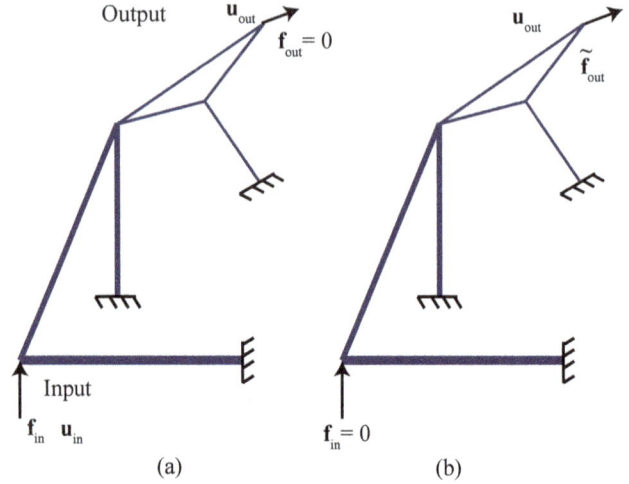

Figure 9. Deriving the Load Transfer matrix for Complaint Mechanisms (Krishnan et al., 2010). (**a**) Output displacement is evaluated for an applied input load (**b**) Output reaction load is evaluated by enforcing the output displacement from (**a**) with no input load. This reaction load is the transferred load.

While this characterization captures the relative compliance between two ports, its usefulness for a systematic synthesis is captured through an important property that enables modularity. Consider the two geometries and their deformed profiles in Fig. 10a and b. These geometries are composed of a beam that acts as an input constraint in series with a beam that connects the input and output. The output in Fig. 10a is constrained by a third beam which is absent in Fig. 10b. It is found that the transferred load evaluated at the output for the two different geometries is exactly equal, implying its independence on the output constraint. This property is true for all geometries consisting of a general input constraint in series with a general transmitter element between the input and the output. The detailed proof of the property is presented in Krishnan et al. (2010). Thus, the fundamental building block for load transfer between two points is identified as a Load-Transmitter Constraint (LTC) set. The implication of this property in a mechanism composed of a number of LTC sets is that the transferred load in each LTC set can be independently evaluated of succeeding ones. This enables modularity in analysis and design of two-port compliant mechanisms. One other observation is that the transferred load is independent of the deformation profile of the transmitter. For example, the transmitter in Fig. 10a is almost fixed-guided (like Fig. 2b), while the transmitter in Fig. 10b deforms as a rigid body.

The simplest of the LTC sets is a compliant dyad as seen in Fig. 10b. It has a beam for a constraint and a beam for a transmitter. Irrespective of the direction of the input force applied the transferred force is always oriented along the axis of the transmitter. If a unit input force is applied perpendicular to the orientation of the input constraint,the output transferred force is given by

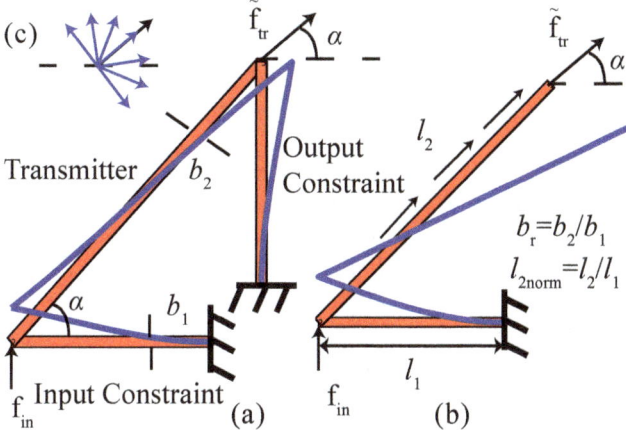

Figure 10. Comparison of two geometries containing **(a)** input constraint beam, transmitter beam and output constraint beam and, **(b)** input constraint beam and transmitter beam alone reveals that the transferred force at the output is independent of the output constraint. **(c)** The output displacement depends on the orientation of the output constraint. In general any direction of the output displacement is permitted ±90° with respect to the direction of transferred force.

$$\tilde{f}_{tr} = \frac{f_{in_y}}{\sin(\alpha)} \tag{3}$$

where α is the inclination of the transmitter with respect to the constraint. Applying an input moment changes the direction of the transferred force along with inducing transferred moment. The transferred force evaluated will then yield

$$\tilde{f}_{tr_x} = \cot(\alpha) f_{in} - \frac{3(l_{2norm}^2 \cos(\alpha) + b_r^3) m_i}{2l_1(l_{2norm} + b_r^3) l_{2norm} \sin(\alpha)}$$

$$\tilde{f}_{tr_y} = f_{in} - \frac{3l_{2norm} m_i}{2l_1(b_r^3 + l_{2norm})}$$

$$\tilde{m}_{tr} = -\frac{b_r^3 m_i}{2(b_r^3 + l_{2norm})} \tag{4}$$

where l_{2norm} is the ratio of the transmitter beam length to the input constraint beam length, l_1 is the length of input constraint beam length, b_r is the ratio of the transmitter beam thickness (in-plane) to the input constraint beam thickness. The output transferred forces ($\tilde{f}_{o_x}$ and $\tilde{f}_{o_y}$) depends upon the input force and input moment. However, the output transferred moment ($\tilde{m}_o$) is dependent on the input moment alone and its direction is opposite to the input moment. Furthermore, from the above equation, the magnitude of the transferred moment is lower than the applied moment.

Though output constraints do not affect the transferred force, they determine the magnitude and direction of output displacement. Shown in Fig. 10c is a semicircular band ±90° with respect to the transferred load. From the positive definiteness of the stiffness or compliance matrix, the output displacement is constrained within this band. Its actual direction is dependent on the output constraint. For example

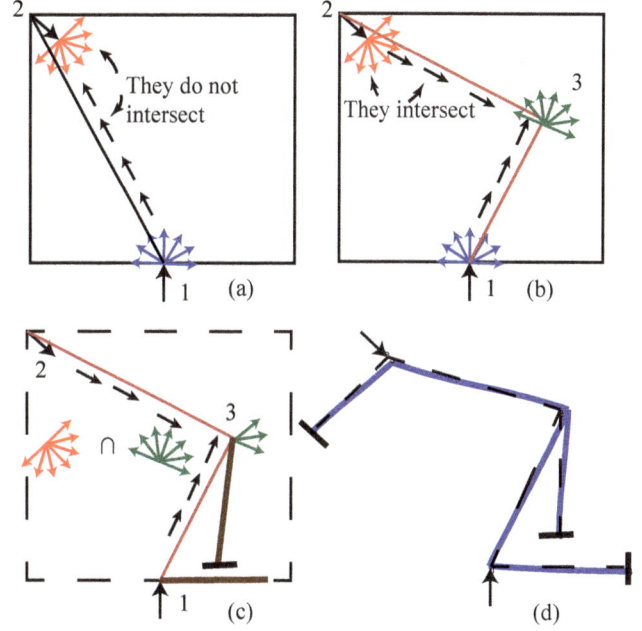

Figure 11. Steps involved in the load flow based conceptual synthesis of two-port compliant mechanisms. **(a)** Kinematic problem specification and the inability of a single load path to solve the problem, **(b)** two load paths with the appropriate load flow directions that meet the kinematic specifications, **(c)** constraints that enforce the load flow directions must be oriented along a truncated band, and **(d)** final mechanism topology and the deformed configuration.

the output constraint beam in Fig. 10a constrains the output to move along the direction determined by the intersection of its degree of freedom with the semicircular band.

So far, we have formulated a representation for relative compliance between two distinct points, identified its unique properties that permit modular analysis and characterized a simple building block, namely a dyad. We will present a simple example of how this technique can be used for systematic synthesis. Consider a problem specification shown in Fig. 11a where force applied at the input is at point "1" in the y-direction and the required displacement is at point "2" at an angle of −45°. This is a nontrivial problem because no direct connection between the input and output will yield the required kinematic specification. This is apparent from the figure as the direction of load flow in the transmitter does not intersect with any of the semicircular band directions at the output. To enable intersection, the problem is decomposed into two load paths, and the direction of load flow in the transmitters is determined. The mechanism topology (i.e. the constraints) will be designed such that the predetermined load flow directions are imposed. The constraints at input point "1" must enable y-direction displacement. A beam shown in Fig. 11c thus acts as the input constraint. The constraint at point "3" must be oriented such that load flow directions are preserved. This can be ensured if the degree of

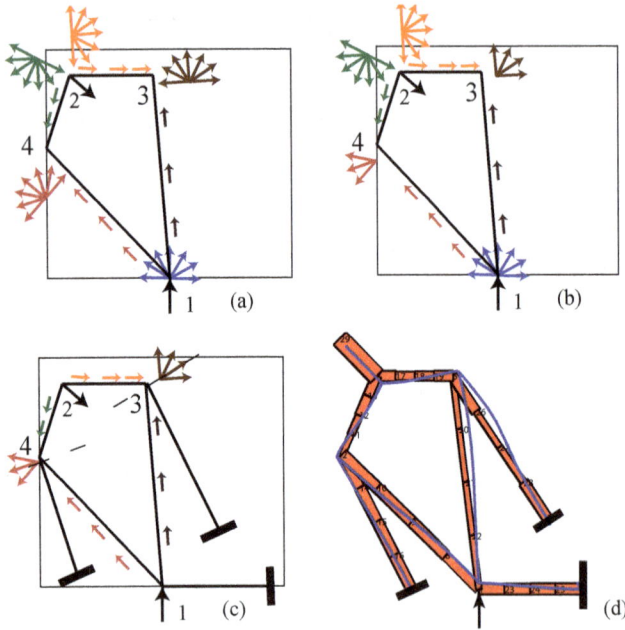

Figure 12. Steps involved in generating conceptual solutions having multiple load paths. (**a**) Kinematic problem specification and planning two parallel load paths between input and output, (**b**) truncated bands at each node, (**c**) constraints that enforce the load flow directions, and (**d**) final mechanism topology and the deformed configuration with widths of transmitters and constraints optimized so that the required deformation is achieved.

freedom direction of the constraint is along a truncated band obtained by the intersection of the bands at points "2" and "3" as shown in Fig. 11c. Furthermore the output constraint at point "2" is a beam that constraints it to move at an angle of −45°. The deformed configuration is shown in Fig. 11d.

The example illustrates the generality of the methodology involving a single load path from input to the output. The same example can be used to demonstrate the use of multiple load paths between the input and the output. This is illustrated in Fig. 12. The required direction of transferred load at point 1 can be obtained by a combination of load paths 1-3-2 and 1-4-2. The net load transferred due to each path add in proportion to the stiffness of the individual paths. For example, if stiffness of path 1-3-2 is greater than 1-4-2 then the former would dominate. If the stiffness of each load path is tuned so that they are equal, then the net load transferred is a vector combination of the individual paths. The constraints are designed in Fig. 12c such that they correspond to one of the truncated band directions. The widths of the constraints and transmitters are optimized such that the output in point 2 moves along a 45° angle.

Thus, a number of conceptual solutions can be generated by planning load paths and constraints using the above method. Comparison of each conceptual solution in terms of stress distribution, stiffness and mechanical efficiency is re-

quired to choose the best solution for a given problem specification.

5   Contributions and future research directions

This article shows how building block methods can be used to generate conceptual designs for compliant mechanisms. The ease with which kinematic specifications are met by systematic combination of simple deformable members, and the ability to quickly obtain alternate solutions highlights the usefulness of this method. Furthermore the lack of numerical complexity and the emphasis on user insight makes this technique excellent for classroom education. It is the representation of compliance that enables this user insight into systematic synthesis. In this article, such a representation and synthesis methodology is reviewed for single port problems, where force displacement relationship is characterized at a single point of interest, and for multi-port problems where relative compliance between two or more ports are characterized.

5.1   Contributions

Representing compliance at a single point is accomplished by decoupling translational and rotational terms thus preserving dimensional consistency. Ability to represent translational compliance as an ellipse and the coupling between translations and rotations as vectors enables insightful quantification of the compliance characteristics. Series and parallel combination of any deformable member is explained as a geometric combination of their individual ellipses and vectors. Thus solving for a given compliance characteristics, which was so far remotely accomplished through optimization is now possible through systematic, yet intuitive methods.

Synthesis of two-port mechanisms have always been considered non-intuitive as the contributions of each member towards the overall mechanism behavior is hard to understand. This is the first attempt towards identifying the functions of each member as a transmitter and a constraint. Representing deformation behavior as load flow enables identifying and thus determining feasible load paths that meets a given kinematic specification. Such a representation enables load path to act as a skeleton for the overall mechanism geometry. As seen in the examples, it is possible to synthesize single continuous load path and multiple parallel load paths for any application with relative ease. This insight and ability to obtain alternate solutions with ease highlights the usefulness of this method.

To summarize, building block method with geometrically insightful compliance representation is a novel synthesis method from first principles. Though this article focused on designing for kinematic specifications alone, the versatility of the method may show promise in designing for strength based considerations and manufacturing limitations.

5.2 Future research directions

The main challenge of compliant mechanism design is obtaining distribution of stresses evenly within all its constituent members. While meeting kinematic specifications alone leads to multiple solutions, it is necessary to evaluate which of these solutions lead towards distributed compliance. Secondly, the effects of large deformations on compliance representations must be studied. In single-port design this translates towards determining the change of the eigen-twist and eigen-wrench parameters with deformation. In two-ports the changes of load flow direction and magnitude must be determined for large deformations. Thus, the future directions in building block methods is towards understanding what constitutes distributed compliance, and formulating strategies in achieving them.

While planar examples were alone presented in this paper, the ideas proposed are as relevant to spatial mechanisms. The development of screw theory based methods (Hopkins and Culpepper, 2010a,b) in designing mechanisms with predetermined degrees of freedom and constraint can be considered as a spatial extension of the single port problem. However in mechanisms with relative deformation between two or more ports that are not in general connected by a rigid body, a combination of screw theory and load flow can be envisioned.

Edited by: J. Herder

References

Ananthasuresh, G. K.: A new design paradigm in microelectromechanical systems and investigations on compliant mechanisms, Ph.D. thesis, Ann Arbor, MI, USA, 1994.

Awtar, S., Slocum, A. H., and Sevincer, E.: Characteristics of beam-based flexure modules, J. Mech. Design, 129, 625–639, 2007.

Canfield, S. L., Chlarson, D. L., Shibakov, A., Richardson, J. D., and Saxena, A.: Multi-objective optimization of compliant mechanisms including failure theories, ASME Conference Proceedings, 48094, 179–190, 2007.

Cappelleri, D.: Flexible automation of micro and meso-scale manipulation tasks with applications to manufacturing and biotechnology, Ph.D. thesis, Philadelphia, PA, USA, 2008.

Cappelleri, D. J., Krishnan, G., Kim, C., Kumar, V., and Kota, S.: Toward the design of a decoupled, two-dimensional, vision-based mu n force sensor, Journal of Mechanisms and Robotics, 2, p. 021010, 2010.

Ciblak, N. and Lipkin, H.: Design and analysis of remote center of compliance structures, Journal of Robotic Systems, 20, 415–427, 2003.

Erdman, A., Sandor, G., and Kota, S.: Mechanism Design: Analysis and Synthesis Volume I, Prentice Hall, Upper Saddle River, New Jersey, 2001.

Harasaki, H. and Arora, J. S.: New concepts of transferred and potential transferred forces in structures, Comput. Methods Appl. Eng., 191, 385–406, 2001.

Hopkins, J. B. and Culpepper, M. L.: Synthesis of multi-degree of freedom, parallel flexure system concepts via freedom and constraint topology (fact) – part i: Principles, Precision Engineering, 34, 259–270, 2010a.

Hopkins, J. B. and Culpepper, M. L.: Synthesis of multi-degree of freedom, parallel flexure system concepts via freedom and constraint topology (fact). part ii: Practice, Precision Engineering, 34, 271–278, 2010b.

Howell, L. L. (Ed.): Compliant Mechanisms, John-Wiley, 2001.

Kim, C.: Functional characterization of a compliant building block utilizing eigentwists and eigenwrenches, Proceedings of the ASME International Design Engineering Technical Conferences and Computers and Information in Engineering Conference, 5–7 August, Brooklyn, NY, USA, 2008.

Kim, C. J.: A conceptual approach to the computational synthesis of compliant mechanisms, Ph.D. thesis, Ann Arbor, MI, USA, 2005.

Kim, C. J., Moon, Y. M., and Kota, S.: A building block approach to the conceptual synthesis of compliant mechanisms utilizing compliance and stiffness ellipsoids, J. Mech. Design, 130, 1–11, 2008.

Kota, S. and Chiou, S.-J.: Conceptual design of mechanisms based on computational synthesis and simulation of kinematic building blocks, Res. Eng. Des., 4, 75–87, doi:10.1007/BF01580146, 1992.

Krishnan, G., Kim, C., and Kota, S.: Load-transmitter constraint sets: Part i – an effective tool to visualize load flow in compliant mechanisms and structures, 2010.

Krishnan, G., Kim, C., and Kota, S.: An intrinsic geometric framework for the building block synthesis of single point compliant mechanisms, Journal of Mechanisms and Robotics, 3, p. 011001, 2011.

Lipkin, H. and Patterson, T.: Geometrical properties of modeled robot elasticity: Part i-decomposition, ASME Design Tech. Conf. and Computers in Engineering Conf., 45, 179–185, 1992.

Saxena, A. and Ananthasuresh, G. K.: A computational approach to the number of synthesis of linkages, J. Mech. Design, 125, 110–118, 2003.

Sclater, N. and Chironis, N. P.: Mechanisms and Mechanical Devices Sourcebook, McGraw-Hill, New York, 2007.

Yin, L. and Ananthasuresh, G. K.: Design of distributed compliant mechanisms, Mechanics based design of structures and machines, 2003.

3

Understanding the drivers for the development of design rules for the synthesis of cylindrical flexures

M. J. Telleria and M. L. Culpepper

Massachusetts Institute of Technology, Cambridge, USA

Correspondence to: M. J. Telleria (mtelleri@mit.edu)

Abstract. Cylindrical flexures (CFs), defined as flexures with only one finite radius of curvature loaded normal to the plane of curvature, present an interesting research direction in compliant mechanisms. CFs are constructed out of a cylindrical stock which leads to geometry, manufacturability, and compatibility advantages. Synthesis rules must be developed to design these new systems effectively. Current knowledge in flexure design pertains to straight-beam flexures or curved flexures loaded along the plane of curvature. CFs present a challenge because their mechanics differ from those of straight beams, and although their modelling has been researched thoroughly it has yet to be distilled into element and system creation rules. This paper uses models and finite element analysis to demonstrate that current design rules for straight-beam flexures are insufficient and inadequate for the design of CF systems. The presented discussion will show that CFs differ both at the element and systems levels, and therefore future research will focus on developing the three components of the building block approach: (i) reworking of element mechanics models to reveal the parameters which cause the kinematics of the curved beam to differ from those of the straight beam, (ii) development of a visual stiffness representation, and (iii) formation of system creation rules.

1 Introduction

Cylindrical Flexures, CFs, are defined as flexure systems with elements that have only one finite radius of curvature and are loaded normal to their plane of curvature. In other words, systems composed of flexure beams that are curved in a single plane. Figure 1 shows a prototype of a particular CF system. The system shown in Fig. 1 is actuated by loading the flexures normal to their plane of curvature. This specific loading condition is presented because it offers the most challenging research aspects and it is the least studied.

Past research has given different names to flexures that fall under this definition. The most applicable definition is Smith's *"hinges of rotational symmetry"*, which he defines as flexures constructed from solids of revolution (Smith, 2000). The flexure shown in Fig. 1 fits within Smith's definition because of its axial symmetry. This definition is expanded to allow CFs to be fractions of a cylinder. The work is scoped by constraining CFs to elements with a single finite radius. Finally the system must have well defined distortions for it

to be classified as a CF. The focus of this paper is to demonstrate: (i) the usefulness of design rules in the design process, (ii) that current design rules for straight-beam flexures are insufficient and inadequate for the design of systems with curved-beam flexures, (iii) the need for future research at the element and systems level, to develop guidelines for the design of CF systems.

1.1 Prior art

Past work that pertains to CFs has focused on two areas: (i) models for curved beams and (ii) analysis of specific CF concepts. Curved beams loaded normal to their plane of curvature have been studied for over 80 yr, leading to closed-form expressions of their mechanics and dynamics (Timoshenko, 1930; Young and Budynas, 2002; Lee, 1969). These models, however, have not been distilled to provide a designer with guidelines as to how to use curved beams to achieve a set of functional requirements. This work will extract, from these analytical models, clear rules for both element and system design. An example of this kind of

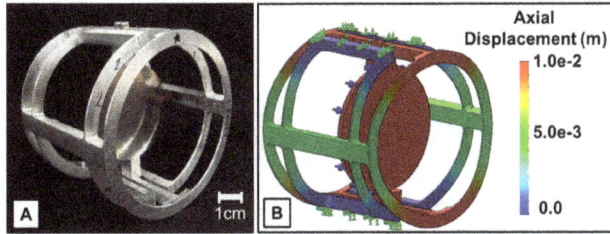

Figure 1. The quad compound-spring cylindrical flexure, CF, example: (**A**) 7075Al prototype and (**B**) FEA model depicting its actuation.

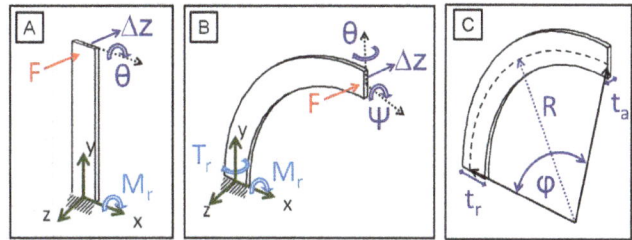

Figure 2. (**A**) Straight-beam mechanics, F is the load on the beam, M_r is the resulting bending moment, Δz is the displacement along the z-axis, and θ is the parasitic rotation about the x-axis. (**B**) Curved-beam mechanics, highlighting the added twist, ψ, and torque, T_r and (**C**) Curved-beam parameters. R represents the radius of curvature, ϕ is the sweep angle, t_r is the thickness of the beam in the radial direction, and t_a is the thickness in the axial direction.

synthesis is the work by Kim et al. (2008), which uses curved beam building blocks (CBB) to create flexure systems. This paper differentiates itself from CBB in that in this case the beams are loaded normal as opposed to parallel to the plane of curvature. This loading condition requires the analysis of the flexures in three dimensions.

The other area of prior art pertains to the analysis of specific systems that fit within the CF definition. Smith (2000) presents detailed analysis on the disc coupling and the rotationally symmetric hinge. These types of analyses have produced useful flexure systems; however there has been little overarching insight developed that could be used to create new CF concepts. In many cases the analysis of these systems has relied on FEA given the lack of knowledge on what parameters determine the system performance. The biggest knowledge gap comes in the form of understanding how to assemble these curved elements to create predictable systems.

The lack of design guidelines restricts the design process. The rapid generation of concepts is limited, since the designer does not have a simple way to predict the general behavior of a system composed of CF elements. In addition optimization is tedious because there is little understanding of the effect of different parameters.

1.2 Advantages of CFs

Cylindrical flexures present geometry, manufacturability, and compatibility advantages over traditional flexure systems. Their axial symmetry may be used to achieve insensitivity to thermal changes, and to decrease the effects of manufacturing and load placement errors. Monolithic systems with a variety of flexure elements can be created out of a single piece of round stock reducing assembly cost and errors.

CF fabrication is facilitated by the availability of accurate round stock. CFs can be manufactured at low cost by using traditional machining methods. The prototype in Fig. 1 was machined using a 4-axis Mazak brand lathe. Other manufacturing methods include: a waterjet with a rotary axis and a 5-axis mill. CF's most attractive quality is their compatibility with rotating applications, laparoscopic tools, optical

systems, and other applications benefiting from their cylindrical geometry.

2 Knowledge gap: need for CF research

This section focuses on explaining how and why CFs behave differently than straight-beam flexures. These differences make current design guidelines inadequate for the effective creation of CF concepts. The variations from straight-beam behavior are illustrated by looking at a curved cantilever beam loaded at its free end. In this case the desired motion is a displacement along the z-axis. All other displacements are defined as parasitic motions.

2.1 Mechanics

Curved flexures have additional complexities over straight beams both at the element and system levels. The curvature of the beam leads to an added rotation and resulting torque. Figure 2 shows that for a given load, F, the flexure will twist, ψ, which leads to a resulting torque, T_r, at the base. Sect. 3.1 presents the proposed process for developing element design rules from current curved-beam models.

The curvature of the beam also leads to challenges in the conceptual assembly of curved-beam flexure systems. The traditional rules for adding elements have to be augmented to include the effects of the added twist and torque. The four-bar linkage and Jones et al.'s (1956) compound rectilinear spring are used to demonstrate the need for additional system creation rules.

A common flexure bearing system is the four-bar linkage, shown in Fig. 3a, where a stage is connected to ground through 2 parallel straight beams or blades. Figure 3a shows that the four-bar linkage's motion can be described as the super position of two deflections which lead to two parasitic motions in addition to the desired displacement along the z-axis. Slocum (1992) gives the equations used to calculate the parasitic pitching motion of the stage, θ_{pitch}, and estimate the

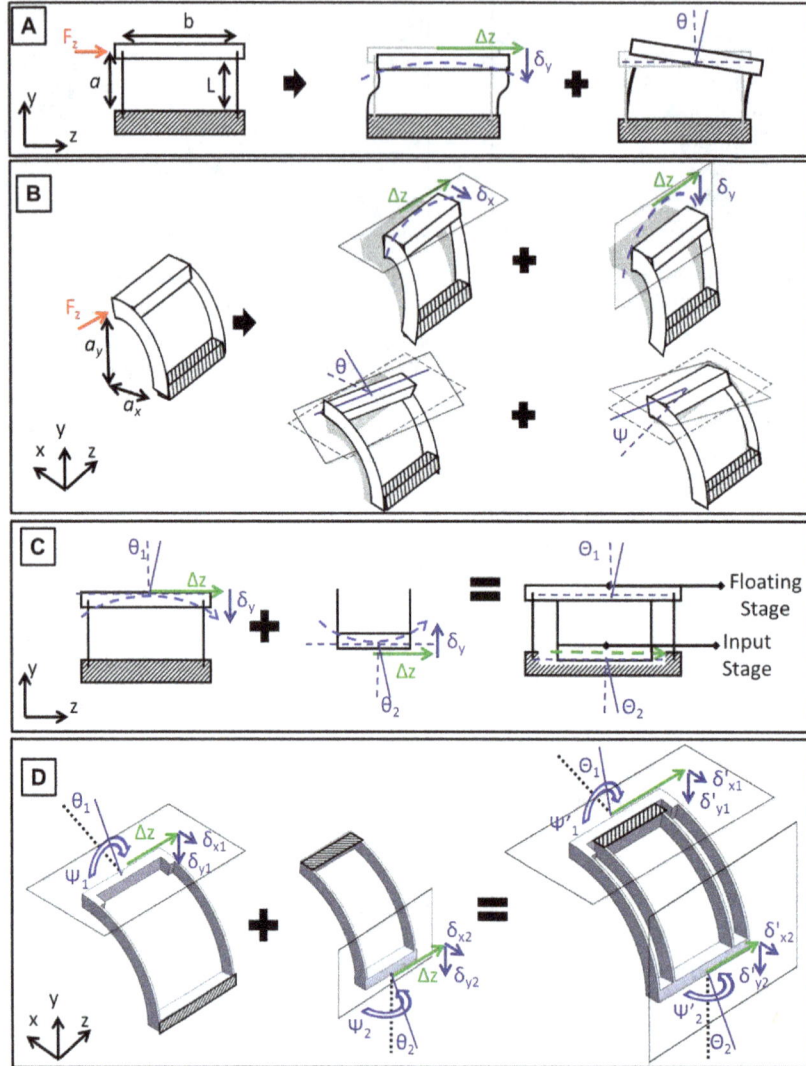

Figure 3. **(A)** Straight four-bar linkage parameters, a indicates the location of F_z relative to ground. Figure shows the three motions associated with F_z applied onto the platform. Desired displacement Δz is indicated with green, while the parasitic motions, θ and δ_y, are shown in blue. **(B)** Curved four-bar linkage parameters and displacements under F_z; a_y and a_x indicate the location of the force relative to ground. The curvature results in two δ parasitic motions, δ_x and δ_y, and an additional rotation of the stage, Ψ. **(C)** Straight compound-spring. Nesting of the two four-bars results in cancellation of the δ parasitic motions. Θ_1 and Θ_2 indicate the rotations of the stages in the nested system. **(D)** Curved compound-spring. Cancellation of the δ motions is not straight forward.

vertical motion, δ_y, presented in Eqs. (1) and (2). In these equations L represents the length of the flexure beam, b is the distance separating the flexures, a is the distance from the load input location to ground, t is the thickness of the beams, and Δz is the desired displacement along the load direction. The parameters are defined in Fig. 3a. The vertical motion of the stage, δ_y, in conjunction with the desired displacement, Δz, will result in the stage following an arc as F_z is applied, as highlighted in Fig. 3a.

$$\theta_{\text{pitch}} = \left(\frac{6(L - 2a) \cdot t^2}{3b^2 L - 2t^2 + 6at^2} \right)\left(\frac{\Delta z}{L} \right), \tag{1}$$

$$\delta_y \approx \frac{(\Delta z)^2}{2L}. \tag{2}$$

Equations (1) and (2) are used by designers to estimate the parasitic motions of a straight four-bar system. These equations have been derived for straight-beam flexures and must be expanded in order to describe a four-bar composed of curved beams. Figure 3b shows that for an input load normal to the plane of curvature, F_z, the motion of the input stage can be decomposed into four deflections which lead to four parasitic motions: θ, Ψ, δ_y and δ_x. As a result of the curvature of the flexures, the CF four-bar the stage will travel in two arcs dictated by Δz, δ_y and δ_x, and it will experience two rotations about its center θ, Ψ. Future research is necessary in order to develop a set of equations that accurately predict the parasitic motions of the four-bar CF system.

The design of precision machines requires that the parasitic motions are well understood and minimized to achieve the desired displacement. Studies have shown that for a four-bar linkage load placement and nesting of systems can be used to reduce the parasitic motions. As Eq. (1) shows θ_{pitch} can be reduced by moving the location of the input load, a. The challenge in the curved version of the four-bar linkage is that the load location is now defined by two coordinates a_x and a_y, as shown in Fig. 3b. Once again the currently available knowledge is insufficient to design a curved-beam flexure system.

Nesting of two four-bar flexure systems, referred to as the compound rectilinear spring, has been used to mitigate the parasitic arcing motion caused by δ_y as shown in Fig. 3c. In the compound-spring the δ_y motion of the input stage is matched by an equal and opposite δ_y of the floating stage resulting in a cancellation of the arcing motion of the input stage. The δ_y motions are mitigated as long as both four-bars are constructed of identical flexure beams, therefore achieving the same Δz displacement and as a result the same magnitude of δ_y. It is important to note that the nesting of the flexures does not remove the parasitic rotations of the stages, θ_1 and θ_2, but these rotations are affected by the nesting and are labelled Θ_1 and Θ_2 in Fig. 3c to reflect this. Figure 3d shows what happens when a compound-spring is created using two curved four-bars. Due to the curvature, the floating and input stages are located on different planes. As a result the cancellation of the δ motions is not straight forward and requires further research; the resulting δ motions will depend on the magnitude of δ_y and δ_x as well as the sweep angle, ϕ. Section 3.3 discusses the proposed research approach for creating system rules that will allow the designer to manage the parasitic motions of CF systems.

2.2 Stress

In order to determine the range of a system, both the displacement to a given load and the resulting stress must be considered. The curvature of the element leads to a torque on the fixed end of the cantilever beam. This torque will affect the resulting stress calculation. Using energy conservation principles it can be identified that an added torque requires that there be a decrease in the original resulting moment. Both of these effects will change the magnitude and distribution of the stress along the beam.

Figure 4 shows that a stress concentration is observed at the base on the inner radius of the curved element. Future work will develop guidelines that will allow the designer to account for the new stress distribution and perhaps find a way to distribute the stress more evenly. Previous work has shown

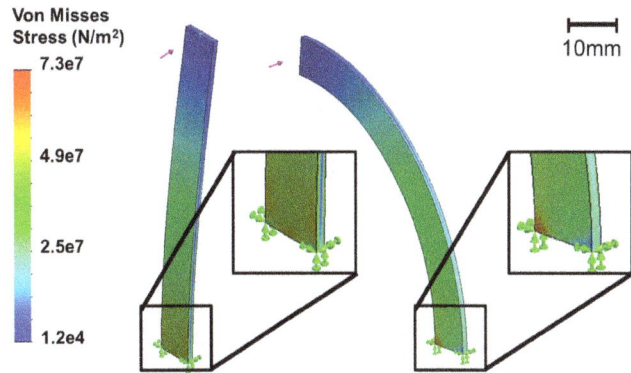

Figure 4. Finite element analysis of two cantilevered beams of the same properties under the same loading force. The stress concentration on the curved beam is highlighted.

contouring of the thickness of the flexure to be a successful way to distribute stress along the length of a beam (Timoshenko, 1930).

3 Research approach and impact

The main three compliant mechanism synthesis approaches are topology synthesis, pseudo-rigid modelling, and building block approach. A building block approach derived from constraint-based design is proposed as the most appropriate synthesis methodology for CFs at this time (Maxwell, 1890; Blanding, 1999; Hale, 1999). The plan is to first develop a full understanding of the parameters that affect the element's behavior. This element then becomes the building block for CF systems. The next step is to develop rules for how these blocks interact when added together. The desired outcome of the research consists of (i) a visual representation that allows the designer to quickly understand how different parameters will affect the behavior of an element, similar to a stiffness ellipsoid (Kim, 2008) and (ii) design guidelines for the generation of CF systems.

A constraint-based design approach has been chosen because it is intuitive for precision engineers. This synthesis approach presents a quick way to understand why a system created from a set of building blocks will behave a certain way. It also makes the rapid concept generation phase of the design process very efficient as the designer can use the stiffness representations and system rules to quickly lay out the elements to achieve a desired performance, while being able to account for external constraints such as manufacturing. The knowledge gathered through the development of these building block rules can then be used to create arguments for the other synthesis methods.

A Curvature Adjustment Factor (ζ) vs. Sweep Angle (φ)

B Curvature Adjustment Factor (ζ) vs. β (Normalized to $\zeta(0.5)$)

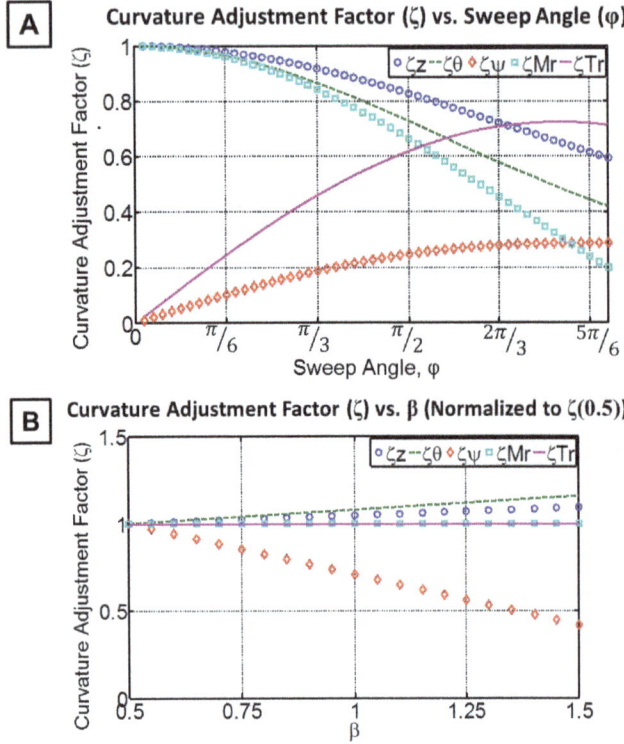

Figure 5. The curvature adjustment factors, ζ, show how the behavior of the curve beam deviates from that of a straight beam as a function of the geometry of the beam. Curvature adjustment factors are obtained from Roark's closed-form solutions. ζ_z corresponds to the z-axis displacement multiplier as shown in Eq. (7), while ζ_θ and ζ_Ψ correspond to the multipliers for the two parasitic motions θ and Ψ, respectively. Finally ζ_{M_r} and ζ_{T_r} represent the curvature adjustment factors for the resulting moment and torque at the base of the beam. Flexure length is held constant by varying the radius of curvature. (**A**) Curvature adjustment factors, ζ, vs. sweep angle, ϕ. The graph shows that as the sweep angle goes to zero the curved beam behaves as a straight beam, ζ_z, ζ_θ, ζ_{M_r} go to 1, and ζ_Ψ and ζ_{T_r} approach zero. While as the sweep angle increases the behavior of the curved beam deviates from that of the straight beam. (**B**) Curvature adjustment factors, ζ, vs. β, β is the ratio of elastic to shear properties of a curved beam as defined by Eq. (4).

3.1 Element mechanics

The first step in the building block approach is to understand the flexure element mechanics. The goal is to identify what parameters affect the element's performance. Roark's equations for curved beams loaded normal to their plane of curvature are used to understand the kinematics of the curved beam and to find the parameters that play an additional role in the motion of the curved beam (Young and Budynas, 2002). Cantilever beams loaded at their free end are first examined. Future work will explore the effect of load location.

Figure 2c shows the different parameters that define the curved flexure element. Roark presents equations for the displacements resulting from a load, F, at the free end of the

beam (Young and Budynas, 2002). The z-displacement, Δz, is the desired motion. This displacement is given by Eq. (3), where R is the radius of the beam, E is the elastic modulus, I, is the second moment of area, and ϕ is the sweep angle. C_{a6}, C_{a9}, C_{a3} are functions of ϕ and β, defined in Roark (Young and Budynas, 2002). Equation (4) defines β as the ratio of elastic to shear properties of the beam, where G is the shear modulus, and k is a torsional stiffness constant. More details on these equations can be found in Roark (Young and Budynas, 2002). Equations (5) and (6) from Roark are used to calculate the two parasitic motion of the cantilever beam, θ and Ψ which are defined in Fig. 2b.

$$\Delta z = \frac{FR^3}{EI} \cdot (C_{a6}\sin(\varphi) - C_{a9}(1 - \cos(\varphi)) - C_{a3}), \quad (3)$$

$$\beta = \frac{E \cdot I}{G \cdot k}. \quad (4)$$

$$\theta = \frac{F \cdot R^2}{E \cdot I}(C_{a6}\cos(\varphi) - C_{a9}\sin(\varphi)) \quad (5)$$

$$\Psi = \frac{F \cdot R^2}{E \cdot I}(C_{a9}\cos(\varphi) + C_{a6}\sin(\varphi)) \quad (6)$$

It is proposed that a more efficient way to look at Eq. (3) is to factor it into the straight-beam equation and a curvature adjustment factor, ζ_z, as shown in Eq. (7). In Eq. (7) the length, L, is the product of R, and ϕ. Equation (8) shows that the curvature adjustment factor is calculated by dividing the displacement given by Roark in Eq. (3) by the straight-beam displacement equation. The curvature adjustment factors for the parasitic motions, ζ_θ and ζ_Ψ, are similarly calculated.

$$\Delta z = \zeta_z \frac{FL^3}{3EI}, \quad L = R \cdot \varphi \quad (7)$$

$$\zeta_z = 3 \cdot (C_{a6}\sin(\varphi) - C_{a9}(1 - \cos(\varphi)) - C_{a3})\big/\varphi^3 \quad (8)$$

This new representation allows the designer to use all previous knowledge of straight-beam behavior and then evaluate what parameters play an additional role in the behavior of the curved beam. The added role of parameters is evaluated by identifying their effect on ζ_z. If ζ_z does not depend on the parameter then that parameter does not have an additional effect, relative to its effect on a straight beam's mechanics. If ζ_z is a function of that parameter then the parameter plays an additional role. Analysing the parameters in Eq. (1) it is found that ζ_z, ζ_θ, and ζ_Ψ depend only on ϕ and β.

The next step is to understand how these two parameters affect the mechanics of the beam. Figure 5a shows how as ϕ approaches zero the curved beam behaves like a straight beam. Then as ϕ increases the behavior deviates. Figure 5b shows the curvature adjustment factor for each of the motions ζ_z, ζ_θ and ζ_Ψ, vs. β. The value of β for a given geometry varies only a small amount with different materials because the elastic and shear modulus are related by the Poisson ratio, which is close to 0.3 for common flexure materials. Having identified the two additional parameters necessary to describe the kinematic behavior of the curved beam, future

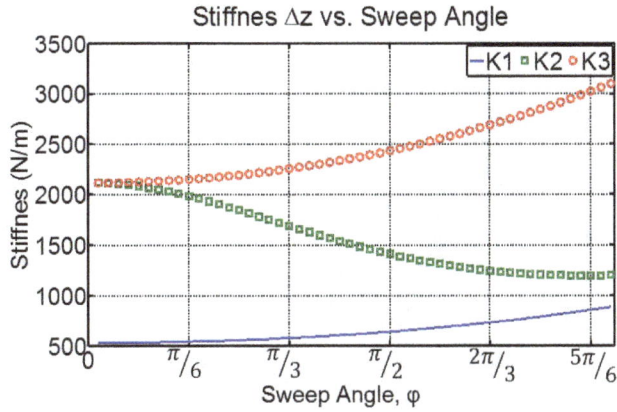

Figure 6. Δz-stiffness vs. sweep angle, ϕ, for three constraint conditions. $K1$ corresponds to a cantilever beam, $K2$ represents a fixed end-slope constrained beam, and $K3$ corresponds to a fixed end-slope and twist constrained beam. Flexure length is held constant by varying the radius of curvature.

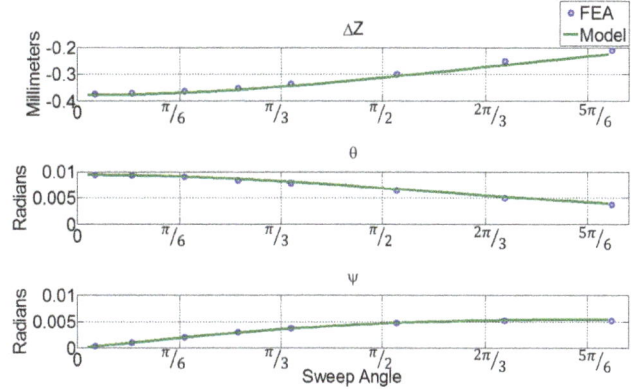

Figure 7. Comparison of the FEA and closed-form equation calculations of the curved beam displacements vs. sweep angle, ϕ, for a cantilever beam, $K1$. Max error between models is 7 %. Δz represents the desired motion along the z-axis of the beam as indicated in Fig. 2b. θ and Ψ correspond to the two parasitic rotations of the beam under a load normal to the plane of curvature. Flexure length is held constant by varying the radius of curvature.

research will analyse how the sweep angle, ϕ, and β affect common flexure performance metrics.

3.2 Stiffness representation

Once the mechanics are understood, a stiffness representation can be developed. This representation will allow the designer to quickly identify how the elements will behave. This can be done using either stiffness or compliance ellipsoids (Kim et al., 2008). A stiffness ellipsoid visually shows the relative magnitude of the stiffnesses of a beam. The Δz-stiffness of the beam depends on the loading conditions and element constraints. The previous analysis focused on cantilevered beams; however, flexures are usually constrained in systems. Therefore, the effect of different constraint conditions on the Δz-stiffness is explored. The change in Δz-stiffness with ϕ is evaluated, because sweep angle is the dominant parameter.

Figure 6 shows the Δz-stiffness vs. ϕ for three different constraint conditions: (i) cantilever beam, $K1$, (ii) slope constrained, $K2$, and (iii) slope and twist constrained, $K3$. The length of the flexures is held constant in this analysis by varying the radius of curvature. The graph for K1 corresponds to the cantilever beam analysed in Fig. 5. From the mechanics analysis $K1$ is expected to increase with ϕ, given Δz decreases with ϕ. $K2$ decreases with ϕ because both Δz and θ decrease with ϕ. This decrease in slope translates to a smaller moment being applied at the free end to achieve the zero slope condition. Finally $K3$ increases with ϕ. This same analysis will be carried out for all stiffnesses leading to the creation of the stiffness ellipsoids.

The presented equations of motion for the curved cantilever beam analysed in Figs. 5 and 6 are corroborated using FEA models. Figure 7 plots the predicted displacements

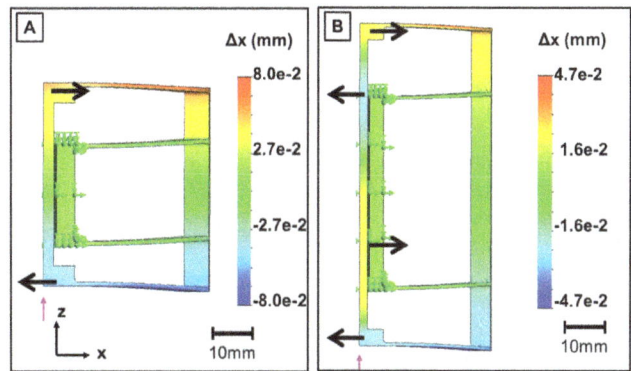

Figure 8. FEA of a curved compound-spring under a 10 N load (**A**) No deformation of input stage (**B**) Deformed input stage as a result of the four-bar twist stiffness, K_ψ, being larger than the stage stiffness.

from both models. The next step will be to compare these models to experimental results. Future research will expand on the kinematic analysis by assembling the 6×6 stiffness matrix for a curved beam.

3.3 System analysis

The system creation guidelines will focus on the interaction between the element building blocks. With these rules the designer will be able to assemble the blocks together to achieve a desired system performance. Three main system creation research areas have been identified so far: (i) the effect of element separation, (ii) the importance of the sweep angle, which leads to the ground and stage of the flexure system to be located on two different planes, and (iii) the effect of load location.

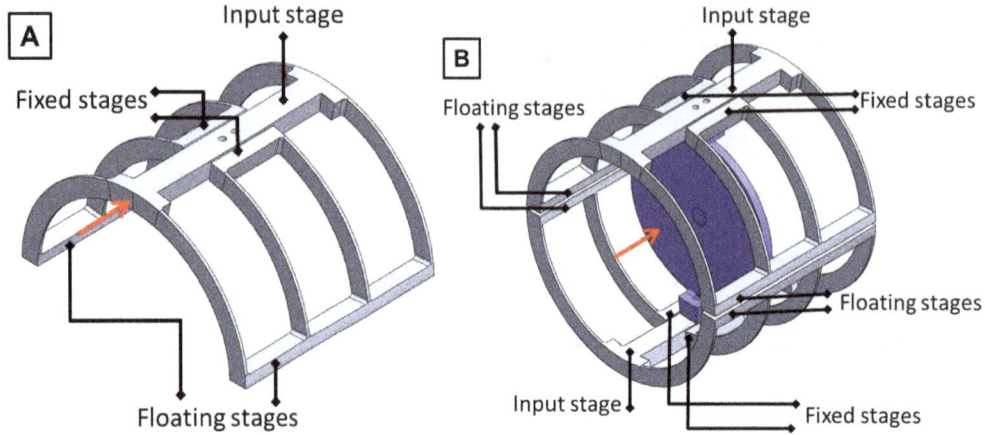

Figure 9. (**A**) Double curved compound-spring. (**B**) Stage driven by two separate double curved compound-springs, the quad compound-spring CF design.

Table 1. Main stage tip and tilt angle in microradians per millimeter of z-axis displacement for the different compound-spring iterations (flexure length is constant for all iterations).

	Tilt angle θ μrad mm^{-1}	Tip angle Ψ μrad mm^{-1}
Flat straight compound	58.5	0.00
90° curved compound	321	101
Double curved compound	0.00	211
Two double curved compound	0.00	5.04

Element spacing plays a critical role in system creation. It is well-established that parasitic rotations of a shaft decrease with the distance between the bearings, b, squared. Equation (1) shows that the pitching motion of a four-bar decreases with b^2 and Eq. (9) presents that the twist stiffness, K_ψ, of a straight four-bar increases with b^2 (Smith, 2000).

$$K_\psi = \frac{EI}{b^2 L}, \tag{9}$$

For straight-beam systems both parasitic motions decrease with b^2. Using only this information may lead the designer to maximize the spacing between the flexure beams. This is not only a problem in terms of weight and volume. It is also found that since the parasitic motions have not been mitigated, when the four-bar's spacing is large enough, K_ψ is greater than the stiffness of the stages and the stages begin to deform, as shown in Fig. 8. Increasing the spacing between the beams is not a complete solution for reducing the parasitic motions.

Load location has been shown to play a critical role in reducing the parasitic motions of a system (Slocum, 1992; Hopkins, 2010). Therefore, future research must establish rules and equations that describe the effect of the load location on the parasitic motions of a curved system. The re-

search approach proposed is to first establish the optimal load location for a curved-beam element and how this position changes with sweep angle. Then to determine how to use center of stiffness rules to find the best actuation point of a CF system.

Section 2.1 showed that a curved four-bar system has four parasitic motions and postulated that in a curved compound-spring the δ motions of the stage are not mitigated through the nesting because they occur on two different planes and have different δ_x and δ_y magnitudes. The question then becomes how do to deal with the additional parasitic motions and the fact that they are occurring on two different planes. The following example is used as a way to demonstrate some of the challenges of CF system creation. The goal of the presented system is to translate a stage along the CF's central-axis with minimal tip and tilt error. To do this the parasitic θ and Ψ motions of the flexure system have to be minimized and the Δz displacement maximized. Precision engineering applications require high accuracy flexure systems; microns of parasitic motion can cause a design to fail.

To improve the performance of the curved version, a double compound-spring is used, as shown in Fig. 9. This design removes the tilt angle, θ, of the input stage by using symmetry without severely over-constraining, given the side/floating stages are free to pitch and tilt. Symmetry can be used once again to reduce the tip angle, Ψ, by driving a stage with two double curved compound-springs as shown in Fig. 9. The displacements of the different iterations are compared to the compound rectilinear spring in Table 1. Finally the side stages have to be connected to achieve a full cylinder in order to have a monolithic system. The challenge is to ensure that the side stages are able to continue to move relative to each other. The stages can be connected through a flexure spring that allows for this relative motion.

4 Conclusions and future work

This paper uses analysis of common flexure mechanisms to demonstrate that current design rules for straight-beam flexures are insufficient for designing effective CFs. The insights that this research has already generated guided the correction of a CF design achieving a 20× reduction in parasitic errors. The result is a new CF concept that is compatible with cylindrical geometries. Continuing research in this area will develop a full set of design insights and a stiffness representation that will enable a designer to create CF systems effectively.

Acknowledgements. This work was supported in part by the National Science Foundation Graduate Fellowship Program and Lincoln Laboratories. Special thanks to MIT's BioInstrumentation lab for the use of their Mazak Lathe.

Edited by: J. Herder

References

Blanding, D. L.: Exact Constraint: Machine Design Using Kinematic Principles, ASME Press, New York, 1999.

Hale, L. C.: Principles and Techniques for Designing Precision Machines, Ph.D. thesis, Massachusetts Institute of Technology, USA, 1999.

Hopkins, J. B.: Design of Flexure-Based Motion Stages for Mechatronic Systems via Freedom, Actuation and Constraint Topologies (FACT), Ph.D. thesis, Massachusetts Institute of Technology, 2010.

Jones, R. V., Phil, D., and Young, I. R.: Some parasitic deflexions in parallel spring movements, Scientific Instruments, 33, 11–15, 1956.

Kim, C. J., Moon, Y., and Kota, S.: A building block approach to the conceptual synthesis of compliant mechanisms utilizing compliance and stiffness ellipsoids, Mech. Design, 022308-1–002308-11, 130, 2008.

Lee, H.: Generalized Stiffness Matrix of a Curved-Beam Element, AIAA Journal Technical Notes, 7, 2043–2045, 1969.

Maxwell, J. C.: General considerations concerning scientific apparatus, The scientific papers of James Clerk Maxwell, edited by: Niven, W. D., Dover Press, (Reprinted from the handbook to the special loan collection of scientific apparatus, 1876), 1890.

Slocum, A. H.: Precision machine design, Society of Manufacturing Engineers, Society of Manufacturing Engineers, Prentice-Hall, Inc., USA, 1992.

Smith, S. T.: Flexures: Elements of elastic mechanisms, CRC Press, USA, 2000.

Timoshenko, S.: Strength of materials, D. Van Nostrand, USA, 1930.

Young, W. C. and Budynas, R. G.: Roark's Formulas for Stress and Strain, McGraw-Hill, Singapore, 2002.

4

From flapping wings to underactuated fingers and beyond: a broad look to self-adaptive mechanisms

L. Birglen

Ecole Polytechnique of Montreal, Department of Mechanical Engineering,
Ecole Polytechnique of Montreal, Montreal, QC, H3T 1J4, Canada

Abstract. In this paper, the author first reviews the different terminologies used in underactuated grasping and illustrates the current increase of activity on this topic. Then, the (probably) oldest known self-adaptive mechanism is presented and its performance as an underactuated finger is discussed. Its original application, namely a flapping wing, is also shown. Finally, it is proposed that the mechanisms currently used in underactuated grasping have actually other applications similarly to the previously discussed architecture could be used for both an underactuated finger and a flapping wing.

This paper was presented at the IFToMM/ASME International Workshop on Underactuated Grasping (UG2010), 19 August 2010, Montréal, Canada.

1 Self-adaptive, adaptive, intelligent, underactuated, differential, or compliant?

Self-adaptive mechanisms as used in underactuated fingers (Birglen et al., 2008) are often confused with other classes of mechanisms because of the lack of a clear definition. According to Gosselin (2006), adaptive mechanical systems are defined as systems *"in which the ability to adapt to new external situations relies strictly on mechanical properties"* (quoted). Although the denomination of "adaptive mechanisms" proposed in the latter reference is absolutely correct, the author of this paper often prefers referring to them as "self-adaptive mechanisms" to avoid confusion. Indeed, the latter expression emphasizes that the adaptation capability refers to the mechanical system itself and does not describe an algorithmic procedure. As noted in Gosselin (2006), adaptive systems are usually associated with computer systems and this has lead to some confusion in underactuated grasping. It should also be noted that "self-adaptive" has already been used by other authors to describe these systems (Rubinger et al., 2001; Carrozza et al., 2004). They are also sometimes referred to as "intelligent" systems (Ulrich, 1988; Gosselin, 2005), which is more vague and therefore, should probably be avoided.

The objective of self-adaptive mechanisms is to delegate part of the control tasks from the electronic board to the mechanical layout of the system itself. Hence, with self-adaptive robotic hands, parts of the possible motions of the fingers have to be uncontrolled electronically. However, these motions must be carefully predicted and studied in order to achieve the desired closing sequence, or else, they can lead to degenerate behaviours (Birglen and Gosselin, 2006b). Since some degrees of freedom (DOF) of these hands are not controlled, they are often referred to as "underactuated" (Laliberté and Gosselin, 1998). Again, this adjective is technically correct but has also lead to some confusion. Underactuation in robotic hands is different from the concept of underactuation usually presented in robotic systems and the differences between both notions should be made clear. An underactuated serial robot is defined as a manipulator with one or more unactuated joints. On the other hand, "underactuated" or self-adaptive fingers use passive elements (the most common of which are springs) in the design of their unactuated joints. Thus, one should rather think of these joints as uncontrollable or passively driven instead of unactuated.

With self-adaptive robotic systems, and conversely to usual underactuated manipulators, the actuation torque (or force) is distributed to each joint of the system. This distribution property is essential and can be related to the Transmission matrix of the linkage (Birglen, 2009) which is characteristic of the type of transmission mechanism required to achieve this distribution. In principle, this distribution

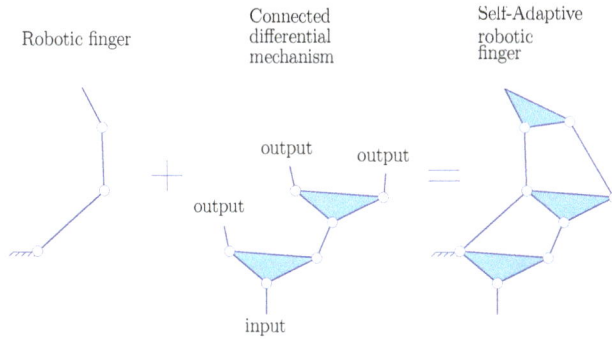

Figure 1. A connected differential mechanism driving a robotic finger.

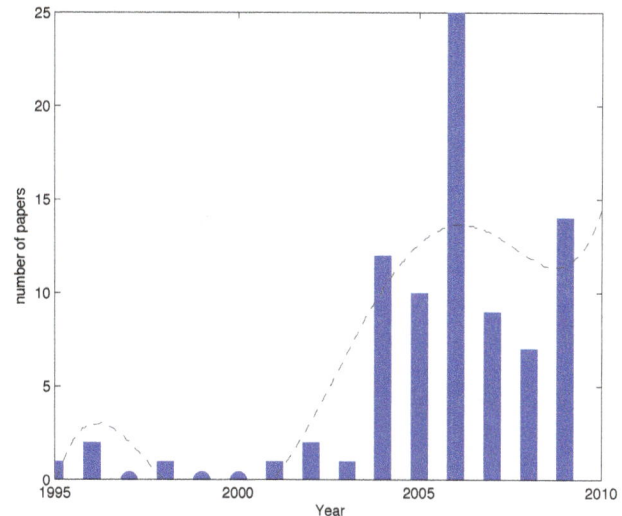

Figure 2. Publication activity in underactuated grasping.

property is similar to the behaviour of differential mechanisms. Let us recall that according to the IFToMM terminology (IFToMM, 1991), a differential mechanism is a two-DOF mechanism that may resolve a single input into two outputs and vice versa. Therefore, if more than two outputs are required – typically three with robotic fingers (for three phalanges) – the simplest approach is to stack multiple differential devices, each stage adding one DOF to the system (Hirose, 1985; Birglen and Gosselin, 2006c). An illustration of this process is given in Fig. 1, the outputs of two differential seesaw mechanisms (middle, presented in Birglen and Gosselin (2006c)) connected in series drive a robotic finger (left) resulting in a well-known architecture of self-adaptive robotic finger (right), proposed in Gosselin and Laliberté (1996) and consequently used in several prototypes (Laliberté et al., 2002).

However, the designs obtained with this method are only a small sample of a vastly larger number (literally thousands) of possible mechanisms with equal or less complexity (Birglen, 2009). Hence, differential mechanisms is also not the best term to describe self-adaptive mechanisms.

Passive elements are the second ingredient to self-adaptation and are used to kinematically constrain the system. Strictly speaking, the inclusion of a specific passive element is not absolutely necessary to achieve self-adaptation since inertial properties can be used as "passive elements" (Birglen, 2009). In this particular case, the resulting linkage is close to the meaning of underactuation commonly found in the literature. However, to the best of the author's knowledge only two prototypes of self-adaptive fingers have importantly relied on dynamic parameters such as inertia, namely from Higashimori et al. (2005); Crisman et al. (1996). Since the most common passive element by far is the spring (preloaded or not), it has been deemed mandatory. This is not true, compliance is not at all necessary. For instance, other passive elements are presented in Birglen (2009).

As mentioned before, underactuation in grasping has lead to some misconception in the past as to which systems it de-

scribed because "underactuation" was associated with underactuated manipulators. This has lead the author to start using "self-adaptive" with an increased frequency. However, this might not be necessary in the coming years as underactuation in grasping is getting more well-known. Indeed, the number of papers published on underactuation in grasping has been steadily increasing in the past few years as illustrated in Fig. 2. The data from this figure have been obtained by searching the following terms on all the available databases from the Engineering Village Database[1]:

(((underactuated) OR (self-adaptive)) AND ((hand*) OR (finger*) OR (gripper*))) WN TI

Note that the duplicates and obviously out of scope entries have been manually removed. The total number of references found with this search is 85. Of course, the results of this search engine are not complete. For instance, well-known prototypes such as the ones discussed by Hirose and Umetani (1978); Townsend (2000); Kyberd et al. (2001) do not appear. Additionally, the author's own database of publications has recently reached 500 entries (excluding patents) on underactuated grasping and related topics. However, this test is representative of the level of activity on this topic and has the advantage of being repeatable by anyone (thus verifiable). The author is pondering the diffusion of his database of publications on a public website similarly to what Prof. Bonev did with the ParalleMIC[2] for parallel mechanisms. One can see that during the last ten years, the number of publications on underactuated grasping exploded and this area of research is now not anecdotal anymore. Nevertheless, it is also obvious that it is not yet a "mainstream" issue in robotics. It is the author's opinion that we are only at the beginning of the

---

[1]http://www.engineeringvillage2.org
[2]http://www.parallemic.org/

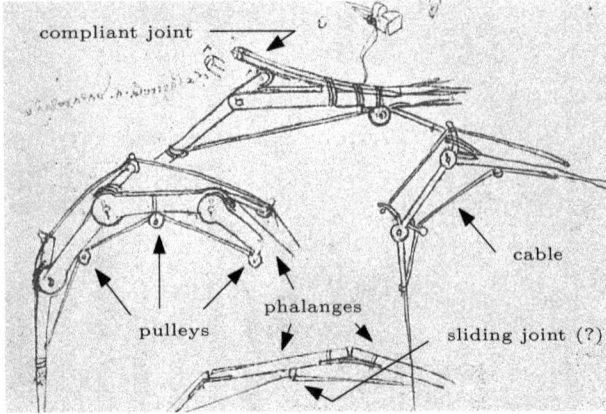

Figure 3. Mechanical wing from Da Vinci's Codex Atlanticus (ca 1496) annotated by the author.

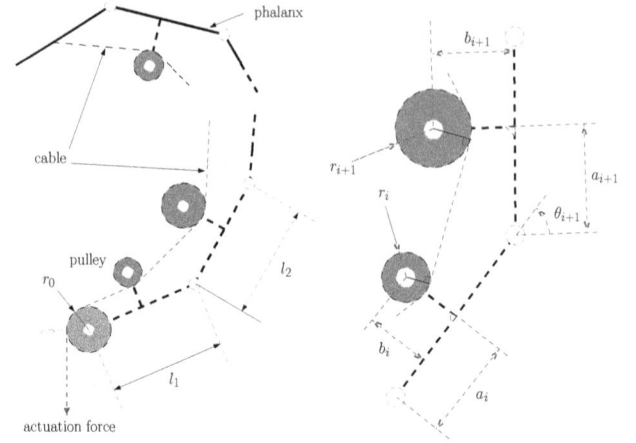

Figure 4. Da Vinci's mechanism.

curve illustrated in Fig. 2 and that a new trend in robotics is emerging for underactuation in grasping is only a small sample of the possible applications of self-adaptive mechanisms. As an example to illustrate this claim, let us consider the oldest self-adaptive linkage known to the author: Da Vinci's articulated wing (Birglen et al., 2008).

2   Da Vinci's linkage

2.1   Introduction

Leonardo Da Vinci proposed in the *Codex Atlanticus* a mechanism illustrated in Fig. 3 based on a cable and pulleys that was intended to drive an artificial wing in one of the fantastic flying machine developed by the Renaissance genius.

It is unknown if such a wing or machine has ever been actually tested let alone built by Da Vinci but it proves the ingenuity of engineers before the discovery of electrical actuators. The mechanism itself consist of several sections, similar to *phalanges*, whose motion was driven by a cable pulled by the human pilot of the machine. This cable was attached to the distal phalanx and ran through the phalanges with the help of pulleys or (maybe) sliding joint. The actual design of this wing bears a striking similarity with an anatomic sketch of the human finger found in another of Da Vinci's codex (see Birglen et al., 2008, for a comparison). More than four hundred years later a *very* similar design was patented for a robotic finger (Rovetta et al., 1982). The first prototype of modern underactuated fingers, namely the Soft-Gripper of Prof. Hirose (Hirose and Umetani, 1978), was also an architecture in which a cable is run through a series of pulleys until the distal end of the finger. Although in that case, the similarity between the two mechanisms ends here since the actual design of the Soft-Gripper is quite different from Da Vinci's architecture. It is nonetheless interesting to note that this first modern prototype which popularized underactua-

tion in grasping among the research community shares many traits with Da Vinci's linkage.

2.2   Grasping

The architecture discussed in this section is shown in Fig. 4, and corresponds to a general model of the drawing found in the *Codex Atlanticus*. Notice that all the pulleys of this linkage can freely rotate around their axes.

The Transmission matrix of this mechanism is Birglen et al. (2008):

$$\mathbf{T} = \begin{bmatrix} 1 & \dfrac{R_1}{r_0} & \dfrac{R_2}{r_0} & \cdots & \dfrac{R_{n-1}}{r_0} \\ \mathbf{0}_{n-1} & & \mathbf{1}_{n-1} & \end{bmatrix} \tag{1}$$

where $\mathbf{0}_{n-1}$ and $\mathbf{1}_{n-1}$ are respectively the null vector and identity matrix of dimension $n-1$. The radius of a pulley equivalent to the i-th transmission stage (illustrated in the right-hand side of Fig. 4) is noted R_i for $i = 1,..,n-1$. The linkage is driven by a base pulley (with a radius r_0) illustrated in the left-hand side of Fig. 4. It can be shown that the equivalent radius of the i-th transmission stage is

$$R_i = r_i + \frac{b_i(rb - al) - (l_i - a_i)(ar + bl)}{a^2 + b^2} \tag{2}$$

with

$$r = r_{i+1} - r_i, \tag{3}$$
$$a = l_i - a_i + a_{i+1}\cos\theta_{i+1} - b_{i+1}\sin\theta_{i+1}, \tag{4}$$
$$b = -b_i + a_{i+1}\sin\theta_{i+1} + b_{i+1}\cos\theta_{i+1}, \tag{5}$$
$$l = \sqrt{a^2 + b^2 - r^2}. \tag{6}$$

When computing this equivalent radius, one must take into account that the points where the cable comes in contact with the pulleys r_i and r_{i+1} is variable and depend on the angle θ_{i+1}. Similarly to the analysis of the *Soft Gripper* presented in (Birglen et al., 2008), the conditions for the forces to become zero are implicit functions that cannot be easily solved.

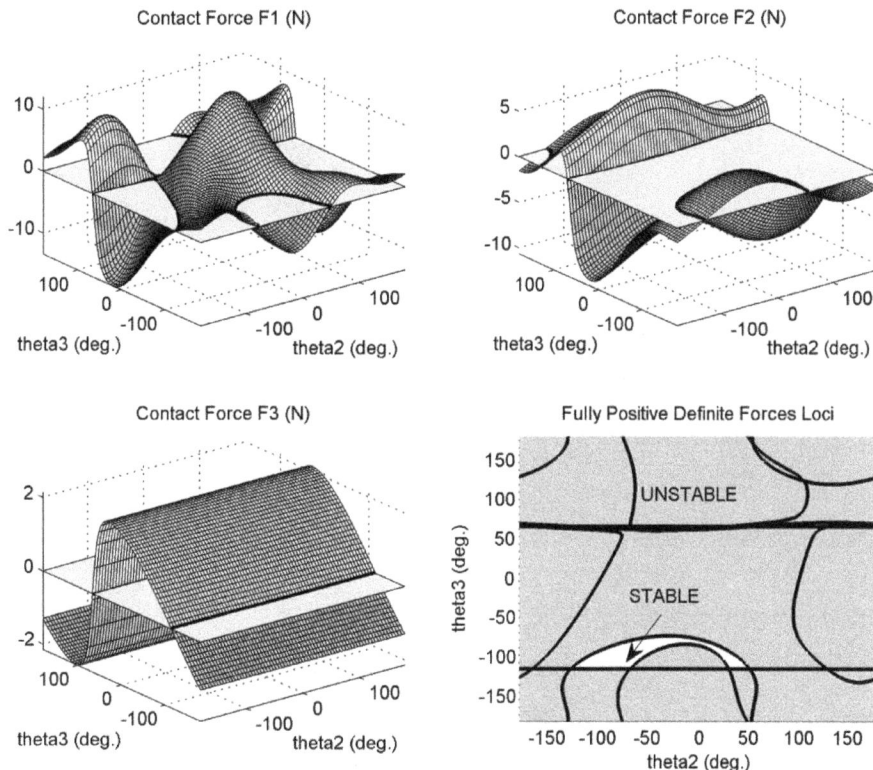

Figure 5. Contact forces and stability loci of Da Vinci's mechanism.

A section of the contact force workspace of the three-phalanx finger based on Da Vinci's drawing is illustrated in Fig. 5 for contact locations at mid-phalanx. The geometric parameters used in this example are directly measured from a copy of the *Codex Atlanticus* drawings. Note that conversely to the analysis of the same architecture found in (Birglen et al., 2008), the parameters used in this paper are these that are illustrated in the lower part of Fig. 3, i.e. with a zero-radius pulley which models a sliding joint. It seemed to be the preferred design of Da Vinci since he used it in several other sketches of his flying machines.

As can be seen in Fig. 5, the fully positive workspace – i.e. the part of it corresponding to the case where all the contact forces are positive or null – of this architecture is rather small. This is especially true when comparing to Prof. Hirose's Soft Gripper. However, as discussed in Birglen and Gosselin (2006b), a fully positive workspace for three-phalanx finger might be actually impossible and more importantly, not really desirable. The analysis of the grasp stability of the finger which aims at predicting if the mobility of a self-adaptive finger will converge to a stable equilibrium or towards ejection of the object from the finger is more meaningful. Yet, this analysis is challenging especially with three-phalanx fingers (Birglen and Gosselin, 2006a). In this paper, focus is placed on the analysis of a simpler two-phalanx version of the finger, using the grasp-state plane. The geometric parameters associated with this design are listed in Table 1.

Table 1. Da Vinci's two-phalanx mechanism geometric parameters.

l_1	l_2	r_0	a_1	b_1	r_1	a_2
1	2.06	1.03	2.08	0	0	.45

The grasp-state plane shows if, for an arbitrary initial contact situation, the finger will be stable or not. And, if not, if the subsequent motion undergone by the finger will converge towards a stable equilibrium or ejection. The initial and final grasp-state associated with the finger under scrutiny are illustrated in Figs. 6 and 7. In the initial grasp-state plane, the areas in gray indicate that at least one contact force is negative which means that a full enveloping grasp with both phalanges is impossible. Then, a sliding motion of the finger along the surface of the object will begin and either reach a situation where the finger is in static equilibrium (resulting in a stable pinch grasp) or lose the object (ejection). In Fig. 7, the areas in gray correspond to ejection while the white zones indicate an eventual static equilibrium. As can be seen, the finger is mostly stable for negative angles of the distal phalanx which is not usually the part of the workspace where the finger is used.

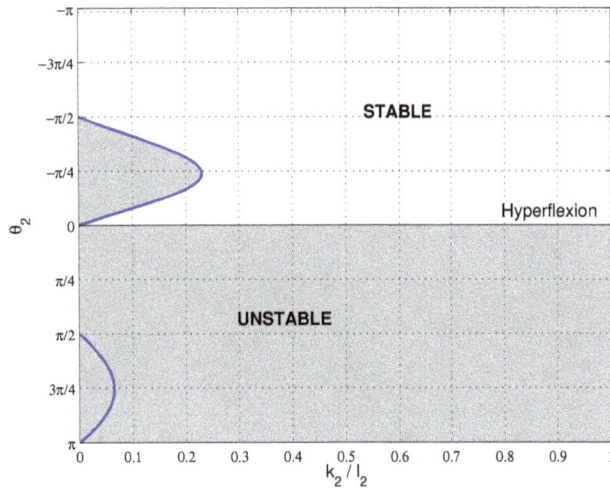

Figure 6. Initial grasp-state plane of Da Vinci's finger.

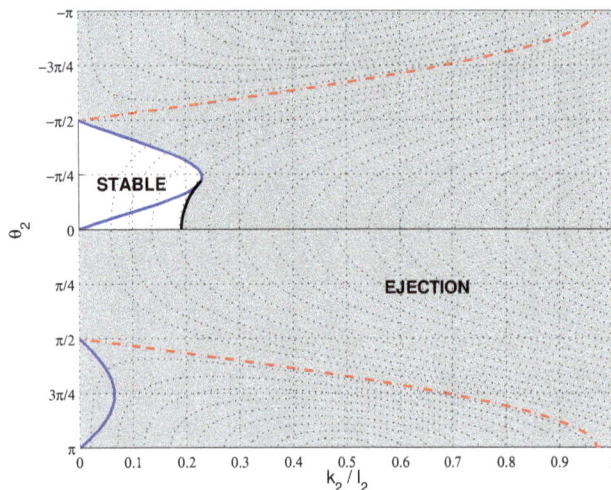

Figure 7. Final grasp-state plane of Da Vinci's finger.

2.3   Flying

From the results of the previous section, we can say that the architecture of Da Vinci makes a poor underactuated finger. Using the same type of transmission, the Soft-Gripper architecture is a far better choice and generally speaking, an excellent gripper. Of course, to be fair, Da Vinci intended his design to be part of a flying machine not a grasping finger so we cannot belittle his invention. For his particular application, Da Vinci also used the shape-adaptation property of the linkage but from an aerodynamic perspective not grasp stability. For a flapping wing, the finger is "in contact" with the air or another fluidic environment. The general motion of a self-adaptive flapping wing constituted of two phalanges is illustrated in Fig. 8. The revolute joint between the two phalanges is equipped with a torsional spring and a mechanical limit. Note that the transmission mechanism achieving this

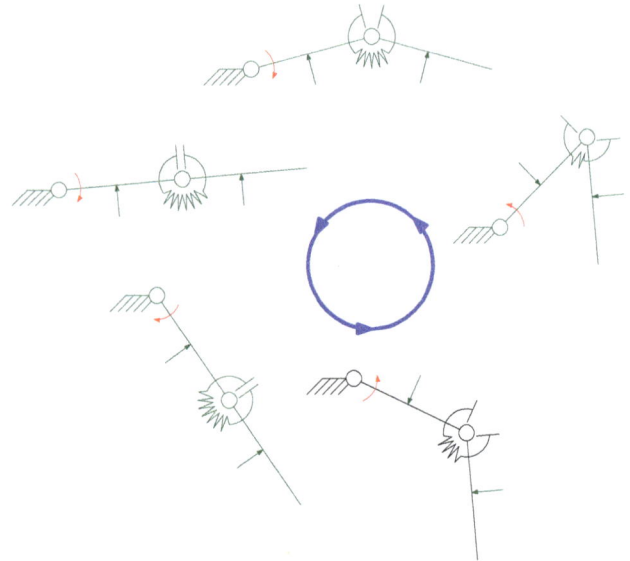

Figure 8.   General motion of a self-adaptive flapping wing, the green curved arrows indicate the resistance of the fluid while the red straight arrow symbolize actuation.

motion is not shown for it can take many form, Da Vinci's being one but any other self-adaptive linkage proposed in the literature could be used. If one looks closely to the sketches made by Da Vinci, it is obvious that the human operator of the machine was to provide the actuation force required to the down stroke of the motion while compliant hinges based on cantilever beams were designed to provide the return actuation (cf. Fig. 3). Thus, the idea of using compliant joints in underactuated fingers must be attributed to Da Vinci. This is particularly humbling to note while this idea is recently actively studied by several authors including the one of this paper (Dollar and Howe, 2006; Boudreault and Gosselin, 2006; Doria and Birglen, 2009).

If considering the values of the contact forces was of interest for a grasping finger, in the case of a flapping wing, the torque distributed by the actuator at the base joint of each phalanx might be a more practical quantity. These torques are obviously linked to the contact forces and they are expressed by another form of the Transmission matrix presented in Birglen (2009). However, to establish the upward force developed by this flapping wing, one needs a model for the air/fluid resistance as a function of the shape of the wing and its velocity. This is also necessary to ensure that the motion of the wing will be optimal. The aerodynamic of a flapping wing is a very complex problem and there are only a few non-numerical models available, e.g. in Madangopal et al. (2004). The characterization of the performance of a flapping wing using a self-adaptive linkage is yet an open issue.

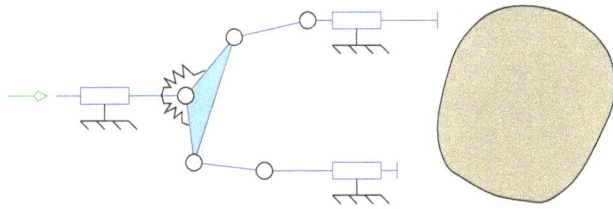

Figure 9. A self-adaptive gripper based on architecture *C234*.

3 Other mechanisms and other applications

The mechanisms synthesized for underactuated grasping do not necessarily have to be used as architectures of robotic fingers or flapping wings using a design based on *phalanges*. Even for grasping, one can use these mechanisms to synthesize new grippers, as illustrated in Fig. 9. The gripper presented in this figure corresponds to the architecture *C234* (Birglen, 2009) and has two-DOF driven by a single actuator. A spring is used as a passive element to constrain the mechanism (it can be used to keep both jaws synchronized during the pregrasping phase), while the leftmost prismatic joint is actuated. A similar mechanism has been proposed to drive two self-adaptive fingers (Birglen and Gosselin, 2006c). Indeed, self-adaptive systems can be connected together to obtain new architectures and this holds for any mechanism properly designed, i.e. not only with differential mechanisms. In the example depicted in Fig. 9, the driven system consists of the two outputs struts with the ground link between them.

Additionally, other designs can be built incrementally from the architectures previously synthesized. Indeed, the known architectures of underactuated fingers are two- or three-DOF mechanisms that can be connected in series or using an arbitrary scheme. Providing that the hypotheses described in Birglen (2009) are satisfied, the resulting architecture will be valid. One can use for instance the motion of the "distal" phalanx of an initial design to drive subsequent phalanges of an extended version of this finger.

Furthermore, the known architectures are not limited to the design of self-adaptive robotic fingers or even grippers but can be extended to a wide range of applications. In Hirose (1985), Prof. Hirose (again) pioneered the design of self-adaptive mechanisms using connected differential mechanisms and proposed to adapt the design he used in the *Soft Gripper* to a wrist-bracing mechanism for a mobile robot. This robot was to be used to navigate inside a pipe and upon attaining its destination, deploy rigid limbs driven by the self-adaptive mechanism, which allowed them to adapt to the internal shape of the pipe, hence providing stable support to engage in maintenance tasks. With all these systems, the only restriction is that the system reacts to an external constrain of its DOF (i.e. a contact).

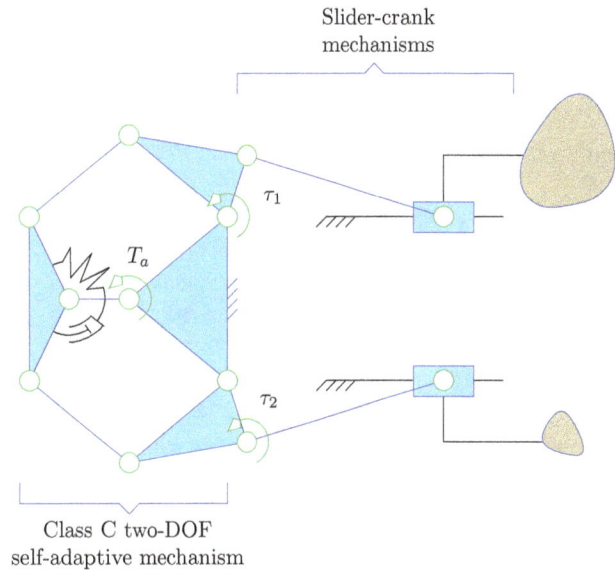

Figure 10. A two-DOF self-adaptive linkage driving two slider-crank linkages.

This restriction itself can be lifted with other types of self-adaptive mechanisms, closely related to the principle of continuously variable transmission. The latter is defined as a technological solution to compensate a variation in the load driven by an actuator axis. Namely, the ability of a mechanical system to continuously change the reduction ratio of its actuator in response to a load variation. These systems are in fact self-adaptive mechanisms too although they have generally only one output, namely the actuator axis. Several mechanisms have been proposed to achieve this capability, mostly based on gears (Hirose et al., 2004; Ishikawa et al., 2000; Fujushima et al., 2001; Takaki and Omata, 2004) (notice that one of these references is from none other than Prof. Hirose again). This principle has yet to be extended to multi-DOF systems but, provided that the actuation distribution condition is satisfied, nothing prevents it. An example of such a mechanism is presented in Fig. 10 where a two-DOF self-adaptive mechanism (the same that was illustrated in Fig. 9) is used to drive two slider-crank linkages. The mechanism is completely symmetrical in order to allow complete revolutions of the joints corresponding to τ_1 and τ_2. The actuation torque is noted T_a. The passive element used here is a spring in parallel with a damper (the choice is arbitrary) which reduces the mechanism into a simple Watt's linkage if $\tau_1 = \tau_2$ (neglecting the dynamics of the linkage itself). If the loads of both slider-crank linkages are different as symbolically illustrated in Fig. 10, the passive element will accommodate this difference by providing another internal mobility. The transmission linkage will still be distributing the actuation torque to both outputs. This accommodation could then be designed to increase the output torque corresponding to the largest load in order to manage the increased load. This is the basic principle behind continuously variable transmission as found in the literature. Of course, in this example, it is much simpler to design the system using gears than linkages.

4 Conclusions

As discussed in this paper, the topic of underactuated (or self-adaptive) mechanisms recently attracted a lot of interest from the research community. During the Renaissance, Da Vinci designed a flapping wing that has been centuries later used in an underactuated finger. This example was used to illustrate that self-adaptive mechanisms do not have a single application and are therefore currently underexploited. The design of ingenious mechanisms is nowadays mainly a lost art due to the ever progressing of electronics. Yet, this skill can be of tremendous interest in modern robotic and mechatronic devices. The author strongly believes that the research community has to put back more focus on the *mecha-* in mechatronics if the best performance are to be obtained or if practical success is required. As a well-known senior researcher once said: *kinematics' not dead* (Merlet, 2000)!

Acknowledgements. The financial support of the Natural Sciences and Engineering Research Council of Canada (NSERC) and the Fonds Québécois de la Recherche sur la Nature et les Technologies (FQRNT) is gratefully acknowledged.

Edited by: J. L. Herder

References

Birglen, L.: Type Synthesis of Linkage-Driven Self-Adaptive Fingers, J. Mechanisms Robotics 1, 021010, doi:10.1115/1.3046139, 2009.

Birglen, L. and Gosselin, C.: Geometric Design of Three-Phalanx Underactuated Fingers, ASME Journal of Mechanical Design, 128, 356–364, 2006a.

Birglen, L. and Gosselin, C.: Grasp-State Plane Analysis of Two-Phalanx Underactuated Fingers, Mech. Mach. Theory, 41, 807–822, 2006b.

Birglen, L. and Gosselin, C.: Force Analysis of Connected Differential Mechanisms: Application to Grasping, Int. J. Robot. Res., 25, 1033–1046, 2006c.

Birglen, L., Laliberté, T., and Gosselin, C.: Underactuated Robotic Hands, Springer, New-York, 2008.

Boudreault, E. and Gosselin, C.: Design of sub-centimetre underactuated compliant grippers, in: 2006 ASME International Design Engineering Technical Conferences, Philadephia, PA, USA, 2006.

Carrozza, M. C., Suppo, C., Sebastiani, F., Massa, B., Vecchi, F., Lazzarini, R., Cutkosky, M. R., and Dario, P.: The SPRING Hand: Development of a Self-Adaptive Prosthesis for Restoring Natural Grasping, Auton. Robot., 16, 125–141, 2004.

Crisman, J. D., Kanojia, C., and Zeid, I.: Graspar: A Flexible, Easily Controllable Robotic Hand, IEEE Robot. Autom. Mag., 3(2), 32–38, 1996.

Dollar, A. M. and Howe, R. D.: A Robust Compliant Grasper via Shape Deposition Manufacturing, IEEE-ASME T. Mech., 11, 154–161, 2006.

Doria, M. and Birglen, L.: Design of an Underactuated Compliant Gripper for Surgery Using Nitinol, Journal of Medical Devices, 3, 011007, doi:10.1115/1.3089249, 2009.

Fujushima, E. F., Nakamoto, H., Damoto, R., and Hirose, S.: Optimal load-sensitive control for mobile robots equipped with continuously variable transmissions, in: Proceedings of the 2001 IEEE International Conference on Robotics and Automation, Vol. 1, Maui, HI, USA, 476–481, 2001.

Gosselin, C.: Mechanically Intelligent Systems: Smart Designs for High-Performance Robotics, in: Proceedings of the 2005 Canadian Congress of Applied Mechanics, Montreal, Canada, 74–84, 2005.

Gosselin, C.: Adaptive Robotic Mechanical Systems: A Design Paradigm, ASME J. Mech. Design, 128, 192–198, 2006.

Gosselin, C. and Laliberté, T.: Underactuated mechanical finger with return actuation, US Patent No. 5 762 390, 1996.

Higashimori, M., Kaneko, M., Namiki, A., and Ishikawa, M.: Design of the 100G Capturing Robot Based on Dynamic Preshaping, Int. J. Robot. Res., 24, 743–753, 2005.

Hirose, S.: Connected Differential Mechanism and its Applications, in: Proceedings of 1985 International Conference on Advanced Robotics, Tokyo, Japan, 319–325, 1985.

Hirose, S. and Umetani, Y.: The Development of Soft Gripper for the Versatile Robot Hand, Mech. Mach. Theory, 13, 351–358, 1978.

Hirose, S., Tibbetts, C., and Hagiwara, T.: Development of X-screw: a load-sensitive actuator incorporating a variable transmission, in: Proceedings of the 1999 IEEE International Conference on Robotics and Automation, Vol. 1, Detroit, MI, USA, 193–199, 2004.

IFToMM: Terminology for the Theory of Machines and Mechanisms, Mech. Mach. Theory, 26, 435–539, 1991.

Ishikawa, Y., Yu, W., Yokoi, H., and Kakazu, Y.: Development of Robot Hands with an Adjustable Power Transmitting Mechanism, Intelligent Engineering Systems Through Neural Networks, 10, 631–636, 2000.

Kyberd, P. J., Light, C., Chappel, P. H., Nightingale, J. M., Whatley, D., and Evans, M.: The design of an anthropomorphic prosthetic hands: A study of the Southampton hand, Robotica, 19, 593–600, 2001.

Laliberté, T. and Gosselin, C.: Simulation and Design of Underactuated Mechanical Hands, Mech. Mach. Theory, 33, 39–57, 1998.

Laliberté, T., Birglen, L., and Gosselin, C.: Underactuation in Robotic Grasping Hands, Japanese Journal of Machine Intelligence and Robotic Control, Special Issue on Underactuated Robots, 4, 77–87, 2002.

Madangopal, R., Khan, Z. A., and Agrawal, S. K.: Biologically inspired design of small flapping wing air vehicles using Four bar mechanisms and Quasi steady aerodynamics, J. Mech. Design, 127, 809–816, 2004.

Merlet, J.: Kinematics' not dead!, in: IEEE International Conference on Robotics and Automation, 2000, Proceedings ICRA'00, Vol. 1, 1–6, 2000.

Rovetta, A., Franchetti, I., and Vicentini, P.: Multi-Purpose Mechanical Hand, US Patent No. 4 351 553, 1982.

Rubinger, B., Fulford, P., Gregoris, L., Gosselin, C., and Laliberté, T.: Self-Adapting Robotic Auxiliary Hand (SARAH) for SPDM Operations on the International Space Station, in: Proceedings of the 6th International Symposium on Artificial Intelligence and Robotics and Automation in Space: i-SAIRAS 2001, Saint-Hubert, QC, Canada, 2001.

Takaki, T. and Omata, T.: Load-Sensitive Continuously Variable Transmission for Robots Hands, in: Proceedings of the 2004 IEEE International Conference on Robotics and Automation, New Orleans, LA, USA, 3391–3396, 2004.

Townsend, W.: The BarrettHand grasper – programmably flexible part handling and assembly, Ind. Robot, 27, 181–188, 2000.

Ulrich, N. T.: Grasping with Mechanical Intelligence, Ph.D. thesis, School of Engineering and Applied Sciences, University of Pennsylvania, Philadelphia, Pennsylvania, 1988.

5

Fully-compliant statically-balanced mechanisms without prestressing assembly: concepts and case studies

G. Chen and S. Zhang

School of Mechatronics, Xidian University, Xi'an, Shaanxi 710071, China

Abstract. The purpose of this paper is to present new concepts for designing fully-compliant statically-balanced mechanisms without prestressing assembly. A statically-balanced compliant mechanism can ideally provide zero stiffness and energy free motion like a traditional rigid-body mechanism. These characteristics are important in design of compliant mechanisms where low actuation force, accurate force transmission or high-fidelity force feedback are primary concerns. Typically, static balancing of compliant mechanisms has been achieved by means of prestressing assembly. However, this can often lead to creep and stress relaxation arising in the flexible members. In this paper two concepts are presented which eliminate the need for prestressing assembly of compliant mechanisms: (1) a weight compensator which employs a constant-force compliant mechanism, (2) a near-zero-stiffness mechanism which combines two multistable mechanisms. In addition to the advantages provided by statically-balanced compliant mechanisms, two other notable features of these statically-balanced mechanisms are their ability to be monolithically fabricated and to return to their as-fabricated position without any disassembly when not in use.

1 Introduction

Compliant mechanisms, which utilize the deflection of flexible segments rather than from articulated joints to achieve their mobility, offer many advantages over their rigid-body counterparts such as decreased part count, increased precision and reduced wear (Howell, 2001). Because a compliant mechanism obtains its motion from deflection of its members, spring-back forces play a significant role in its input and output capabilities. Hence the study of the force-deflection characteristics (Jutte and Kota, 2008) of a compliant mechanism is among the most important methods of understanding its behavior.

Compliant mechanisms can be roughly divided into four categories depending on their force-deflection characteristics: spring-like compliant mechanisms (Trease et al., 2005), multistable compliant mechanisms (Oh and Kota, 2009; Gerson et al., 2008; Chen et al., 2011b, 2010), constant-force compliant mechanisms (Nahar and Sugar, 2003; Lan et al., 2010), and statically-balanced compliant mechanisms (Herder and van den Berg, 2000; Gallego and Herder, 2010). Statically-balanced compliant mechanisms may be consid-

ered a constant-force compliant mechanism with zero force input. The design approaches of statically balanced mechanisms may be used to design constant-force mechanisms (Gallego and Herder, 2010).

The concept of a statically balanced compliant mechanism was first introduced and studied by Herder and van den Berg (2000). A statically balanced compliant mechanism, which utilizes energy-release elements to balance the energy stored in the flexible segments of the mechanism, maintains neutral equilibrium throughout its range of motion (Hoetmer et al., 2009; Gallego and Herder, 2010). Like their traditional rigid body counterparts, these mechanisms provide energy free motion with zero accompanying stiffness – a fact particularly useful in cases where accurate force transmission and high-fidelity force feedback are of primary concern in compliant mechanisms (Hoetmer et al., 2009).

Most of the statically balanced compliant mechanisms rely on prestressing to achieve static balancing (Trease and Dede, 2011; Powell and Frecker, 2005; Hoetmer et al., 2009). However implementation difficulty, creep and stress relaxation of flexible members are the challenges associated with prestressing. Accurate prestressing is difficult to achieve in practice, particularly in micromechanical devices due to the small available operating space (Tolou et al., 2010). Moreover, creep and stress relaxation arising from prestressing may dramatically deteriorate the performance of the mechanism

(Howell, 2001). In contrast, less work has been done on statically-balanced mechanisms without prestressing assembly, although they receive increasing attention. For example, Stapel and Herder (2004) presented a feasibility study of a fully compliant statically balanced grasper in 2004, and Tolou et al. (2010) presented two concepts of fully-compliant statically-balanced compliant micro mechanisms and validated them through simulation in 2010.

In this paper two novel concepts of fully-compliant statically balanced compliant mechanisms are proposed which eliminate the need for prestressing assembly altogether. The first concept is a weight compensator using a constant-force compliant mechanism, and the second is a near-zero-stiffness mechanism based on combination of two multistable mechanisms. Two design cases are presented to demonstrate the feasibility of the concepts. In addition to the advantages such as energy efficiency, accurate force transmission and actuator effort reduction provided by statically balanced compliant mechanisms (Radaelli et al., 2010), they exhibit two other key features: they are able to be monolithically fabricated, and they return to their as-fabricated position without any disassembly when not in operation.

The rest of this paper is organized as follows: Section 2 presents a gravity compensator based on a fully compliant constant-force mechanism. Section 3 studies how to achieve static balancing by combining two multistable mechanisms. Section 4 provides concluding remarks.

2 Gravity compensation using constant-force mechanisms

In this section, we take humanoid robot as an example to illustrate gravity compensation.

Because the weight of a humanoid robot can seriously degrade its dynamic performance and result in energy-inefficient operation (Wongratanaphisan and Cole, 2008), gravity compensation is often employed to eliminate or minimize the effects of gravity. The resulting gravity compensated robot is a typical example of a statically balanced mechanism. Because the effect of gravity on the body is constant in the vertical direction, we propose a concept of using a constant-force compliant mechanism to support and compensate for the weight, as illustrated in Fig. 1. This is possible because a constant-force compliant mechanism is a mechanism that produces a constant output force over a range of motion (Nahar and Sugar, 2003).

2.1 Design of constant-force mechanism for gravity compensation

It is observed that the fully compliant end-effector shown in Fig. 1a (Chen et al., 2009b) exhibits a constant-force behavior. Based on the pseudo-rigid-body model (PRBM) shown

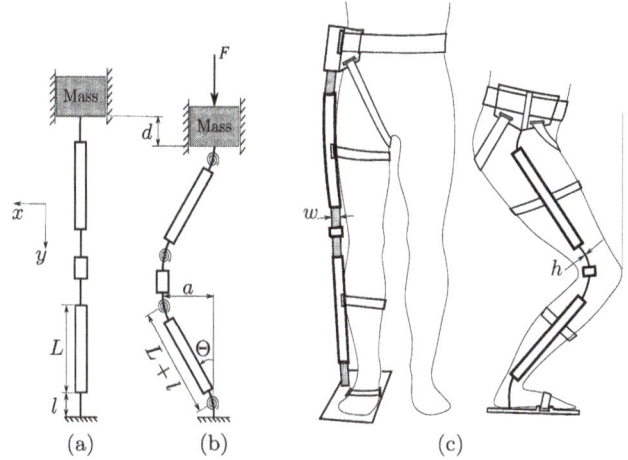

Figure 1. (a) Schematic diagram of a constant-force mechanism and a mass to be balanced, (b) the corresponding pseudo-rigid-body model (PRBM), and (c) an implementation example for humanoid robot.

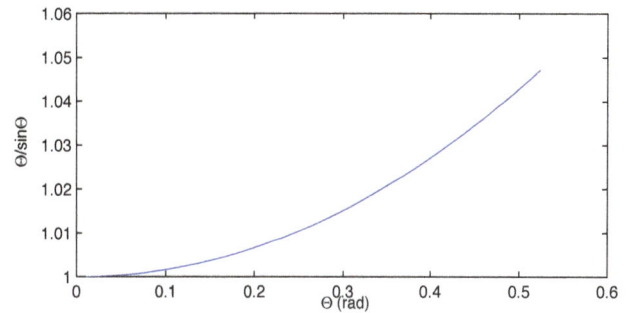

Figure 2. $\Theta/\sin\Theta$.

in Fig. 1b, its force-deflection relationship is given as (Chen et al., 2009b)

$$F = \frac{2K\Theta}{(L+l)\sin\Theta} = \frac{2K}{L+l} \cdot \frac{\Theta}{\sin\Theta} = \frac{4K\arccos\left[1 - \dfrac{d}{2(L+l)}\right]}{\sqrt{4(L+l)d - d^2}} \quad (1)$$

where, $K = EI/l$, Θ is the pseudo-rigid-body angle, $L+l$ the length of the pseudo-rigid-body link, E the Young's modulus of the material, and I the moment of inertia of the small-length flexural pivots. Note that the term $[2K/(L+l)]$ in Eq. (1) is constant for a specific design, and the change of $(\Theta/\sin\Theta)$ is very small over the range of $0 < \Theta \leq 30°$ (less than 5 %), as illustrated in Fig. 2. This indicates that the mechanism acts as a near constant-force, gravity balancing mechanism, and Eq. (1) can be simplified as

$$F \approx \frac{2K}{L+l} \quad (2)$$

In this paper a gravity compensator is referred to as a passive mechanical device that counteracts the gravity of a humanoid robot body to improve its dynamic performance

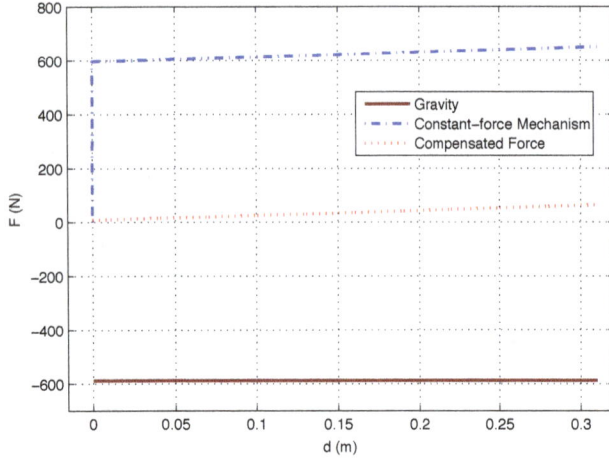

Figure 3. The force-displacement characteristics of the gravity compensated robot. Note that the force-displacement curve for the constant-force mechanism is plotted using Eq. (1) instead of its simplified version, Eq. (2).

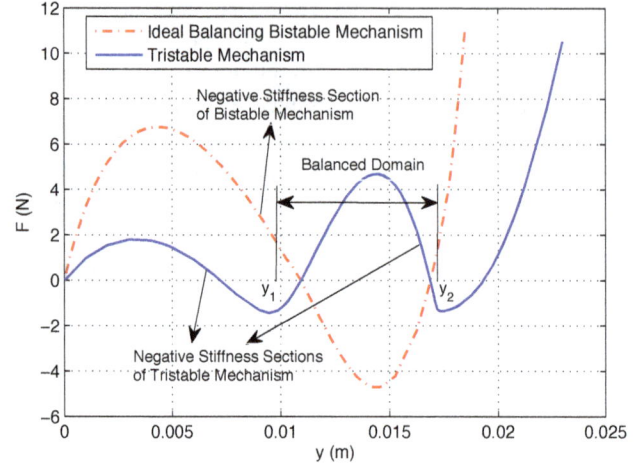

Figure 4. Force-displacement characteristics of an ideal balancing bistable mechanism.

and energy efficiency. The gravity compensator based on the constant-force mechanism is designed to satisfy the following requirement:

$$G = N_p F = \frac{2 N_p K}{L + l} \qquad (3)$$

where N_p is the number of flexible segment sets in parallel.

2.2 Case study

This subsection presents an example in order to demonstrate the design procedure of a gravity compensator for a humanoid robot using a constant-force compliant mechanism, as shown in Fig. 1c. We first assume the following: $G = 60 \times 9.8$ N, $N_p = 2$, $L = 0.6$ m, $l = 0.02$ m and $E = 2.07 \times 10^{11}$ Pa (stainless steel). The torsional stiffness of the small-length flexure pivots, K, can be calculated from Eq. (3) ($K = 91.14$ Nm in this example). Then the moment of inertia for the flexural pivots is determined using $K = EI/l$. As a result, $w = 0.02$ m and $h = 0.00175$ m can satisfy the design (w and t are the width and thickness of the cross-section of the small-length pivots respectively). From Fig. 3 we can see that the gravity is well compensated for by the constant force mechanism, with the maximum error less than 10 % for d from 0 to 0.3 m.

3 Static balancing using multistable compliant mechanisms

3.1 Multistable compliant mechanism as negative stiffness building block

The deflections of the flexible members often produce positive stiffness in a compliant mechanism. Positive stiffness

means that the resulting deflection is in the same direction as the external applied force, like the spring-back effect of a linear spring. If a component exhibiting negative stiffness is introduced to the mechanism, the appropriate counterbalancing of positive and negative stiffness in the mechanism can achieve a state of zero-stiffness, i.e., static balancing (Hoetmer et al., 2009).

A bistable mechanism possesses two distinct stable equilibrium positions within its range of motion. At each of these positions the mechanism can maintain stability without power input and, if exposed to a small disturbance while in a stable position, tends to return to the same stable position. It can be seen from Fig. 4 that, for a bistable mechanism, one portion of the force-deflection curve exhibits negative stiffness. Thus Tolou et al. (2010) employed bistable mechanisms as the negative stiffness building blocks in their design of statically-balanced compliant micro mechanisms.

In general, a multistable mechanism with n stable equilibrium positions offers $(n-1)$ negative stiffness sections and n positive stiffness sections on its force-deflection curve (e.g., as shown in Fig. 4, a tristable mechanism possessed two negative stiffness sections and three positive stiffness sections), and as such is a potential building block for synthesis of statically balanced mechanisms. Therefore, we propose a concept for designing fully-compliant statically-balanced mechanisms which combines two fully compliant multistable mechanisms.

There has been a large amount of work done on multistable mechanisms, including tristable mechanisms (Chen et al., 2009a,b), quadristable mechanisms (Han et al., 2007), and synthesis approaches for multistable mechanisms (Oh and Kota, 2009; Gerson et al., 2008; Chen et al., 2011b). A brief summary of the work on compliant multistable mechanisms can also be found in Chen et al. (2010). Thus this concept may easily be expanded to include these as building blocks

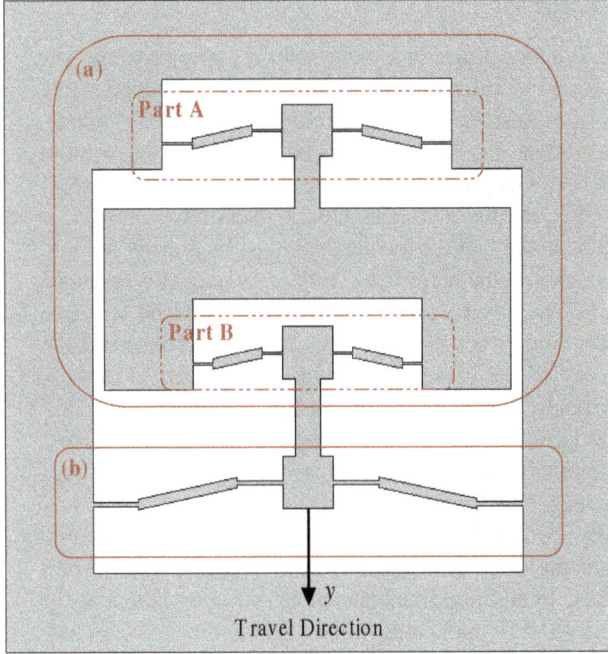

Figure 5. Schematic diagram of a fully-compliant statically-balanced mechanism consisting of two multistable building blocks, i.e., a tristable mechanism and a bistable mechanism. (a) The tristable mechanism consisting of two different bistable mechanisms (i.e., Part A and Part B) connected in series, and (b) the balancing bistable mechanism.

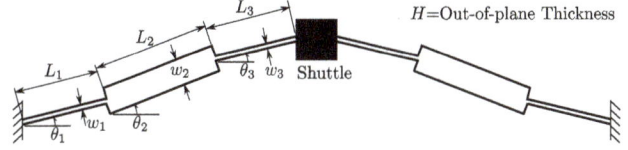

Figure 6. Parameters of bistable mechanism. It should be noted that three bistable mechanisms of this kind are employed in the statically-balanced mechanism shown in Fig. 5, i.e., Part A, Part B and the balancing bistable mechanism.

Table 1. Design values of the tristable mechanism (consisting of two bistable mechanisms connected in series, i.e., Part A and Part B) and the balancing bistable mechanism (see Fig. 6 for the parameters).

Parameter	Part A of Tristable Mechanism	Part B of Tristable Mechanism	Balancing Bistable Mechanism
E	1.4×10^9	1.4×10^9	1.4×10^9
H	6 mm	6 mm	6 mm
L_1	12 mm	8 mm	18 mm
θ_1	0°	0°	0°
w_1	0.6 mm	0.8 mm	1.68 mm
L_2	24 mm	20 mm	40 mm
θ_2	12°	12°	11.1°
w_2	4 mm	4 mm	4 mm
L_3	12 mm	8 mm	18 mm
θ_3	0°	0°	0°
w_3	0.6 mm	0.8 mm	1.68 mm

in the design of more complex statically-balanced mechanisms. In the following subsection a case study is presented to demonstrate the principle of using multistable mechanisms to achieve static balance.

3.2 Case study

In this subsection we present a fully-compliant statically-balanced mechanism combining a tristable mechanism with a bistable mechanism.

To begin our design example, we first chose a specific tristable mechanism and then found a balancing bistable mechanism through an optimization algorithm. The tristable mechanism employed in this example consists of two bistable mechanisms (i.e., Part A and Part B) of different load thresholds connected in series (Oh and Kota, 2009), as shown in Fig. 5a. The design parameters of each bistable mechanism are shown in Fig. 6 and listed in Table 1. Figure 4 gives the force-deflection characteristics of the tristable mechanism (shown as a blue solid line), which is achieved using nonlinear finite element analysis (the detailed finite element modeling approach for this type of bistable mechanism can be found in Cherry et al. (2008)).

We suppose there exists an ideal balancing bistable mechanism which can counterbalance the given tristable mechanism to a maximum balanced domain. Figure 4 plots the force-deflection characteristics of such an ideal bistable mechanism (the red dash-dot curve) and the corresponding balanced domain (from y_1 to y_2). As illustrated in Fig. 4, in this ideal balanced domain, the positive stiffness section of the tristable mechanism is compensated by the corresponding part of the negative stiffness section of the bistable mechanism, while its negative stiffness section is compensated by the corresponding positive stiffness section of the bistable mechanism. As the shuttle of the balancing bistable mechanism moves along the y-axis until Part A of the tristable mechanism switches to its second stable equilibrium position, Part B of the tristable mechanism and the balancing bistable mechanism balance each others stiffness, which results in static balancing of the whole mechanism.

The design of a balancing bistable mechanism hinges upon matching its force-deflection properties with those of the ideal balancing bistable mechanism in the balanced domain, which may be formulated as an optimization problem as follows:

$$\text{Min IAR} = \int_{y_1}^{y_2} |F_r(y) - F_i(y)| dy \quad (4)$$

where IAR is the integral of the absolute compensation residual, $F_r(y)$ and $F_i(y)$ are the force-deflection characteristics

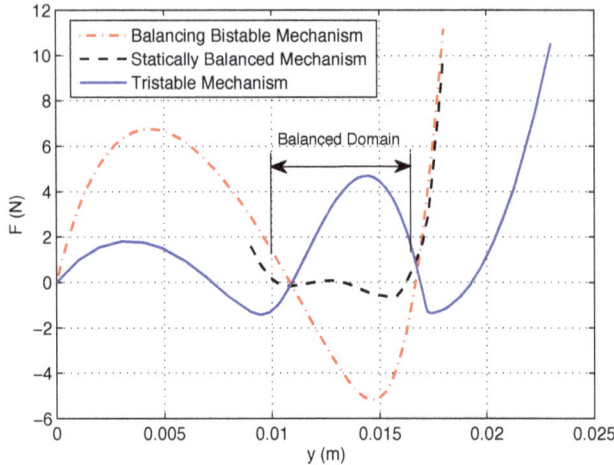

Figure 7. Force-displacement characteristics of the optimized balancing bistable mechanism.

of the candidate and the ideal bistable mechanisms, respectively.

A primary optimization search was conducted using a particle swarm optimizer (PSO) (Chen et al., 2011a) integrated with ANSYS (ANSYS was used to solve the force-deflection characteristics of each candidate bistable mechanism). For simplicity, we assume $L_1 = L_3$, $w_1 = w_3$, $w_2 = 4\,\text{mm}$, $\theta_1 = \theta_3 = 0$ and $H = 6\,\text{mm}$ for the balancing bistable mechanism, thus reducing the number of parameters to be optimized to 4 (i.e., L_1, L_2, w_1 and θ_2). In the implementation of PSO, the swarm size is set to 40 and the maximum number of iterations to 200. The optimization process is briefly described as follows:

– Step 1: Randomly initialize 40 particles in the search space, with each particle corresponding to a candidate bistable mechanism.

– Step 2: Evaluate the fitness of each bistable mechanism using Eq. (4).

– Step 3: Update the swarm and generate the next generation particles according to the rules of PSO.

– Step 4: If the current iteration number is less than 200, go to Step 2. Otherwise, the optimization is stopped and the bistable mechanism with the best fitness value is selected as the balancing mechanism for the tristable mechanism.

The optimized parameters of the balancing bistable mechanism are given in Table 1. The resulting force-deflection curves of the tristable and balancing bistable mechanisms (as well as the unbalanced residual force) are plotted in Fig. 7 around the balanced domain.

3.3 Discussion

Figure 7 demonstrates that the system is approximately balanced for a specific range of motion around the second stable equilibrium position of the tristable mechanism. Of course, the results can be further improved by conducting a more comprehensive optimization on the balancing bistable mechanism (e.g., taking other design parameters into account). It should be noted that the mechanism must be prestressed into the balanced domain for it to work as a statically balanced device. Statically-balanced mechanisms might be sensitive to fabrication errors (Tolou et al., 2011). We observed the sensitivity of the balancing bistable mechanism to fabrication errors by assuming an error of 5 % is caused during fabrication in both w_1 and w_3. Results from our finite element calculations indicate that this causes force-deflection errors of less than 7 % deviated from the ideal. This principle may be easily extended to the static balancing of two multistable mechanisms, each with more than three stable equilibrium positions. In addition, the concept has especial potential for usage in MEMS applications due to its ability to be monolithically fabricated.

4 Conclusions

In this paper we have presented two novel concepts which may be used to eliminate the need for prestressing assembly through the use of fully-compliant statically-balanced compliant mechanisms. The first concept utilizes constant-force mechanisms in order to achieve gravity compensation; the second is based upon the combination of two multistable mechanisms in order to achieve static balancing over a certain range of motion (having especial potential for usage in MEMS applications). Each of these concepts has been demonstrated with a case study in which the principles and equations used to formulate them have been shown to perform with good accuracy. Two additional key benefits of these mechanisms is that they can be monolithically fabricated, and they return to their as-fabricated positions without disassembly when they are not in use.

Acknowledgement. The authors gratefully acknowledge the financial support from the National Natural Science Foundation of China under Grant No. 50805110, the Scientific Research Foundation for the Returned Overseas Chinese Scholars under Grant No. JY0600100401, and the Fundamental Research Funds for the Central Universities under No. JY10000904010.

Edited by: J. A. Gallego Sánchez

References

Chen, G., Wilcox, D. L., and Howell, L. L.: Fully compliant double tensural tristable micromechanisms (DTTM), J. Micromech. Microeng., 19, 025011, 1–8, 2009a.

Chen, G., Aten, Q. T., Zirbel, S., Jensen, B. D., and Howell, L. L.: A tristable mechanism configuration employing orthogonal compliant mechanisms, Trans. ASME, J. Mechan. Robot., 2, 014501, 1–6, 2009b.

Chen, G., Gou, Y., and Yang, L.: Research on Multistable Compliant Mechanisms: The State of the Art, Proceedings of the 9th International Conference on Frontiers of Design and Manufacturing (ICFDM 2010), 17–19 July 2010, Changsha, China, 1–5, 2010.

Chen, G., Xiong, B., and Huang, X.: Finding the optimal characteristic parameters for 3R pseudo-rigid-body model using an improved particle swarm optimizer, Precis. Eng., 35(3), 505–511, 2011a.

Chen, G., Gou, Y., and Zhang, A.: Synthesis of compliant multistable mechanisms through use of a single bistable mechanism, Trans. ASME J. Mechan. Design, 133(8), 081007, doi:10.1115/1.4004543, 2011b.

Cherry, B. B., Howell, L. L., and Jensen, B. D.: Evaluating three-dimensional effects on the behavior of compliant bistable micromechanisms, J. Micromech. Microeng., 18, 095011, 1–10, 2008.

Gallego, J. A. and Herder, J. L.: Criteria for the static balancing of compliant mechanisms, Proceedings of the ASME Design Engineering Technical Conferences & Computers and Information in Engineering Conference, Montreal, Canada, 15–18 August 2010, DETC2010-28469, 2010.

Gerson, Y., Krylov, S., Ilic, B., and Schreiber, D.: Large displacement low voltage multistable micro actuator, IEEE 21st International Conference on Micro Electro Mechanical Systems, 463–466, 2008.

Han, J. S., Muller, C., Wallrabe, U., and Korvink, J. G.: Design, simulation, and fabrication of a quadstable monolithic mechanism with X- and Y-directional bistable curved beams, Trans. ASME J. Mechan. Design, 129, 1198–1203, 2007.

Herder, J. L. and van den Berg, F. P. A.: Statically balanced compliant mechanisms (SBCM's): an example and prospects, Proceedings of the ASME Design Engineering Technical Conferences & Computers and Information in Engineering Conference, 10–14 September 2000, Baltimore, Maryland, DETC2000/MECH-14144, 2000.

Hoetmer, K., Herder, J. L., and Kim, C. J.: A building block approach for the design of statically balanced compliant mechanisms, Proceedings of the ASME Design Engineering Technical Conferences & Computers and Information in Engineering Conference, San Diego, California, USA, 30 August–18 September 2009, DETC2009-87451, 2009.

Howell, L. L.: Compliant Mechanisms, Wiley-Interscience, New York, NY, 2001.

Jutte, C. V. and Kota, S.: Design of nonlinear springs for prescribed load-displacement functions, Trans. ASME J. Mechan. Design, 130, 081403, 1–10, 2008.

Lan, C.-C., Wang, J.-H., and Chen, Y.-H.: A compliant constant-force mechanism for adaptive robot end-effector operations, Proceedings of the 2010 IEEE International Conference on Robotics and Automation, Anchorage Convention District, 3–8 May, An-chorage, Alaska, USA, 2131–2136, 2010.

Morsch, F. M. and Herder, J. L.: Design of a generic zero stiffness compliant joint, Proceedings of the ASME Design Engineering Technical Conferences & Computers and Information in Engineering Conference, Montreal, Canada, 15–18 August 2010, DETC2010-28351, 2010.

Nahar, D. R. and Sugar, T.: Compliant constant-force mechanism with a variable output for micro/macro applications, Proceedings of the IEEE International Conference on Robotics and Automation, 1, 318–323, 2003.

Oh, Y. S. and Kota, S.: Synthesis of multistable equilibrium compliant mechanisms using combinations of bistable mechanisms, Trans. ASME J. Mechan. Design, 131, 021002, 1–11, 2009.

Powell, K. M. and Frecker, M. I.: Method for optimization of a nonlinear static balance mechanism, with application to ophthalmic surgical forceps, Proceedings of the ASME 2005 Design Engineering Technical Conferences and Computers and Information in Engineering Conference, 24–28 September 2005, Long Beach, California USA, DETC2005-84759, 2005.

Radaelli, G., Gallego, J. A., and Herder, J. L.: An energy approach to static balancing of systems with torsion stiffness, Proceedings of the ASME Design Engineering Technical Conferences & Computers and Information in Engineering Conference, Montreal, Canada, 15–18 August 2010, DETC2010-28071, 2010.

Rosenberg, E. J., Radaelli, G., and Herder, J. L.: An energy approach to a 2DOF compliant parallel mechanism with self-guiding statically-balanced straight-line behavior, Proceedings of the ASME Design Engineering Technical Conferences & Computers and Information in Engineering Conference, Montreal, Canada, 15–18 August 2010, DETC2010-28447, 2010.

Stapel, A. and Herder, J. L.: Feasibility study of a fully compliant statically balanced laparoscopy grasper, Proceedings of the ASME International Design Engineering Technical Conferences and Computers and Information in Engineering Conference, 28 September–2 October 2004, Salt Lake City, Utah, USA, DETC2004-57242, 635–643, 2004.

Tolou, M., Henneken, V. A., and Herder, J. L.: Statically-balanced compliant micro mechanisms (SB-MEMS): concepts and simulation, Proceedings of the ASME Design Engineering Technical Conferences & Computers and Information in Engineering Conference, Montreal, Canada, 15–18 August 2010, DETC2010-28406, 2010.

Tolou, M., Estevez, P., and Herder, J. L.: Collinear-type statically balanced compliant micro mechanism (SB-CMM): experimental comparison between pre-curved and straight beams, Proceedings of the ASME Design Engineering Technical Conferences & Computers and Information in Engineering Conference, 28–31 August 2011, Washington, DC, USA, DETC2011-47678, in press, 2011.

Trease, B. and Dede, E.: Statically balanced bompliant four-Bar mechanism, ME-599 Design Project Final Report, 2011.

Trease, B. P., Moon, Y.-M., and Kota, S.: Design Of Large-Displacement Compliant Joints, Trans. ASME J. Mechan. Design, 127, 778–788, 2005.

Wongratanaphisan, T. and Cole, M. O. T.: Analysis of a gravity compensated four-Bar linkage mechanism with linear spring suspension, Trans. ASME J. Mechan. Design, 130, 011006, 1–8, 2008.

6

Gantry crane control of a double-pendulum, distributed-mass load, using mechanical wave concepts

W. O'Connor and H. Habibi

UCD School of Mechanical and Materials Engineering, UCD Belfield, Dublin 4, Ireland

Correspondence to: H. Habibi (hossein.habibi@ucdconnect.ie)

Abstract. The overhead trolley of a gantry crane can be moved in two directions in the plane. The trolley is attempting to control the motion of a suspended, rigid-body, distributed mass load, supported by a hook, modelled as a lumped mass, in turn connected to the trolley by a light flexible cable. This flexible system has six degrees of freedom, four variables describing the flexible, hanging load dynamics and two (directly controlled) input variables for the trolley position. The equations of motion are developed and the crane model is verified. Then a form of wave-based control (WBC) is applied to determine what trolley motion should be used to achieve a reference motion of the load, with minimum swing during complex manoeuvres. Despite the trolley's limited control authority over the complex, flexible 3-D dynamics, WBC enables the trolley to achieve very good motion control of the load, in a simple, robust and rapid way, using little sensor information, with all measurements taken at or close to the trolley.

1 Introduction

In a typical problem of crane control, the challenge is to achieve controlled motion of the load, simultaneously moving it to follow a desired trajectory while actively controlling the swing. The system can be described as under-actuated: only the trolley is directly controlled, and it must indirectly control the swinging load at the far end of the cable. Performance can be measured under various headings, including minimization of sway during motion or on arrival at target, tracking desired motion paths, accurate repositioning payloads in target within the shortest possible time, maximum repetition rate, and safety (Abdel-Rahman et al., 2003). Even with no external disturbances, the trolley motion can cause significant payload pendulation, especially when the dynamics are more complex. Also cranes are inherently lightly damped (Todd et al., 1997). Finally, the load can change significantly, in ways which the controller may not know in advance. All these factors add to the control challenge.

Experienced crane drivers can acquire considerable skill, both in steering a load along a desired trajectory and in damping the swinging, especially on arrival at target. The trajectory might require guiding a heavy load safely around or between obstacles, within a factory, on a building site, or between a dock and a ship's hold. But developing the driving skills takes time and is expensive. Also even highly trained drivers can make mistakes. Furthermore drivers will typically err on the side of safety at the cost of longer manoeuvre times. Often drivers will also rely on at least one assistant (a person) to guide the load, especially for final positioning. One view of improving automatic control, therefore, is to try to understand what experienced crane drivers do intuitively, defining when to accelerate and to decelerate, and for planar trolleys, where and when to change direction and how quickly. But fully understanding what a human operator does is far from easy, and implementing this in a robust automatic control system is a further challenge.

This paper is about automating the process of moving the gantry of a crane to control the motion of a load when the load is dynamically complex. Specifically the paper is about a technique to solve this challenge based on the idea of mechanical waves. The technique is shown to have many advantages over existing approaches, including robustness to changes in the system dynamics, known or unknown; ease of implementation; and general applicability.

The reference inputs are requested motions to the system, that is, desired controlled motions of the load. These inputs could be desired crane positions or velocities over time. (Specifying one, of course, implicitly specifies the other.) Here it is assumed that the reference inputs are position control requests for the load motion over time. Often the problem is one of getting the load to track a desired trajectory. Many papers measure performance by the tracking ability of their control strategies (Sun et al., 2013; Kim and Singhose, 2010; Manning et al., 2010; Neupert et al., 2010; Forest et al., 2001; Vaughan et al., 2011).

In published work to date, the load and cable dynamics are frequently modelled as a simple or (at most) a double pendulum system moving in a plane. Abdel-Rahman et al. (2003) in a crane review article report that "most control strategies for this class of crane assume a planar gantry crane, utilize planar, linear models, and assume that the crane path, external forces, and control effort are all planar". The simple pendulum model considerably simplifies the dynamics, giving a swing frequency which is independent of the load mass. The simplification is not always appropriate however. Some researchers have recently modelled the loads as 3-D pendula to which various control techniques have been applied. However they generally assume a single, lumped-mass, load (Al-Garni et al., 1995; Sun et al., 2013; August et al., 2010; Yang and Yang, 2006; Antić et al., 2012; Cheng-Yuan et al., 2006; Maghsoudi et al., 2012; Chen et al., 2005; Schulze and Chang, 2010; Zhong, 2011). The most dynamically advanced developments in crane controllers in 3-D assume point mass loads, or at most, a rod-like body hanging from the trolley.

Some papers report work on controlling double-pendulum cranes (Kim and Singhose, 2006; Kim and Singhose, 2010; Masoud and Nayfeh, 2003; Sawodny et al., 2002; Manning et al., 2010; Tanaka and Kouno, 1998; Cheng-jun et al., 2009; Singhose and Towel, 1998; Kenison and Singhose, 1999; Dan and Li, 2008) but their simulations or experiments are planar rather than spatial. Manning et al. (2010), for example, present a dynamic model of bridge cranes with distributed-mass loads as a planar double pendulum. This work is an example of the use of input shapers. In general, the input shaper design depends on knowing the natural frequencies of the flexible system to be controlled. See Kenison and Singhose (1999) for example.

The authors found no research which considered the control of a 3-D double-pendulum crane involving rigid body dynamics, so this aspect of the work is considered novel. Also novel is the application of wave ideas to this control problem, although it had previously been applied to a simple, one-degree of freedom gantry crane leading to a robust control performance (O'Connor, 2003). In this work it is shown that "wave-based control", or WBC, can be applied successfully to controlling more complex dynamical systems in 3-D, such as controlling the double-pendulum load, retaining many of the advantages that WBC demonstrated in controlling simpler systems.

The paper treats the load as a distributed mass, with translational and swing rotational inertia effects about two axes. Furthermore, this distributed load is assumed to be hanging from a hook of significant mass, about which the load is free to swing. The hook in turn is modeled as a lumped mass, so that its 3-D translational inertia effects can be taken into account. The trolley is taken to move in the horizontal plane, with motion controllers for two perpendicular axes. Three simplifying assumptions or restrictions are made. Firstly, the effects of cable hoisting are not considered. Secondly, the mass of the cable is assumed to be negligible. Thirdly, spin rotation of the load (as opposed to swinging in either planar direction) is neglected. (The authors are confident that WBC can easily be adapted to work without these assumptions, but the present paper assumes them to make the flexible dynamics more manageable.) The entire system has 6 degrees of freedom, four of which are determined by the system dynamics and two of which are controlled, input variables. Figure 1 shows a schematic of the system model and the variables used as coordinates in the dynamic model.

As a control technique, WBC has been successfully applied to various flexible mechanical systems. It sees the actuator motion (in this case the motion of the trolley) as launching a disturbance, or mechanical wave, into the flexible system, while responding to waves coming back from the system, usually trying to absorb them (O'Connor, 2003, 2007; O'Connor et al., 2009; O'Connor and Fumagalli, 2009; O'Connor and McKeown, 2008; McKeown, 2009). This launching and absorbing are considered to be happening simultaneously. These notional motion waves have DC components (or net displacement components), which, on passing thought the flexible system, from actuator to tip and back again, leave behind the desired net displacement, while simultaneously controlling vibrations. The control system decides on the launch wave net displacement, usually setting it to half the reference displacement. The returning motion wave from the system is measured and is added to the launch wave, and this combination forms the input to the trolley motion controllers.

The returning-wave component has two important effects. Firstly, it causes the trolley to absorb the vibration (that is, reduce the load swing) both during the manoeuvre and on arrival at the final position. Secondly, in the absence of significant external, disturbing forces, the net motion associated with the absorbing motion will exactly equal the net motion of the launch wave, so that it makes up the second half of the reference displacement. It will be shown that this fairly simple idea produces rapid, robust control, which is easy to implement, of low order, and which does not depend on having a precise system model, or ideal actuator response, or position sensing at the load.

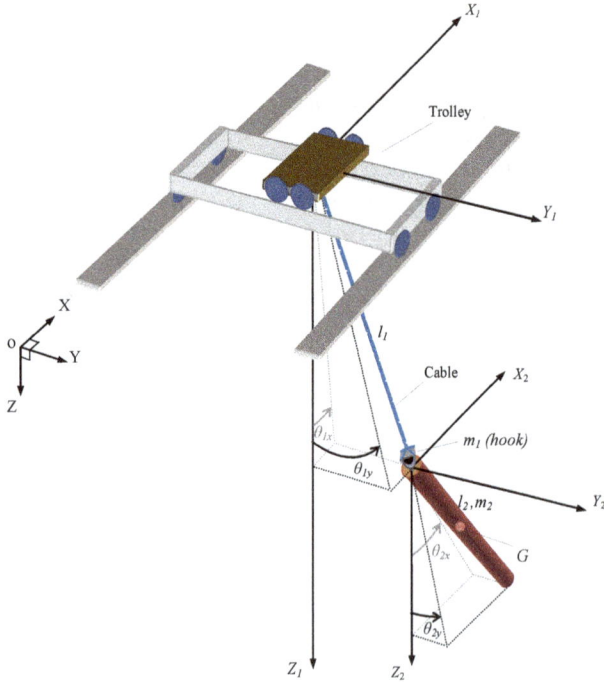

Figure 1. Representation of the 3-D double pendulum gantry crane.

2 3-D model of system dynamics

Figure 1 represents the entire physical system as modeled, including trolley, rails, cable, hook and hanging rigid body payload.

The trolley can move in both x and y directions simultaneously, so that it can follow an arbitrary path in the $X-Y$ plane in response to input signals to the trolley position subcontrollers. The reference system with coordinates $X_1Y_1Z_1$ is attached to the trolley, with Z_1 vertically downwards. The hook is connected to the trolley by a light, flexible cable, with the cable mass considered negligible in comparison with that of the hook and the load. The hook is modeled as a point mass (m_1), the kinetic energy of which is therefore purely translational. Also since cable hoisting is ignored, the cable length is taken as a constant, l_1, so the exact position of the hook can be determined by the angle of the cable in space. This angle is measured by two variables, θ_{1x} and θ_{1y}, which are the projections of the cable onto the $X-Z$ plane and $Y-Z$ planes, respectively. A second reference frame with coordinates system $X_2Y_2Z_2$, parallel to the first, has its origin at the hook, and is used to describe the swing rotation of the load, again using two angles, θ_{2x} and θ_{2y}, which are projections of the load angle onto the $X-Z$ and $Y-Z$ planes, respectively.

The instantaneous location of hook with respect to an inertial (space) coordinate system is given by

$$\begin{cases} x_H = x_T + l_1 \sin\theta_{1x} \\ y_H = y_T + l_1 \cos\theta_{1x}\sin\theta_{1y} \\ z_H = l_1 \cos\theta_{1x}\cos\theta_{1y} \end{cases} \tag{1}$$

where x_T and y_T are the trolley's position in the inertial reference frame. The position of the mass centre, G, of the swinging load is given by

$$\begin{cases} x_G = x_T + l_1 \sin\theta_{1x} + \frac{l_2}{2}\sin\theta_{2x} \\ y_G = y_T + l_1 \cos\theta_{1x}\sin\theta_{1y} + \frac{l_2}{2}\cos\theta_{2x}\sin\theta_{2y} \\ z_G = l_1 \cos\theta_{1x}\cos\theta_{1y} + \frac{l_2}{2}\cos\theta_{2x}\cos\theta_{2y} \end{cases} \tag{2}$$

The equations of motion are obtained from Lagrange's equation

$$\frac{\mathrm{d}}{\mathrm{d}t}\left(\frac{\partial L}{\partial \dot{q}_i}\right) - \left(\frac{\partial L}{\partial q_i}\right) = 0 \tag{3}$$

The generalized coordinates, q_i, are here taken as the four independent variables $\theta_{1x}, \theta_{1y}, \theta_{2x}$ and θ_{2y}. The trolley position variables x_T and y_T are considered as input variables, used to control the attached flexible system, so four equations of motion are required. If m_1 is the mass of the hook and m_2 the mass of the load, total potential energy is

$$U = m_1 g l_1 \left(1 - \cos\theta_{1x}\cos\theta_{1y}\right) + m_2 g$$
$$\left[l_1\left(1 - \cos\theta_{1x}\cos\theta_{1y}\right) + \frac{l_2}{2}(1 - \cos\theta_{2x}\cos\theta_{2y})\right] \tag{4}$$

where g is the acceleration due to gravity. The total kinetic energy may describe as

$$T = \frac{1}{2}m_1\vec{v}_H\vec{v}_H + \frac{1}{2}m_2\vec{v}_G\vec{v}_G + \frac{1}{2}\tilde{\omega}\,[I_G]\,\tilde{\omega} \tag{5}$$

where $\vec{v}_H$ is the hook velocity, $\vec{v}_G$ is the velocity of point G, and $\tilde{\omega}$ is the total angular velocity of the load. The linear velocities are the derivatives of Eqs. (1) and (2), which may be expressed in the form

$$\vec{v}_H : \begin{cases} \dot{x}_H = \dot{x}_T + l_1\dot{\theta}_{1x}\cos\theta_{1x} \\ \dot{y}_H = \dot{y}_T - l_1\dot{\theta}_{1x}\sin\theta_{1x}\sin\theta_{1y} + l_1\dot{\theta}_{1y}\cos\theta_{1x}\cos\theta_{1y} \\ \dot{z}_H = -l_1\dot{\theta}_{1x}\sin\theta_{1x}\cos\theta_{1y} - l_1\dot{\theta}_{1y}\cos\theta_{1x}\sin\theta_{1y} \end{cases} \tag{6}$$

$$\vec{v}_G : \begin{cases} \dot{x}_G = \dot{x}_H + \frac{l_2}{2}\dot{\theta}_{2x}\cos\theta_{2x} \\ \dot{y}_G = \dot{y}_H - \frac{l_2}{2}\dot{\theta}_{2x}\sin\theta_{2x}\sin\theta_{2y} + \frac{l_2}{2}\dot{\theta}_{2y}\cos\theta_{2x}\cos\theta_{2y} \\ \dot{z}_G = \dot{z}_H - \frac{l_2}{2}\dot{\theta}_{2x}\sin\theta_{2x}\cos\theta_{2y} - \frac{l_2}{2}\dot{\theta}_{2y}\cos\theta_{2x}\sin\theta_{2y} \end{cases} \tag{7}$$

To quantify the rotational kinetic energy of the load requires its angular velocity $\tilde{\omega}$ and moment of inertia tensor, $[I_G]$, to be expressed in a common coordinate system. The load inertia tensor is most conveniently expressed using a coordinate system $X_{2c}Y_{2c}Z_{2c}$ fixed to the load and along its principal axes, so that

$$[I_G] = \begin{bmatrix} I_{xx} & 0 & 0 \\ 0 & I_{yy} & 0 \\ 0 & 0 & I_{zz} \end{bmatrix}_{X_{2c}Y_{2c}Z_{2c}} \tag{8}$$

If so, $\tilde{\omega}$ should also be expressed in the same coordinate system. As shown in Fig. 2, the coordinate system $X_{2a}Y_{2a}Z_{2a}$

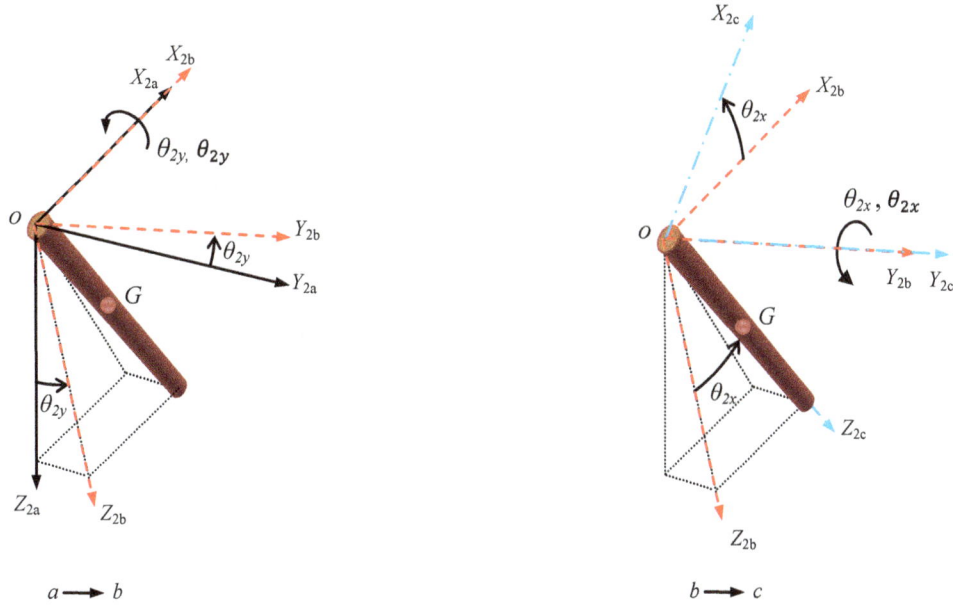

Figure 2. Representation of successive rotations to reach the body coordinates of the hanging load.

moves through two rotations about the main axes to become aligned to the load body axis $X_{2c}Y_{2c}Z_{2c}$. The angular velocity ($\tilde{\omega}$) according to Fig. 2 is defined as

$$\tilde{\omega} = -\dot{\theta}_{2y}\hat{i}_{2b} + \dot{\theta}_{2x}\hat{j}_{2c} \qquad (9)$$

The transformation from $X_{2b}Y_{2b}Z_{2b}$ to $X_{2c}Y_{2c}Z_{2c}$ or $T_{2b\to 2c}$ is

$$T_{(2b\to 2c)} = \begin{bmatrix} \cos\theta_{2x} & 0 & -\sin\theta_{2x} \\ 0 & 1 & 0 \\ \sin\theta_{2x} & 0 & \cos\theta_{2x} \end{bmatrix} \qquad (10)$$

So finally $\tilde{\omega}$ can be expressed in $X_{2c}Y_{2c}Z_{2c}$ as

$$\tilde{\omega} = \begin{bmatrix} -\dot{\theta}_{2y}\cos\theta_{2x} \\ \dot{\theta}_{2x} \\ -\dot{\theta}_{2y}\sin\theta_{2x} \end{bmatrix}_{X_{2c}Y_{2c}Z_{2c}} \qquad (11)$$

The rotational kinetic energy of the hanging load, T_ω, is then expressible as

$$T_\omega = \frac{1}{2}\tilde{\omega}\,[I_G]\,\tilde{\omega} = \frac{1}{2}I_{xx}(\dot{\theta}_{2y})^2(\cos\theta_{2x})^2 + \frac{1}{2}I_{yy}(\dot{\theta}_{2x})^2$$
$$+ \frac{1}{2}I_{zz}(\dot{\theta}_{2y})^2(\sin\theta_{2x})^2 \qquad (12)$$

Substituting Eqs. (6), (7) and (12) into Eq. (5) gives the total kinetic energy of system. Then this T from Eq. (5) minus U from Eq. (4) gives the Lagrangian, $L = T - U$ to be used in Eq. (3), with q_i equal, in turn, to each of the four angles θ_{1x}, θ_{1y}, θ_{2x} and θ_{2y}, giving four equations of motion, which after

simplification become

$$\left(m_1 l_1^2 + m_2 l_1^2\right)\ddot{\theta}_{1x} + (m_1 + m_2)\,l_1\ddot{x}_T\cos\theta_{1x} - (m_1 + m_2)\,l_1\ddot{y}_T$$
$$\sin\theta_{1x}\sin\theta_{1y} + (m_1 + m_2)l_1^2(\dot{\theta}_{1y})^2\cos\theta_{1x}\sin\theta_{1x} + \frac{1}{2}m_2 l_1 l_2\ddot{\theta}_{2x}$$
$$\cos\theta_{1x}\cos\theta_{2x} - \frac{1}{2}m_2 l_1 l_2(\dot{\theta}_{2x})^2\cos\theta_{1x}\sin\theta_{2x} + \frac{1}{2}m_2 l_1 l_2\ddot{\theta}_{2x}$$
$$\sin\theta_{1x}\sin\theta_{2x}\cos(\theta_{1y} - \theta_{2y}) + \frac{1}{2}m_2 l_1 l_2\left[(\dot{\theta}_{2x})^2 + (\dot{\theta}_{2y})^2\right]\sin\theta_{1x}$$
$$\cos\theta_{2x}\cos(\theta_{2y} - \theta_{1y}) + \frac{1}{2}m_2 l_1 l_2\ddot{\theta}_{2y}\sin\theta_{1x}\cos\theta_{2x}\sin(\theta_{2y} - \theta_{1y})$$
$$+ m_2 l_1 l_2\dot{\theta}_{2x}\dot{\theta}_{2y}\sin\theta_{1x}\sin\theta_{2x}\sin(\theta_{1y} - \theta_{2y}) + (m_1 + m_2)\,g l_1$$
$$\sin\theta_{1x}\cos\theta_{1y} = 0 \qquad (13)$$

$$(m_1 + m_2)\,l_1^2(\cos\theta_{1x})^2\ddot{\theta}_{1y} - 2(m_1 + m_2)\,l_1^2\dot{\theta}_{1x}\dot{\theta}_{1y}\sin\theta_{1x}\cos\theta_{1x}$$
$$+ (m_1 + m_2)\,l_1\ddot{y}_T\cos\theta_{1x}\cos\theta_{1y} + \frac{1}{2}m_2 l_1 l_2\ddot{\theta}_{2y}\cos\theta_{1x}\cos\theta_{2x}$$
$$\cos(\theta_{1y} - \theta_{2y}) + \frac{1}{2}m_2 l_1 l_2\left[(\dot{\theta}_{2x})^2 + (\dot{\theta}_{2y})^2\right]\cos\theta_{1x}\cos\theta_{2x}$$
$$\sin(\theta_{1y} - \theta_{2y}) + \frac{1}{2}m_2 l_1 l_2\ddot{\theta}_{2x}\cos\theta_{1x}\sin\theta_{2x}\sin(\theta_{1y} - \theta_{2y})$$
$$- m_2 l_1 l_2\dot{\theta}_{2x}\dot{\theta}_{2y}\cos\theta_{1x}\sin\theta_{2x}\cos(\theta_{1y} - \theta_{2y}) + (m_1 + m_2)$$
$$g l_1\sin\theta_{1y}\cos\theta_{1x} = 0 \qquad (14)$$

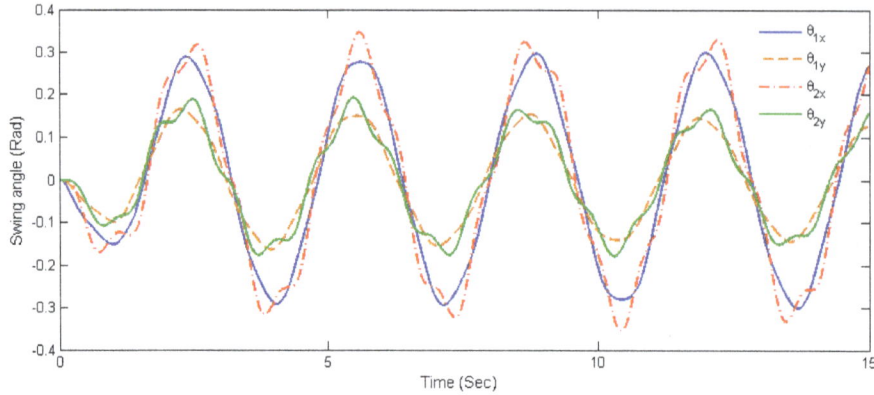

Figure 3. Dynamic response of 3-D double pendulum gantry crane with 4 degree of freedom.

Table 1. Values chosen for the 3-D double pendulum system.

m_1 (kg)	m_2 (kg)	l_1 (m)	l_2 (m)
1	50	1	2

$$\left(m_2\frac{l_2^2}{4} + I_{yy}\right)\ddot{\theta}_{2x} + \frac{1}{2}m_2l_2\ddot{x}_T\cos\theta_{2x} - \frac{1}{2}m_2l_2\ddot{y}_T\sin\theta_{2x}$$

$$\sin\theta_{2y} + \frac{1}{2}m_2l_1l_2\ddot{\theta}_{1x}\cos\theta_{1x}\cos\theta_{2x} - \frac{1}{2}m_2l_1l_2(\dot{\theta}_{1x})^2$$

$$\sin\theta_{1x}\cos\theta_{2x} + \frac{1}{2}m_2l_1l_2\ddot{\theta}_{1x}\sin\theta_{1x}\sin\theta_{2x}\cos\left(\theta_{1y}-\theta_{2y}\right)$$

$$+\frac{1}{2}m_2l_1l_2\left(\dot{\theta}_{1x}\right)^2 + \left(\dot{\theta}_{1y}\right)^2\cos\theta_{1x}\sin\theta_{2x}\cos(\theta_{1y}-\theta_{2y}) + \frac{1}{2}$$

$$m_2l_1l_2\ddot{\theta}_{1y}\cos\theta_{1x}\sin\theta_{2x}\sin\left(\theta_{1y}-\theta_{2y}\right) - m_2l_1l_2\dot{\theta}_{1y}\dot{\theta}_{1x}\sin\theta_{1x}$$

$$\sin\theta_{2x}\sin\left(\theta_{1y}-\theta_{2y}\right) + m_2\frac{l_2^2}{4}\left(\dot{\theta}_{2y}\right)^2\cos\theta_{2x}\sin\theta_{2x} + (I_{xx}-I_{zz})$$

$$\left(\dot{\theta}_{2y}\right)^2\cos\theta_{2x}\sin\theta_{2x} + m_2g\frac{l_2}{2}\sin\theta_{2x}\cos\theta_{2y} = 0 \qquad (15)$$

$$\left(m_2\frac{l_2^2}{4}(\cos\theta_{2x})^2 + I_{xx}(\cos\theta_{2x})^2 + I_{zz}(\sin\theta_{2x})^2\right)\ddot{\theta}_{2y} - m_2\frac{l_2^2}{2}\dot{\theta}_{2x}\dot{\theta}_{2y}$$

$$\cos\theta_{2x}\sin\theta_{2x} + \frac{1}{2}m_2l_2\ddot{y}_T\cos\theta_{2x}\cos\theta_{2y} + \frac{1}{2}m_2l_1l_2\ddot{\theta}_{1x}\sin\theta_{1x}$$

$$\cos\theta_{2x}\sin\left(\theta_{2y}-\theta_{1y}\right) + \frac{1}{2}m_2l_1l_2\left\{\left(\dot{\theta}_{1x}\right)^2 + \left(\dot{\theta}_{1y}\right)^2\right\}\cos\theta_{1x}\cos$$

$$\theta_{2x}\sin\left(\theta_{2y}-\theta_{1y}\right) + \frac{1}{2}m_2l_1l_2\ddot{\theta}_{1y}\cos\theta_{1x}\cos\theta_{2x}\cos\left(\theta_{1y}-\theta_{2y}\right)$$

$$-m_2l_1l_2\dot{\theta}_{1x}\dot{\theta}_{1y}\sin\theta_{1x}\cos\theta_{2x}\cos\left(\theta_{2y}-\theta_{1y}\right) + (I_{zz}-I_{xx})$$

$$\dot{\theta}_{2x}\dot{\theta}_{2y}\sin\left(2\theta_{2x}\right) + m_2\frac{l_2}{2}g\cos\theta_{2x}\sin\theta_{2y} = 0 \qquad (16)$$

These four, highly coupled, equations of motion capture the full system dynamics. In the modeling, no small-angle approximations were made to keep the model accurate even for large swing angles. These four equations can be integrated numerically from given initial conditions to describe the time evolution of the system. The trolley motion components are considered as inputs, defining $\ddot{x}_T$ and $\ddot{y}_T$ in these equations. As an example, Fig. 3 shows the behavior of the system for an arbitrary planar movement of the trolley with no damping

or control action. The chosen system parameters (which can be arbitrarily chosen) are given in Table 1.

While the input motion of the trolley takes no longer than two seconds, the system keeps swinging indefinitely, with multiple frequency components. Alternatively, rather than by trolley motion, the system can be set in motion by giving it initial angular displacements and/or velocities, with the trolley stationary. If desired this can be done in such a way that the subsequent motion corresponds to the mode shapes, at each of four natural frequencies.

With the crane model behaving as expected, the WBC ideas are now developed and used to control the swinging load by controlling the trolley motion.

3   Wave-based approach to crane control

From a control perspective, the trolley is a single actuator attempting to control a flexible system of relatively complex dynamics. The system is under-actuated, with more degrees of freedom than actuators. The actuator does not act directly on the load position and orientation, but must work through the intervening flexible dynamics, of cable and hook, to try to achieve a target motion of the load. The actuator motion is in two perpendicular directions, the load can swing in 3-D, and the motion components are strongly coupled.

The control method adopted here, wave-based control, is a generic approach, which does not depend on having an accurate system model. It uses feedback, but the feedback measurements are taken not at the system output (here, the load position and orientation) but at the actuator (the trolley, in this case). Thus the actuator and sensing are collocated, with

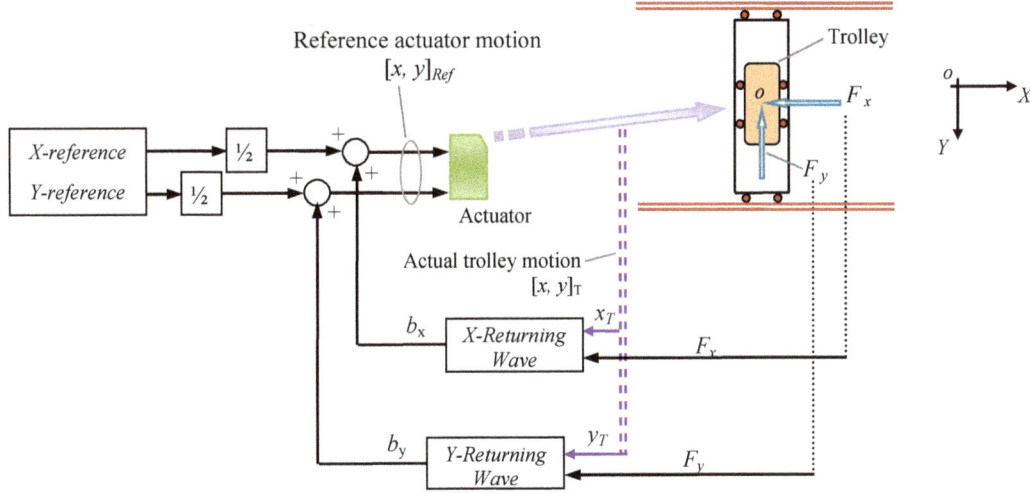

Figure 4. WBC plan for gantry crane with planar-moving trolley.

the consequent stability advantages. To date WBC has been used successfully to control 1-D rectilinear lumped flexible systems, 2-D flexible mass-spring arrays either beam-like or arranged in a grid, laterally flexing manipulators, and a simple pendulum gantry crane as already mentioned (O'Connor, 2003).

In the current case, there are two inputs actuating the trolley motion in two orthogonal directions. So the trolley should be given two reference displacements, in the x and y directions, and two corresponding returning waves, b_x and b_y, to achieve wave absorption in the two directions under WBC. Figure 4 illustrates a version of a general scheme for positioning the trolley, along with the payload, to a target position in the plane.

Here the returning displacement waves are defined and measured as

$$b_x = \frac{1}{2}\left[x_T - \frac{1}{Z_x} \int_0^t F_x \, dt \right] \tag{17}$$

$$b_y = \frac{1}{2}\left[y_T - \frac{1}{Z_y} \int_0^t F_y \, dt \right] \tag{18}$$

where x_T and y_T are the trolley position coordinates, F_x and F_y are the horizontal components in the x and y directions of the cable force measured at the trolley, and Z_x, Z_y are impedance terms. References O'Connor (2003) and McKeown (2009) outline how such expressions for the returning waves can be developed. Here we simply note the following two features.

First, for rest-to-rest motion, from time $t = 0$ to some final time t, as the initial and final momenta are zero, the force integrals must be zero. So the final values of b_x and b_y will be half the trolley displacements, or $1/2 x_T$ and $1/2 y_T$. Note that

this result holds regardless of the values of the impedances Z_x and Z_y. The second observation is that while F_x and F_y are changing, the effect of adding b_x and b_y to the trolley's motion is to make the trolley act as a viscous damper with damping coefficient Z_x, Z_y in response to the cable forces.

The values of impedance are not critical to the control scheme. In this work both impedances were set to

$$Z = (m_1 + m_2) \sqrt{\frac{g}{(l_1 + l_2)}} \tag{19}$$

Variations in the values of Z cause small variations in the transient part of the responses. So Z can be used as a parameter with which to fine-tune the transient, for example to improve a specific performance measure (e.g. rise time, overshoot, or settling time), as appropriate for a given application, invariably at the cost of a slight degradation of some other transient performance measure (although always retaining the zero steady-state error).

The force components F_x and F_y should ideally correspond to the horizontal components of the cable tension, including the dynamic effects of the acceleration of the load mass. For most purposes, however, they can be approximated by assuming that the cable tension is equal to the load weight, and then F_x and F_y from Fig.1 can be taken as

$$F_x = \frac{-(m_1 + m_2)\, g \tan\theta_{1x}}{\cos\theta_{1y}} \tag{20}$$

$$F_y = -(m_1 + m_2)g \tan\theta_{1y} \tag{21}$$

These approximations were used in obtaining the results below.

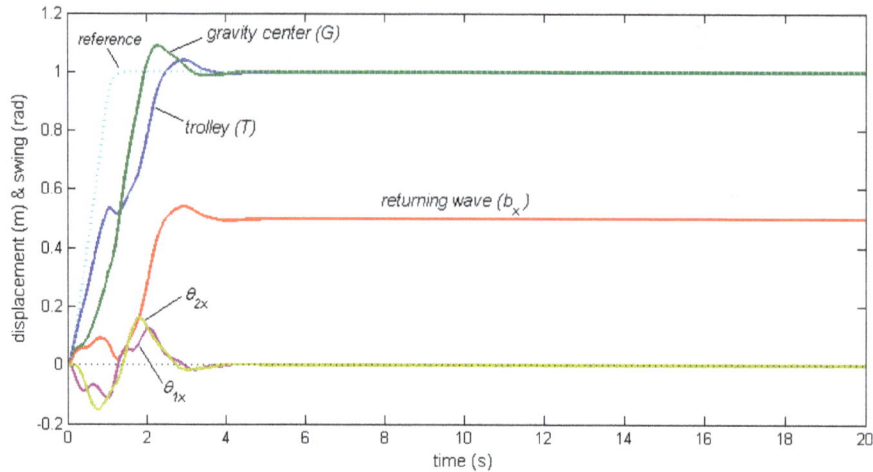

Figure 5. Gantry crane response to a 1 m displacement in x direction using WBC.

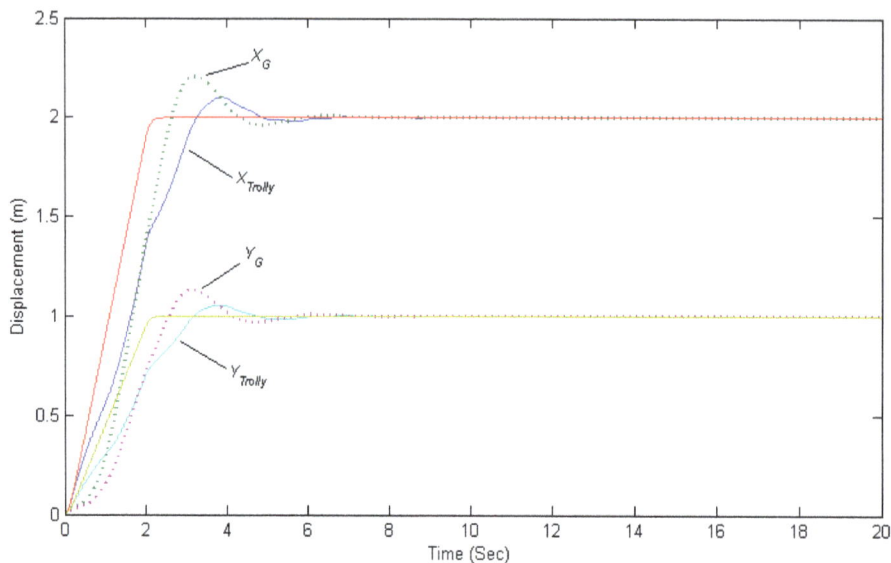

Figure 6. Response to planar trolley motion under WBC (sub G: mass centre of distributed load).

4   Results

In the first manoeuvre the reference is a simple ramp (or constant velocity) displacement of one meter in the x direction, with no motion in the y direction. Figure 5 depicts the response of both the trolley and the centre of mass of the payload, G, under the control system of Fig. 4. Also the swinging angles of the cable (in both directions) and of the load due to this excitation are displayed in radians. The trolley can be seen to settle quickly at the target displacement with an initial overshoot of less than 10 %. The centre of mass of the hanging load, at 2 m from the trolley, comes to rest rapidly, with little swing as shown by θ_{1x} and θ_{2x}, and with an overshoot of about 10 %. Clearly the suspension swinging dies out soon after the trolley reaches the target. Also shown is the returning wave, b_x, which provides the swing absorption

and settles at half the target displacement. (In this case there is no b_y as all the motion is in the x direction.)

In the next example, the reference input is a simultaneous combination of a ramp up to 2 m in the x direction and a ramp of a different slope up to one meter in the y direction. Figure 6 shows the response. The trolley comes to rest gently with no steady-state error while quickly absorbing all oscillations. Figure 7 shows how the swing angles of the double pendulum are actively damped. As the model has no damping built in, the damping is being achieved entirely by the trolley motion which is simultaneously moving the system to the target displacement as seen in Fig. 6. The periodic time of the lowest mode of vibration is of the order of 4 s. After 10 s, that is within two and a half times the periodic time after arrival at the target, all oscillations have been reduced from the

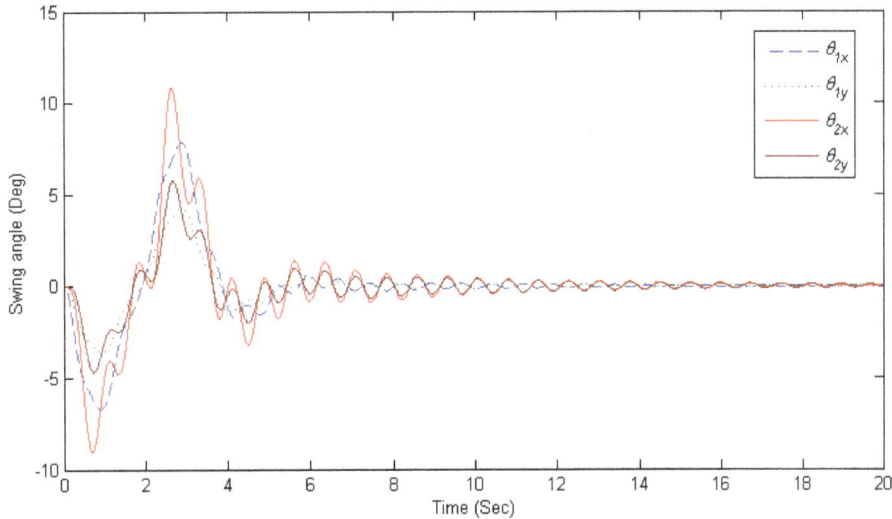

Figure 7. Swing angles for manoeuver of Fig. 6, showing active damping.

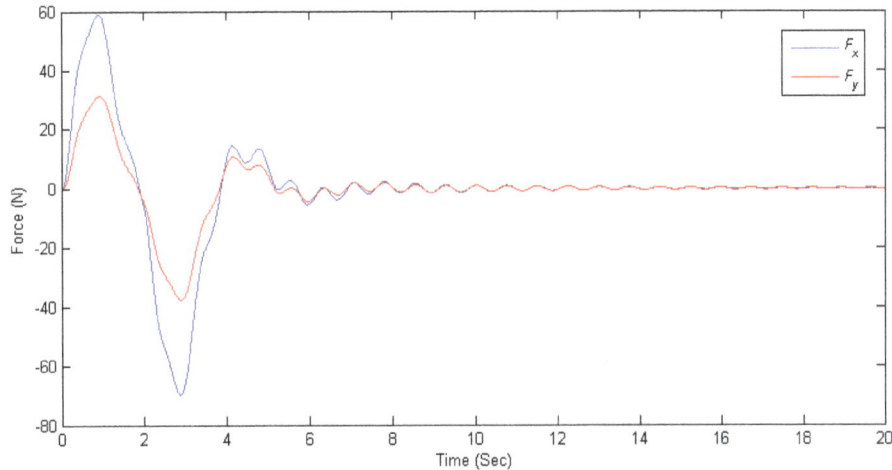

Figure 8. Reduction of forces acting on trolley through WBC process.

maximum value of about 10 degrees to less than 2 degrees and they are decaying steadily.

Figure 8 shows the horizontal force components between trolley and cable for the same manoeuvre. The main accelerations and decelerations occur within 4 s, and after about 6 s the force amplitudes diminish to less than 3 % of the initial peaks. Again WBC is using the forces to combine position control and active vibration damping very effectively.

Figure 9 shows the effect of choosing different values of the impedance parameter Z, as the only control parameter to be tuned, for the single-input manoeuvre of Fig. 5. The reference impedance $Z = Z_{eq}$ is as in Eq. (19), which is the value used to obtain the results presented above. Despite a 12-fold range in impedance values, the responses are good for all cases, showing a stable response, rapid transit and zero steady-state error in the final position. The best choice of Z will depend on the priorities in the desired response.

For example, perhaps a good compromise between minimum overshoot and shortest settling time is when $Z_x = (0.75)Z_{eq}$. For $Z_x \geq 2Z_{eq}$ the various responses become almost indistinguishable (except around the half-way point). More could have been added for other values of Z_x, but they would have fallen on top of the curves shown. On the other hand for low values of impedance, say $Z_x < (0.5)Z_{eq}$, the trolley has a slow transient and slow convergence to the target position.

The robustness of the control response to variations in Z also indirectly illustrates the robustness of WBC to changes in the system under control, whether these are known or unknown, whether modeled or not. For example, if incorrect values of masses or lengths are assumed in using Eq. (19), or these parameter change during operations, the control system still copes well.

In producing these results, the trolley was assumed to have ideal dynamics, that is, that it reproduces exactly and

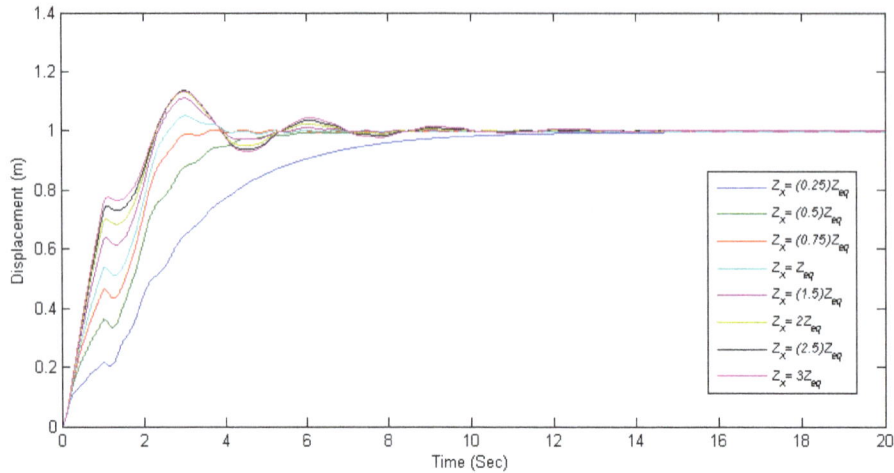

Figure 9. Trolley response to 1 m x direction input with different Z values in the WBC.

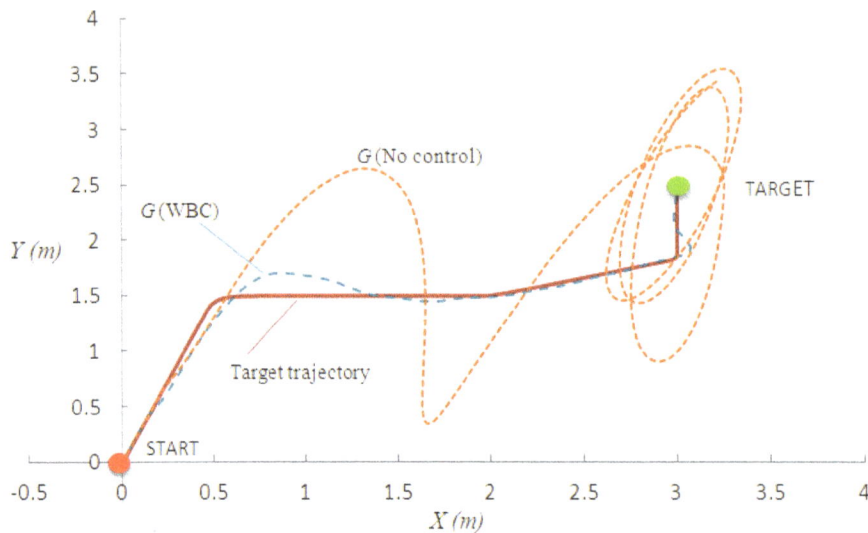

Figure 10. Trajectory tracking, plan view, showing the paths of the mass centre of the load.

immediately the motion requested by the WBC system of Fig. 4. However the system has also been tested with realistic trolley dynamics, where the trolley response shows some dynamic delay in achieving the requested motion. The control system still works well, with comparable results to those presented, provided that the steady state trolley error is zero. This robustness to the trolley's dynamic performance can be explained in part because the measurements used in the WBC control system, including Eqs. (17, 18) come *after* the trolley (see Fig. 4), using the values of position, x_T, y_T, and forces, F_x, F_y, actually achieved and experienced by the trolley.

Finally, in addition to the point-to-point manoeuvres above, input tracking and obstacle avoidance are considered. Figure 10 shows a plan view of a desired input trajectory, and the resulting path of the mass centre of the load, G, both un-

der WBC and with no control. As can be seen, under WBC the tracking is very satisfactory.

5 Conclusions

The double-pendulum, distributed mass, gantry crane model assumed in this work has non-trivial dynamics, and represents a considerable advance on the simple pendulum model often used. A model was developed to capture these dynamics in three dimensions. The 2-D trolley motion has limited control authority over the 3-D suspended system. This paper explores how well a simple version of WBC can work to achieve load position control while damping the swinging in 3-D. The control strategy does not require details of the system model, and all the required measurements are taken at the trolley. The control law has one tuning parameter, a mechanical impedance term, whose value is not critical, but can be used to fine-tune the transient response as desired.

The results illustrate the power, simplicity, effectiveness and robustness of the control approach. For point to point manoeuvres in the plane, WBC proves very effective. It also performs well in trajectory tracking, often required for obstacle avoidance.

Future work will extend the approach to cranes with cable hoisting, to tower cranes in which the trolley moves on a rotating arm, and to jib or luffing cranes in which the arm rotates in the vertical plane. Initial results suggest that the same WBC strategy can be extended successfully to all such cases.

Edited by: A. Müller

References

Abdel-Rahman, E. M., Nayfeh, A. H., and Masoud, Z. N.: Dynamics and Control of Cranes: A Review, J. Vib. Control, 9, 863–908, 2003.

Al-Garni, A. Z., Moustafa, K. A. F, and Javeed Nizami, S. S. A. K.: Optimal control of overhead cranes, Control Eng. Pract., 3, 1277–1284, 1995.

Antić, D., Jovanović, Z., Perić, S., Nikolić, S., Milojković, M., and Milošević, M.: Anti-Swing Fuzzy Controller Applied in a 3D Crane System, Engineering, Technology & Applied Science Research (ETASR), 2, 196–200, 2012.

August, W., Ren, J., Notheis, S., Haase, T., Hein, B., and Wörn, H.: 3D Pendulum Swinging Control by an Industrial Robot Manipulator, 1–7, in: Proceeding of: ISR/ROBOTIK 2010, Proceedings for the joint conference of ISR 2010 (41st Internationel Symposium on Robotics) und ROBOTIK 2010 (6th German Conference on Robotics), 7–9 June 2010, Munich, Germany, 2010.

Chen, H., Gao, B., and Zhang, X.: Dynamical Modelling and Nonlinear Control of a 3D Crane, International Conference on Control and Automation (ICCA2005), Budapest, Hungary, 1085–1090, June 2005.

Cheng-jun, D., Ping, D., Ming-lu, Z., and Yan-fang, Z.: Double Inverted Pendulum System Control Strategy Based On Fuzzy Genetic Algorithm, in: Proceedings of the IEEE International Conference on Automation and Logistics Shenyang, China, 1318–1323, August 2009.

Cheng-Yuan, C., Kou-Cheng, H., Kuo-Hung, C., and Guo-En, H.: An Enhanced Adaptive Sliding Mode Fuzzy Control for Positioning and Anti-Swing Control of the Overhead Crane System, IEEE International Conference on Systems, Man, and Cybernetics, Taipei, Taiwan, 992–997, October 2006.

Dan, Y. and Li, Z.: The Structure of HSIC System and Its Application on Arbitrary Switch Control of Double Pendulum, in: Proceedings of the 7th World Congress on Intelligent Control and Automation, Chongqing, China, 2810–2815, June 2008.

Forest, C., Frakes, D., and Singhose, W.: Input-Shaped Control of Gantry Cranes: Simulation and Curriculum Development, The 18th ASME DETC Biennial Conference on Mech. Vib. and Noise, 2001.

Kenison, M. and Singhose, W.: Input Shaper Design for Double-Pendulum Planar Gantry Cranes, in: Proceedings of the 1999

EEE International Conference on Control Applications, Kohala Coast-Island of Hawai'i, Hawai'i, USA, 539–544, August 1999.

Kim, D. and Singhose, W.: Reduction of Double-Pendulum Bridge Crane Oscillations, The 8th International Conference On Motion And Vibration Control (MOVIC 2006), Atlanta, GA, USA, 300–305, 2006.

Kim, D. and Singhose, W.: Performance Studies Of Human Operators Driving Double-Pendulum Bridge Cranes, Control Eng. Pract., 18, 567–576, 2010.

Maghsoudi, M. J., Mohammed, Z., Pratiwi, A. F., Ahmad, N., and Husain, A. R.: An Experiment for Position and Sway Control of a 3D Gantry Crane, The 4th International Conference on Intelligent and Advanced Systems (ICIAS2012), 497–502, 2012.

Manning, R., Clement, J., Kim, D., and Singhose, W.: Dynamics and Control of Bridge Cranes Transporting Distributed-Mass Payloads, J. Dyn. Syst.-T. ASME, 132, 014505, doi:10.1115/1.4000657, 2010.

Masoud, Z. N. and Nayfeh, A. H.: Sway Reduction on Container Cranes Using Delayed Feedback Controller, Nonlinear Dynam., 34, 347–358, 2003.

McKeown, D. J.: Wave based Control of Elastic Mechanical Systems, Ph.D. Thesis, Department of Mechanical Engineering, University College Dublin, Ireland, 189 pp., 2009.

Neupert, J., Arnold, E., Schneider, K., and Sawodny, O.: Tracking and anti-sway control for boom cranes, Control Eng. Pract., 18, 31–44, 2010.

O'Connor, W. J.: A Gantry Crane Problem Solved, J. Dyn. Syst.-T ASME, 125, 569–576, 2003.

O'Connor, W. J.: Wave-Based Analysis and Control of Lump-Modeled Flexible Robots, IEEE T. Robot., 23, 342–352, 2007.

O'Connor, W. J. and Fumagalli, A.: A Refined Wave-Based Control Applied to Nonlinear, Bending, and Slewing Flexible Systems, J. Appl. Mech., 76, 041005, doi:10.1115/1.3086434, 2009.

O'Connor, W. J. and McKeown, D. J.: A new approach to modal analysis of uniform chain systems, J. Sound Vib., 311, 623–632, 2008.

O'Connor, W. J., Ramos, F., McKeown, D. J., and Feliu, V.: Wave-based control of non-linear flexible mechanical systems, Nonlinear Dynam., 57, 113–123, 2009.

Sawodny, O., Aschemannb, H., and Lahres, S.: An automated gantry crane as a large workspace robot, Control Eng. Pract., 10, 1323–1338, 2002.

Schulze, T. and Chang, T. N.: Zero Vibration Position Control of a Spherical Pendulum for Control Systems Demonstration, American Control Conference, Marriott Waterfront, Baltimore, MD, USA, 738–743, July 2010.

Singhose, W. E. and Towel, S. T.: Double-Pendulum Gantry Crane Dynamics and Control, in: Proceedings of the 1998 IEEE International Conference on Control Applications Trieste, Italy, 1205–1209, September 1998.

Sun, N., Fang, Y., and Zhang, X.: Energy coupling output feedback control of 4-DOF underactuated cranes with saturated inputs, Automatica, 49, 1318–1325, 2013.

Tanaka, S. and Kouno, S.: Automatic measurement and control of the attitude of crane lifters; Lifter-attitude measurement and control, Control Eng. Pract., 6, 1099–1107, 1998.

Todd, M. D., Vohra, S. T., and Leban, F.: Dynamical measurements of ship crane load pendulation, in: Oceans 97 MTS/IEEE: Conference Proceedings, Hailfax, Canada, 2, 1230–1236, 1997.

Vaughan, J., Karajgikar, A., and Singhose, W.: A Study of Crane Operator Performance Comparing PD-Control and Input Shaping, American Control Conference, San Francisco, CA, USA, 545–550, June 2011.

Yang, J. H. and Yang, K. S.: Adaptive Control for 3-D Overhead Crane Systems, in: Proceedings of the 2006 American Control Conference, Minneapolis, Minnesota, USA, 1832–1837, June 2006.

Zhong, B.: Load's 2-Degree of Freedom Swing Angle Model and Dynamic Simulation for Overhead Crane or Gantry Crane, Energy Procedia, 11, 1217–1223, 2011.

7

Design and principles of an innovative compliant fast tool servo for precision engineering

H. Li[1,*], **R. Ibrahim**[2], and **K. Cheng**[2]

[1]Department of Mechanical Design and Manufacturing, College of Engineering,
China Agricultural University, Beijing 100083, China
[2]School of Engineering and Design, Brunel University, Uxbridge, UB8 3PH, UK
[*]currently at: Brunel University, Uxbridge, UB8 3PH, UK

Abstract. The paper presents an innovative design of fast tool servo (FTS) by combining compliant mechanism (CM) and precision engineering (PE) so as to meet the stringent requirements of a one degree of freedom (1DOF) compact FTS. The design requirements of the FTS include: (a) 1 nm resolution, (b) 10–20 μm of range, (c) first natural frequency of over 1400 Hz, (d) compatibility with holding different diamond cutting tools, and (e) without fatigue of the compliant mechanism. Then FEA is used to evaluate the static and dynamic performance of the FTS. The preliminary results show that both the static and dynamic performances are matched to the design objectives.

1 Introduction

The fast tool servo (FTS) plays important roles in precision turning of free-form surfaces with a diamond cutting tool. Some compliant mechanisms (CMs) have been applied in FTS because of their advantages such as no coulomb friction and no backlash (Howell, 2001). A piezoelectric actuator was often used in FTS because of its high resolution and wide bandwidth actuation. Many researchers studied the design and synthesis of the CMs (Li et al., 2010; Yu et al., 2010a, 2010b). But they seldom considered the actual application in precision engineering (PE). On the contrary some FTSs have been designed taking little account of the design principles of CMs (Cuttino et al., 1999; Gan et al., 2007; Huo and Cheng, 2008; Kim and Kim, 2003; Kim et al., 2004, 2009; Kouno, 1984; Noh et al., 2009; Spotts et al., 2004; Tian et al., 2009). Therefore, there is a need for a comprehensive and robust design method combining with CMs and PE to meet the stringent requirements of ultra precision machining.

Kouno (1984) designed a FTS CM that consists of two parallel diaphragms based on the parallel-spring concept but the stiffness of $70 \, \text{N} \, \mu\text{m}^{-1}$ of the CM consumed more energy and reduced the stroke. Cuttino (1999) used two annular flexures in his FTS but the design bandwidth is only 100 Hz. The CM with two sets of parallel circle notch flexure hinges was introduced by Gan et al. (2007), Huo and Cheng (2008), Kim and Kim (2003), Kim et al. (2004, 2009), Tian et al. (2009). Kim and Kim (2003) and Kim et al. (2004) designed a FTS giving a stroke and natural frequency of 7.5 μm and 100 Hz respectively. In order to increase the stroke of the FTS, a flexure mechanism amplifier was embedded on the CM and the stroke was increased to 432 μm but the stiffness decreased (Kim et al., 2009). Gan et al. (2007) pointed out that the dimensions of a CM were determined by the static stiffness along three directions and the stiffness of the CM of his FTS along the motion direction was $4.29 \, \text{N} \, \mu\text{m}^{-1}$; the mass of the rigid part, the stroke and the natural frequency were 170 g, 4.6 μm and 69 Hz respectively. Tian et al. (2009) discovered the contact condition between the actuator and the rigid part of a CM was affected by a low stiffness; the maximum stress analyzed by using FEM was less than the yield limit of the material; the best stiffness was $13.1 \, \text{N} \, \mu\text{m}^{-1}$ by considering the contact condition and the stroke reduction; the mass of the rigid part could be reduced to a certain level to get the desired high dynamic frequency, 1122 Hz. Huo and Cheng (2008) put forward that the higher stiffness and first frequency of the CM reduce effectiveness of the piezoelectric actuator stoke. Both the theoretical calculation and FEA analysis of the CM were used to decide the dimension of the CM, the static stiffness and the first natural frequency. Therefore, the design method of the CM in FTS is not clear till now.

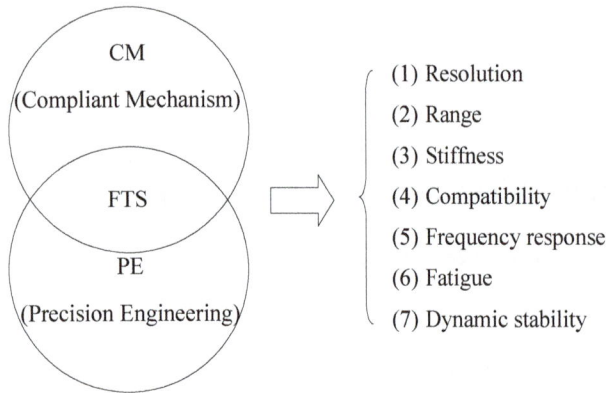

Figure 1. FTS is in the intersection of CM and PE and its design requirements are displayed.

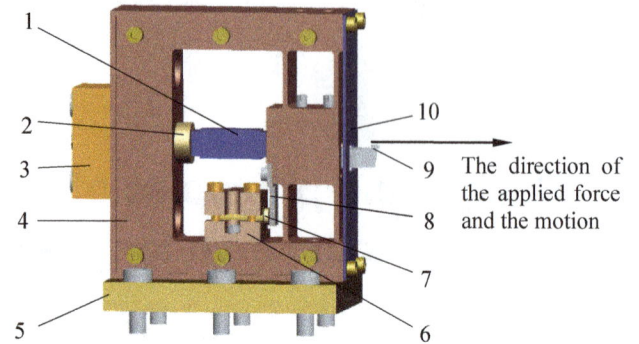

Figure 2. A schematic view of the FTS. The main parts concludes: (1) the piezoelectric actuator, (2) the preload screw, (3) the lock screw base, (4) the CM structure, (5) the base, (6) the sensor holder, (7) the sensor, (8) the sensor plate, (9) the cutting tool, and (10) the front cover.

This paper presents the design criteria and method of a CM to predetermine the dimensions of the CM in design stage. The CM in a FTS should satisfy the requirements of resolution, working range, higher first natural frequency, compatibility with holding various cutting tools and non-fatigue. Then an innovative design of a 1DOF linear FTS and its design principles are developed. FEA is used to evaluate the performance of the design in detail and preliminary results are obtained and discussed. Finally the conclusions are outlined.

2   Conceptual design and principles

A typical FTS consists of four major parts. They are a CM holding a cutting tool, a piezoelectric actuator, a displacement sensor and a control system. Therefore, the design of a FTS should be considered from both CM and PE. So the FTS is the intersection of CM and PE shown in Fig. 1. Based on the understanding of the FTS, there are seven critical parameters or issues that should be considered; they are resolution, range, stiffness, compatibility, frequency response, fatigue and dynamic stability.

An innovative design of FTS is designed to hold various diamond cutting tools and to perform turning operations as shown in Fig. 2. It consists of several parts: a CM structure, a piezoelectric actuator, a preload screw, a lock screw, a diamond cutting tool, a sensor, two covers on both sides, a front cover, a base, etc. The CM can be manufactured by wire-cut EDM to avoid machining errors.

Piezoelectric actuator has high resolution, high stiffness and high frequency response. Therefore a piezoelectric stacked actuator is used for larger stroke. Three different lengths of the stacked actuators which are 18 mm (PPA10M), 28 mm (PPA20M), and 48 mm (PPA40M) from Cedrat Technologies SA can be used in this research project. Different preload screws with corresponding length for each of the piezoelectric stacked actuators are used to house the actua-

tors in the same CM. Spring steel material has been chosen in taking into account of greater Young's modulus.

The CM structure is symmetric and its rigid part can only move in one direction (axial direction), without movement along the other two directions (horizontal and vertical ones). The rigid part can hold various diamond cutting tools. The tool tip should move in the central line of the CM structure.

The working frequency of the FTS, f_{com}, is no more than the natural frequency, f_{com0}, of the CM. f_{com0} is decided by the mass of the rigid part of the CM structure, m_s, and the stiffness of the axial stiffness of the CM, K_{axial}, as expressed as

$$f_{com0} = \frac{1}{2\pi} \sqrt{\frac{K_{axial}}{m_s}} \geq \frac{f_{com}}{S_f} \tag{1}$$

where S_f is the safety factor of the frequency response of the FTS.

The smaller the mass of the rigid part is, the higher the natural frequency becomes. However, the lower the axial stiffness is the lower the natural frequency turns out. While at the same time, a high stiffness will reduce the range of the FTS and consumes more energy. Therefore, two criteria should be considered when the natural frequency of the CM is decided. One criterion is the mass of the rigid part should be reduced to minimum. Another is the appropriate axial stiffness of the CM should be chosen.

The stiffness in other two directions, which are vertical stiffness, $K_{vertical}$, and horizontal stiffness, $K_{horizontal}$, should be evaluated. They should be adequate in order that the tool tip will have very small displacement in the two directions.

Each beam bears maximum cyclic loading at both ends. The life cycle should be calculated to determine whether failure will occur.

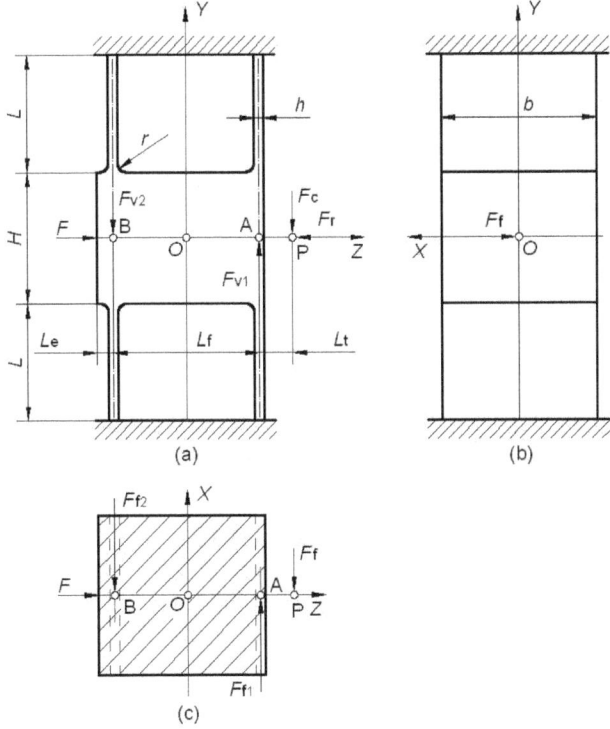

(a) (b)

(c)

Figure 3. Three orthographic projections of the CM: (**a**) front view; (**b**) side view; (**c**) top view. The reference system (the origin O and the axes X, Y, and Z), the critical dimensions (L, H, b, h, L_e, L_f, L_t), the force components (the actuating force F, the cutting force F_c, the feed force F_f, and the thrust force F_r at the tool tip P; the reaction force F_{v1}, F_{f1} at the centre point A of the front set of beams, the reaction force F_{v2}, F_{f2} at the centre point B of the rear set of beams) are also displayed.

3 Design analysis

3.1 The design of the CM

The CM consists of two sets of parallel beams; each set of beams is symmetric and fixed both ends with a rigid part and the base respectively. Figure 3 shows the three orthographic projections of the CM; the rigid part holds the tool.

The critical dimensions of the CM are shown in Table 1. The thickness (h), width (b) and length (L) of the beams play an important role in determining the stiffness of three directions. The beams also should be designed in fillet or radius to prevent the distribution of stress concentration. The width of the rigid part is the same as that of the beam.

A Cartesian coordinate has been defined. XOZ is in the horizontal symmetrical centre plane of the CM; XOY is in the vertical symmetrical plane of the CM. The tool tip, P, is located in Z-axis. The set of beams near the tool tip is called the front set of beams and the other one is the rear set of beams. The centric point of the front set of beams and the rear set of beams are A and B, respectively. When the tool is cutting a work piece, the force applied on the cutting tip is

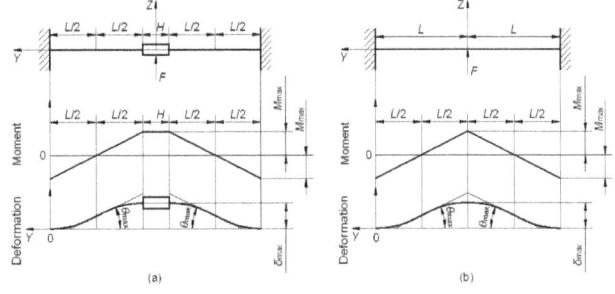

Figure 4. The moment, the deformation, and the rotated angle of the CM from (**a**) conceptual design and (**b**) simplified design. The maximum moment M_{max}, the maximum deformation δ_{max}, and the maximum rotated angle θ_{max} of the CM are also displayed.

divided into three ones; they are cutting force, F_c, feed force, F_f and thrust force, F_r. The actuating force the actuator applies on the rigid part, F, is in the Z-axis direction.

The axial stiffness is much smaller than its vertical and horizontal stiffness. When an axial force is applied to the rigid part in Z-axis the rigid part will move; each beam will bear an axial shear and a tensile stress along its whole length, and bear a moment at both ends of each beam.

3.2 Stiffness

3.2.1 Axial stiffness

Each set of the beams of the CM, which we call the designed beam, can be regarded as a beam, which we call the simplified beam, with both fixed ends and a central load as shown in Fig. 4 because all the beams in both the designed beam and the simplified beam bear the same shear stress, tensile stress and moment.

The axial stiffness of the CM can be calculated according to the linear-elastic beam theory. For a beam with both fixed ends and a centre load, F_{axial}, the axial displacement, δ_{axial} at the centre of the beam, i.e., the axial displacement of the rigid part is expressed as (Howell, 2001; Noh et al., 2009)

$$\delta_{axial} = \frac{F_{axial}(2L)^3}{192EI} \qquad (2)$$

where E is Young's Modulus and I is moment of inertia which is equal to $\frac{bh^3}{12}$.

So we can get

$$F_{axial} = 2Eb\left(\frac{h}{L}\right)^3 \delta_{axial} \qquad (3)$$

Because there two sets of beams with both fixed ends and a centre load, the axial stiffness of the CM can be obtained

$$K_{axial} = 4Eb\left(\frac{h}{L}\right)^3 \qquad (4)$$

Table 1. Critical dimensions of the CM.

Item	Definition	Value/mm
h	The thickness of the beam	1.7
b	The width of the beam	20
L	The length of the beam	20
H	The height of the rigid part	26
L_f	The distance between the two sets of beams in horizontal direction	18
L_e	The distance between the end of the rigid part and the rear set of beams in horizontal direction	6
L_t	The distance between the tip of the tool and the front set of beams in horizontal direction	13.85
r	The fillet of the CM	1.5

3.2.2 Horizontal stiffness

The horizontal stiffness of the two sets of beams with both fixed ends and a centre load are calculated respectively and they are equal. They are both calculated by using the same method used in calculating axial stiffness (see Eq. 3) except the moment of inertia is $\frac{b^3 h}{12}$. The horizontal stiffness is expressed as below

$$K_{\text{horizontal}} = 2Eh\left(\frac{b}{L}\right)^3 \qquad (5)$$

3.2.3 Vertical stiffness

The vertical stiffness of the two sets of beams with both fixed ends are also calculated respectively and they are equal as well as. Assumed the applied vertical force is F_{vertical}. The lengths of each beam in one set of beams are equal. The forces applied on each beam are equal but one beam is compressed and another is extended. Therefore, the forces applied on each beam are $F_{\text{vertical}}/2$. The equation is obtained as

$$E = \frac{\sigma}{\varepsilon} = \frac{F_{\text{vertical}} L}{2 A_b \Delta L} \qquad (6)$$

where ΔL is the length of extension, A_b is the sectional area of the beam and it is equal to bh.

Transforming Eq. (6), the vertical stiffness is expressed as

$$K_{\text{vertical}} = \frac{F_v}{\Delta L} = \frac{2EA_b}{L} = \frac{2Ebh}{L} \qquad (7)$$

3.2.4 Comparison of the vertical and the horizontal stiffness

Because the FST is used for both face turning and outline turning the vertical stiffness should be equal to the horizontal stiffness. Letting Eq. (5) is equal to Eq. (7), we can obtain $b = L$ which means the length and the width of the beam are equal.

3.2.5 Comparison of the axial stiffness with the vertical/horizontal stiffness

Both the vertical stiffness and the horizontal stiffness should be high enough to ensure that the displacement of the tool tip remains very small in both the vertical and the horizontal directions. Comparing Eqs. (5) and (4), the ratio of vertical stiffness to axial stiffness is expressed as

$$\frac{K_{\text{horizontal}}}{K_{\text{axial}}} = \frac{1}{2}\left(\frac{b}{h}\right)^2 \qquad (8)$$

3.2.6 Stiffness of the FTS

The actuator and the CM are in parallel connection. Therefore the stiffness of the FTS is equal to $K_{\text{FTS}} = K_{\text{act}} + K_{\text{axial}}$, where K_{act} is the stiffness of the actuator.

3.3 Natural frequency

The mass of the rigid part, which is approximately equal to a cube whose volume is the outline of the rigid part, is obtained as following equation

$$m_s = \rho(L_e + h + L_f)bH \qquad (9)$$

where ρ is the density of the material, L_e the distance between the end of the rigid part and the rear set of beams in horizontal direction, L_f the distance between the two sets of beams in horizontal direction, H the height of the rigid part

The volume of the FTS should be compact. Therefore the length of the beam, L, or the width of the beam, b, are determined and the axial stiffness is obtained. According to Eq. (1), the natural frequency of the CM can be calculated.

3.4 Force and deformation analysis

3.4.1 Vertical force analysis and deformation analysis

The vertical forces applied to the two sets of beams, respectively, are

$$F_{v1} = \frac{L_t + L_f}{L_f} F_c \qquad (10)$$

$$F_{v2} = \frac{L_t}{L_f} F_c \quad (11)$$

where L_t is the distance between the tip of the tool and the front set of beams in horizontal direction.

Correspond displacements according to $K_{horizontal}$ respectively are

$$\delta_{vi} = \frac{F_{vi}}{K_{vertical}} \quad (i=1,2) \quad (12)$$

The reaction force and deformation in the front set of beams are bigger than those in the rear set of beams, respectively.

3.4.2 Horizontal force analysis and deformation analysis

The horizontal forces applied to the two sets of beams, respectively, are

$$F_{h1} = \frac{L_t + L_f}{L_f} F_f \quad (13)$$

$$F_{h2} = \frac{L_t}{L_f} F_f \quad (14)$$

Correspond displacements according to $K_{horizontal}$ respectively are

$$\delta_{hi} = \frac{F_{hi}}{K_{horizontal}} \quad (i=1,2) \quad (15)$$

The reaction force and the deformation in the front set of beams are bigger than those in the rear of beams, respectively. The moment in the front set of beams bigger than that in the rear set of beams. The maximum moment locates at both ends of the beam and its value is

$$M_{h1} = \frac{F_{h1}L}{8} \quad (16)$$

3.4.3 Axial force and deformation analysis

The maximum moment occurs at both ends of each beam is shown as

$$M_{axial\,max} = \frac{\frac{F_{axial}}{2} 2L}{8} = \frac{F_{axial}L}{8} \quad (17)$$

where the maximum stress is calculated as below

$$\sigma_{axial\,max} = \frac{M_{axial\,max}C}{I} = \frac{\frac{F_{axial}L}{8}\frac{h}{2}}{\frac{bh^3}{12}} = \frac{3F_{axial}L}{4bh2} \quad (18)$$

Substituting Eq. (3) into Eq. (18), the results is as following

$$\sigma_{axial\,max} = \frac{M_{axial\,max}C}{I} = \frac{\frac{F_{axial}L}{8}\frac{h}{2}}{\frac{bh^3}{12}} = \frac{3Eh}{2L^2}\delta_{axial} \quad (19)$$

3.5 Fatigue analysis

The modified Basquin equation is used to determine the life cycle till failure; the life cycle is calculated as below (Spotts et al., 2004)

$$N = \left(\frac{\sigma_R}{A}\right)^{\frac{1}{B}} \quad (20)$$

where σ_R is the completely equivalent reverse stress, which can be determined by A and B as following

$$\sigma_R = \frac{K_f \sigma_r \sigma_u}{\sigma_u - \sigma_{avg}} \quad (21)$$

$$A = \frac{\sigma_e}{10^{6B}} \quad (22)$$

$$B = \frac{\log(\sigma_e) - \log(0.9\sigma_u)}{3} \quad (23)$$

where K_f is the fatigue stress concentration factor, σ_r the range stress, σ_u the ultimate strength, σ_{avg} the average stress, σ_e the endurance limit.

4 Preliminary results and discussion

The material properties of the spring steel are: Young's modulus E of 210 GPa; the Poisson's ratio, v, of 0.3; the density, ρ, of 7.85×10^{-3} g mm^{-3}; the ultimate strength, σ_u, of 1274 MPa; the endurance limit, σ_e, of 700 MPa. The fatigue stress concentration factor K_f is 1.4.

All the dimensions of the FTS are determined by above detailed analysis. The dimensions of the CM are shown in Table 1. The mass of the rigid part m_s is approximately 104.9074 g. The first natural frequency is 1578 Hz. The safety factor S_f is assumed 1.05.

The applied forces are assumed: cutting force F_c of 10 N; feed force F_f of 10 N; the preload axial force of 20.635 N; the maximum axial force of 206.346 N.

4.1 Design analysis results

Some critical parameters of the FTS are obtained; the axial stiffness K_{axial} is 10.3173 N μm^{-1}; both the horizontal stiffness $K_{horizontal}$ and the vertical stiffness $K_{vertical}$ are 0.714 N nm^{-1}.

The horizontal force applied to the front set of beams F_{h1} and that in the rear set of beams F_{h2} are 17.69 N and 7.69 N, respectively. The vertical force applied to the front set of beams F_{v1} and that in the rear set of beams F_{v2} are 17.69 N and 7.69 N, respectively. The horizontal displacements of the end of front set of beams δ_{h1} and that in the rear set of beams δ_{h2} are 24.78 nm and 10.78 nm, respectively. The vertical displacements of the end of front set of beams δ_{v1} and that in the rear set of beams δ_{v2} are 24.78 nm and 10.78 nm, respectively.

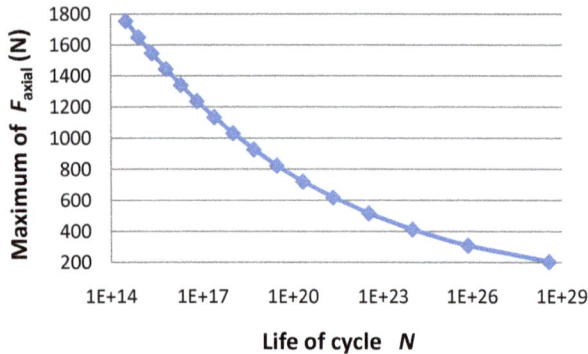

Figure 5. Relationship between the maximum axial force and the life cycle.

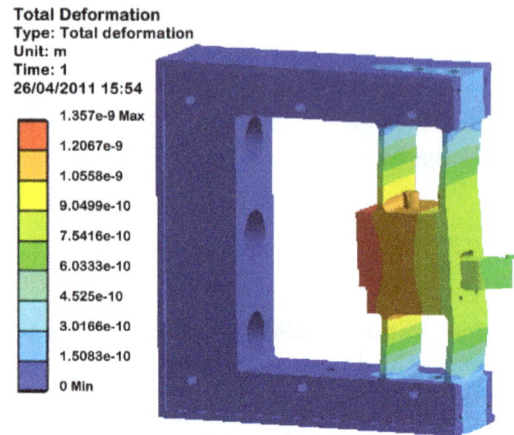

Figure 6. Deformation of the CM structure only subjected to gravity.

The relationship between the maximum axial force and the life cycle N is shown in Fig. 5. It clearly shows the life cycle is more than 10^{28} when the maximum axial force is 206.346 N. Therefore, fatigue failure will not occur.

4.2 FEA results

In order to verify the results of the design analysis method, FEM is used to doing static analysis of the CM structure including the cutting tool and its locking bolts.

The first static analysis was to investigate the influence of gravity on the tool tip and gave a value of 1.3 nm as shown in Fig. 6.

The first reason for finding the value for static analysis is to evaluate the displacement from gravitational force error when the machine is in static conditions. The second reason was finding the weakest point of the FTS structure which will make the greatest effect during machining in term of further modification and model updating. From that point of view, the modal analysis was performed under the same conditions. Figure 7 shows the modes shape of the desktop machine including 1st mode 1634 Hz (actuation in Z-axis, i.e., axial direction), 2nd mode 6947 Hz (twisting, left-right in Y-axis), 3rd mode, 8792 Hz (twisting and bending) and 4th mode 12 305 Hz (twisting in X-axis).

Figure 8 shows the shear stress analysis in X-, Y- and Z-directions. Both in the directions of X and Y 10 N are applied to investigate the shear area. In the Z-direction, the applied force is 200 N for investigating the stiffness of the CM. It clearly shows in Fig. 8 that the maximum shear stress occurs around the hinges where the beams connected with the rigid part or the base. The red colour presents where the maximum shear stress occurs.

The displacement in Z-direction when force is applied from 0 to 200 N, increased by 10 N, is analyzed. The relationship between the displacement and the force in the Z direction shows in Fig. 9. It can be observed that the relationship between them is in the linear manner. The gradient of the line; the axial stiffness of the CM in the axial direc-

tion is $11.979\,\mathrm{N\,\mu m^{-1}}$; the correlation coefficient is 1.0; the Y-intercept is $-0.0248\,\mathrm{N}$.

4.3 Discussion

1. The first natural frequency of the CM analyzed by FEA is close to the detailed analysis. The FTS can work at no more than 1400 Hz which is less than 10 % of the first natural frequency 1634 Hz.

2. Both the vertical and horizontal stiffness are greater than the axial stiffness and their displacements are very small when applied cutting force and feed force. This kind of design of the CM assures the 1DOF motion of the tool tip and the manufacturing accuracy.

3. The smaller the length of the tool tip protruding from the rigid part of the CM is, the smaller the displacement of the rigid part in horizontal and vertical directions becomes. The high rigidity of the tool is necessary.

4. Fatigue should not occur in the FTS during the analysis.

5   Conclusions

A comprehensive and robust design method of compliant fast tool servo, which has a 1DOF translational motion, for precision engineering is presented. The stiffness in three directions and the first natural frequency of the CM can be predetermined. The reaction forces and maximum deformation analysis of the CM can be calculated. The results of FEA are matching to that of the design requirements. The fatigue of the CM is predicted and it should work without failure.

The author would like to report the experimental cutting trial results by using the fast tool servo in alignment with the FEA-based simulation and analysis as an extension in the future.

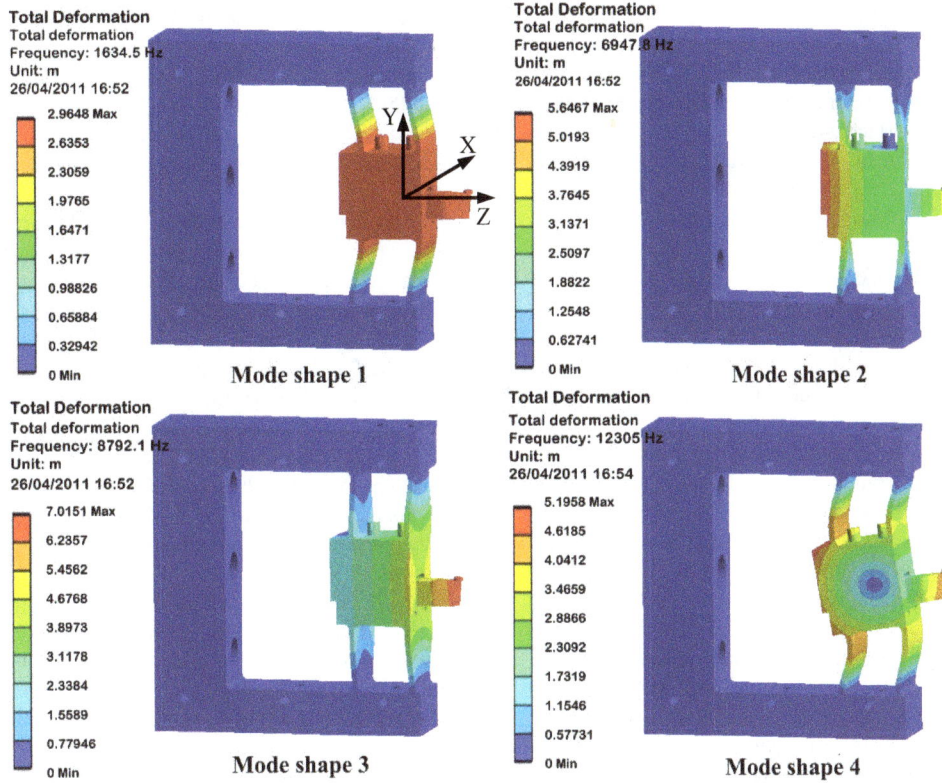

Figure 7. The first four natural frequencies and mode shapes.

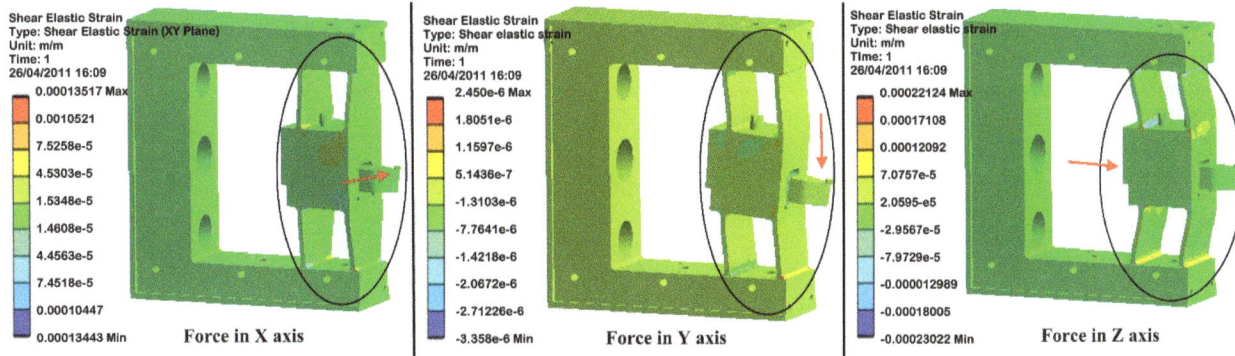

Figure 8. The shear stain and stress area in X-, Y- and Z-direction applied force.

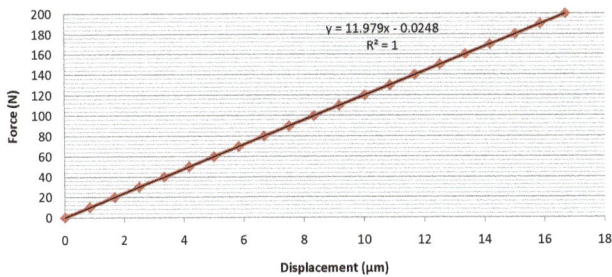

Figure 9. Stiffness experiment in Z-axis.

Acknowledgements. The authors would like to thank Paul Yates and Chao Wang at Brunel University for helpful discussions and the partial support of The National High-tech R&D Program (863 Program, SQ200802LS1479459) for funding the visiting project at Brunel University.

Edited by: N. Tolou

References

Cuttino, J. F., Miller, A. C., and Schinstock, D. E.: Performance optimization of a fast tool servo for single-point diamond turning machines, IEEE-ASME T. Mech., 4, 169–179, 1999.

Gan, S. W., Lim, H. S., Rahman, M., and Watt, F.: A fine tool servo for global position error compensation for a miniature ultra-precision lathe, Int. J. Mach. Tool. Manu., 47, 1302–1310, 2007.

Howell, L. L.: Compliant Mechanisms, John Wiley & Sons, Inc., New York, USA, 2001.

Huo, D. and Cheng, K.: A dynamics-driven approach to the design of precision machine tools for micro-manufacturing and its implementation perspectives, in: Proceedings of ImechE Part B: Journal of Engineering Manufacture, 222, 1–13, 2008.

Kim, H. S. and Kim, E. J.: Feed-forward control of fast tool servo for real-time correction of spindle error in diamond turning of flat surfaces, Int. J. Mach. Tool. Manu., 43, 1177–1183, 2003.

Kim, H. S., Kim, E. J., and Song, B. S.: Diamond turning of large off-axis aspheric mirrors using a fast tool servo with on-machine measurement, J. Mater. Process. Tech., 146, 349–355, 2004.

Kim, H. S., Lee, K. I., Lee, K. M., and Bang, Y. B.: Fabrication of free-form surfaces using a long-stroke fast tool servo and corrective figuring with on-machine measurement, Int. J. Mach. Tool. Manu., 49, 991–997, 2009.

Kouno, E.: A fast response piezoelectric actuator for servo correction of systematic errors in precision machining, Annals of the CIRP, 33, 369–372, 1984.

Li, S. Z., Yu, J. J., Pei, X., Su, H. J., Hopkins, J. B., and Culpepper, M. L.: Type synthesis principle and practice of flexure systems in the framework of screw theory Part III: numerations and type synthesis of flexure mechanisms, in: Proceeding of the ASME 2010 international design engineering technical conferences & computers and information in engineering conference IDETC/CIE 2010, Montreal, Quebec, Canada, 15–18 August 2010, DETC2010-28963, 2010.

Noh, Y. J., Nagashima, M., Arai, Y., and Gao, W.: Fast positioning of cutting tool by a voice coil actuator for micro-lens fabrication, International Journal of Automation Technology, 3, 257–262, 2009.

Spotts, M. F., Shoup, T. E., and Hornberger, L. E.: Design of Machine Elements, 8th Edn., Pearson Prentice Hall, New Jersey, 2004.

Tian, Y., Shirinzadeh, B., and Zhang, D.: A flexure-based mechanism and control methodology for ultra-precision turning operation, Precis. Eng., 33, 160–166, 2009.

Yu, J. J., Li, S. Z., Pei, X., Su, H. J., Hopkins, J. B., and Culpepper, M. L.: Type synthesis principle and practice of flexure systems in the framework of screw theory Part I: general methodology, in: Proceeding of the ASME 2010 international design engineering technical conferences & computers and information in engineering conference IDETC/CIE 2010, Montreal, Quebec, Canada, 15–18 August 2010, DETC2010-28783, 2010a.

Yu, J. J., Pei, X., Li, S. Z., Su, H. J., Hopkins, J. B., Culpepper, M. L.: Type synthesis principle and practice of flexure systems in the framework of screw theory Part II: numerations and synthesis of complex flexible joints, in: Proceeding of the ASME 2010 international design engineering technical conferences & computers and information in engineering conference IDETC/CIE 2010, Montreal, Quebec, Canada, 15–18 August 2010, DETC2010-28794, 2010b.

8

Compliant mechanisms for an active cardiac stabilizer: lessons and new requirements in the design of a novel surgical tool

L. Rubbert[1], **P. Renaud**[1], **W. Bachta**[2], **and J. Gangloff**[1]

[1]LSIIT, Université de Strasbourg-CNRS, Strasbourg, France
[2]ISIR, Université Pierre et Marie Curie-CNRS, Paris, France

Abstract. In this paper, three aspects of the use of compliant mechanisms for a new surgical tool, an active cardiac stabilizer, are outlined. First, the interest of compliant mechanisms in the design of the stabilizer is demonstrated with *in vivo* experimental evaluation of the efficiency of a prototype. We then show that the specific surgical constraints lead to the development of compliant mechanisms, with the design of new original mechanical amplifiers. Finally, the requirements in the design of stabilizers exhibiting a higher level of integration are outlined. Novel architectures and design procedures are actually needed, and we introduce an exploratory study with a proof-of-concept designed using the combination of ant colony optimization and classical pseudo rigid body modeling. Relative errors in the estimation of the displacement do not exceed 5 %. The proposed design method constitutes an interesting approach that may be applied more generally to the design of compliant mechanisms.

1 Introduction

Compliant mechanisms exhibit higher accuracy and compactness than conventional mechanisms. They can consequently contribute to the development of Minimally Invasive Surgery (MIS) procedures, high accuracy procedures that require small size instruments. The use of compliant devices also simplifies the sterilization process (Rebello, 2004), and the suppression of lubrication improves surgical compatibility. Surgical tools for organ manipulation (Awtar et al., 2010), grasping (Frecker et al., 2005a) or tissue cutting (Frecker et al., 2005b) have therefore been proposed previously.

In the Cardiolock project, we investigate the design of an active cardiac stabilizer, a new surgical tool for heart surgery. In this context, this article has three aims. First, we show that the field of heart surgery can benefit from the field of compliant mechanisms by outlining the efficiency of a compliant-based active cardiac stabilizer with *in vivo* experimental re-

sults. Second, we show that the surgical context introduces design requirements that can lead to contributions in the field of compliant mechanisms. An original compliant mechanism is introduced, whose architecture is indeed directly derived from the surgical necessities. Finally, we propose an exploratory study on the design of a highly integrated active stabilizer: a proof-of-concept is introduced with a design methodology based on ant colony optimization and pseudo rigid body modeling.

The paper is organized as follows. The principle of active stabilization and the efficiency of the use of a compliant mechanism for active stabilization are presented in Sect. 2. The design of an original compliant mechanism based on the surgical requirements is introduced in Sect. 3, with the use of kinematic singularities to synthesize an amplification mechanism. Then, in Sect. 4, a design methodology is proposed to assess the level of integration that can be achieved in order to design an active stabilizer. A very recent proof-of-concept is experimentally evaluated, before concluding in Sect. 5.

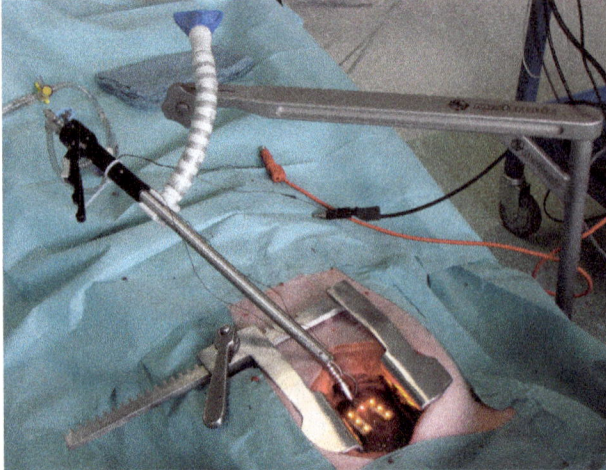

Figure 1. Passive stabilizer during *in vivo* evaluation on pig.

2 Active stabilization for cardiac surgery

In the field of heart surgery, one of the most common interventions is Coronary Artery Bypass Grafting (CABG). In order to improve the supply of blood to the myocardium, arterial or vein grafts are connected to the coronary arteries. The most delicate part of a CABG is the suture of the grafts to the coronary arteries. These arteries have a diameter of 1–2 mm, and more than 10 knots must be performed around each artery. For the patient, the best approach is to perform such a procedure on a beating heart, i.e. without the use of an external heart-lung machine, and with a minimally invasive approach. Only small incisions are then made to insert the surgical tools and an endoscope through trocars.

Today, in that situation surgeons use a stabilizer, a device that aims at locally immobilizing the beating heart surface during the suturing process. In a MIS context, this stabilizer is composed of fingers applied on the heart and a long shaft (Fig. 1) that is inserted through the subxiphoid process, at the base of the sternum, to reach the area of interest on the heart surface.

From a mechanical point of view, one can easily understand that the forces developed by the heart, in the order of 5 N (Bachta et al., 2008), cause the deformation of the mechanism. The current commercial device (Fig. 1) exhibits displacements at its tip that exceed the required accuracy, as outlined by surgeons (Cattin et al., 2004; Lemma et al., 2005). Displacements of the stabilizer tip should remain in the order of 0.1 mm, with respect to the suturing task described earlier.

As a consequence, we have proposed in the Cardiolock project (Bachta et al., 2007) the development of an active device to improve the stabilization accuracy and to allow CABG to be performed satisfactorily on a beating heart with MIS. The main idea is to use the endoscope introduced during the surgery (Fig. 2) to observe the presence of any deflection of the stabilizer and compensate, in real time, for the

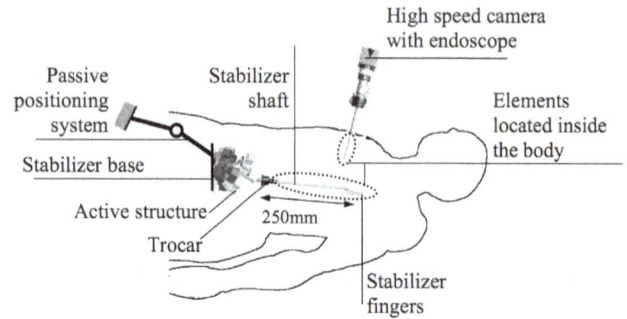

Figure 2. Principle of insertion of the active stabilizer.

displacement of the area of interest. Compliant mechanisms obviously appear of interest when combined with piezoelectric actuation: high dynamics can be obtained without any backlash. Thus the Cardiolock principle relies on the use of a compliant mechanism controlled by a piezoelectric actuator using the information given by the endoscope.

3 Compliant mechanisms for a new surgical tool

3.1 Design requirements

The stabilization task consists of compensating, in the presence of the heart force, for displacements in the order of 1 mm (Bachta et al., 2008). The main source of displacement of the stabilizer tip is the bending of the shaft. As a consequence, two directions of compensation at the stabilizer tip are needed. The displacement along the shaft axis is not significant. Forces exerted by the heart on the stabilizer have been assessed experimentally (Bachta et al., 2008). The amplitude of the force is in the order of 5 N in the direction perpendicular to the stabilizer shaft.

The stabilizer is in interaction with the heart at its tip, but also with the trocar that allows its insertion into the patient's body. As a consequence, the movement of the trocar due to respiratory motion creates forces on the stabilizer body in addition to the heart forces. A simple way to avoid these additional perturbations increasing the complexity of the stabilizer control is to consider that the compliant mechanism and the actuator are located outside the body and allow the stabilizer shaft to rotate with respect to the trocar. This simultaneously simplifies the sterilization of the device, since the active elements are located outside the patient's body. Revolute joints cannot be located at the trocar position, which means that the stabilizer must exhibit a Remote Center of Motion (RCM).

3.2 Design of a first device: Cardiolock 1

Cardiolock 1 is a first prototype that has been designed to assess the efficiency of the active stabilization approach during *in vivo* experiments. Experiments on pigs cannot be

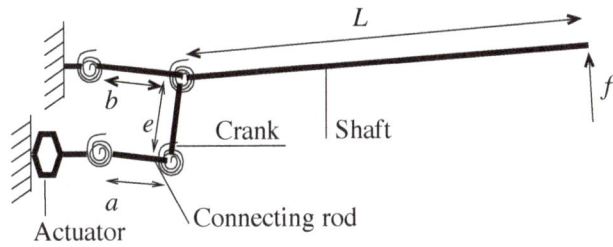

Figure 3. PRBM model of the device.

Figure 4. CAD view of Cardiolock 1, an active stabilizer with 1 DOF.

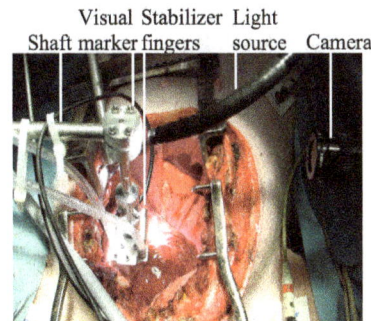

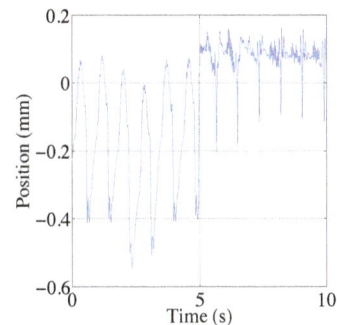

performed with a MIS approach for anatomical reasons. The RCM constraint was therefore not taken into account and compensation is performed only in one direction.

An actuator with integrated amplification structure (Cedrat Technologies APA120ML) is considered. Its translational movement is converted using a slider-crank system (Fig. 3). The stabilizer shaft is connected to the crank, the actuator is the slider which is connected to the crank by means of a connecting rod. The mechanism synthesis has been carried out using a Pseudo Rigid Body Model (PRBM, Howell, 2001). A torsion spring is used to represent each compliant joint and an additional joint is introduced to represent the flexibility of the shaft. The achievable displacement can be easily derived from Fig. 3 (Bachta et al., 2007). The stiffness of each compliant joint is described with the analytical model of symmetric right-circular profile joints (Howell, 2001) and the model from Pham and Chen (2002) gives the maximum stresses in the joints. The parameters to be determined are the geometrical parameters (a,b,e) and the minimum thickness t of the compliant joints. A non linear optimization was achieved to determine the best set of parameters. The optimization criterion is the size of the device. The compensation condition as well as the maximum admissible stresses in the joints are considered as non linear constraints. The CAD view of the system corresponding to the result of the optimization is introduced in Fig. 4. Further evaluation of the PRBM accuracy by comparison with simulations using Finite Element Analysis (FEA) showed the relevance of the model. The stabilizer tip maximum displacement is correctly described by the PRBM as well as its stiffness.

3.3 Experimentation – stabilization efficiency

The stabilizer tip displacement given by the PRBM could be reached within 5 % with laboratory experiments. The device stiffness and the eigenfrequencies are however significantly different from the values estimated with FEA: eigenfrequencies are for instance lowered by more than 13 %. Further analysis demonstrated that the assembly of the mechanism, and especially the connection of the connecting rod to the actuator lowers the device's performances. Even though the elements are rigidly connected, the behavior of the device remains very sensitive to the quality of assembly.

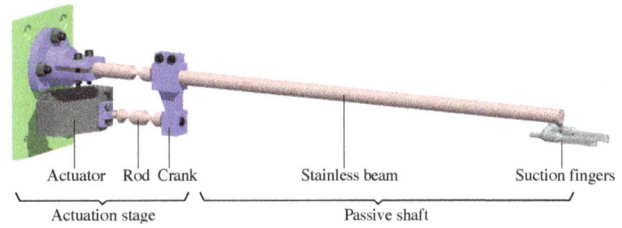

Figure 5. Cardiolock 1 prototype during *in vivo* experiments (left) and corresponding recorded residual displacement (right).

After the implementation of the stabilizer control, *in vivo* experiments have been performed. The stabilizer tip is positioned above the sternum of a 35 kg pig, with a custom end-effector to reach the beating heart. A high-speed camera (Dalsa CAD-6, 333 fps) is used to detect the stabilizer tip displacement. It is positioned in front of the stabilizer tip to follow a marker attached to the stabilizer tip. Because of the anatomy of the animal, a custom end-effector is added to access to the heart surface.

The position of the stabilizer tip measured with the camera is shown in Fig. 5. During the active stabilization, after 6 s on the graph, the standard deviation of the position of the stabilizer tip is equal to 35 microns. The device efficiency is high enough to confirm the benefit of using a compliant mechanism in combination with vision to develop a new surgical tool, an active stabilizer.

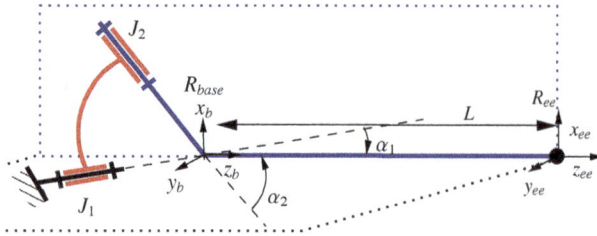

Figure 6. Kinematic scheme of Cardiolock 2.

4   A new compliant device from surgical requirements: Cardiolock 2

4.1   Stabilizer kinematics

As introduced in Sect. 3.1, the design objective is to provide the stabilizer with 2 DOF in rotation, with respect to the RCM located at the trocar. Parallel architectures can be considered, with spatial arrangements (Gosselin and Angeles, 1989; Gregorio, 2004), sometimes with the use of spherical joints, but they seem very difficult to manufacture as compliant devices. The experimental evaluation of Cardiolock 1 has also shown that the complexity of the design has to be limited, in particular with a small number of elements in the assembly to minimize sources of flexibility. As a consequence, a simple serial mechanism is considered as represented in Fig. 6 (Bachta et al., 2009). Displacements around the represented configuration can be obtained by the rotation of joints J_1 and J_2, with a decoupling of the two displacements that simplifies the stabilizer control.

The displacement of the stabilizer tip is a function of the angles α_1 and α_2, the length L being constrained by the medical requirements. From a dynamic point of view, parameters α_1 and α_2 should be minimized to get a compact structure with lower inertias. This implies the development of compliant joints actuated with piezoelectric actuators that provide large rotations.

4.2   From kinematic singularities to mechanical amplifiers

Piezoelectric actuators provide linear motions that need to be converted into rotations, that we want to maximize. Parallel mechanisms can exhibit a behavior that corresponds to that situation, in the vicinity of the so-called parallel singularities (Gosselin and Angeles, 1990). Even though such configurations are rarely considered (Stoughton and Arai, 1992; Ranganath et al., 2006), we have thus proposed (Bachta et al., 2009) consideration of a parallel mechanism to design the compliant mechanism that would convert the actuator displacement into a rotation. The parallel mechanism at the origin of our amplifier is represented in Fig. 7. Parallel singularity corresponds to the situation $\epsilon = 0$.

Considering the points A_1 and A_3 as immobilized and an actuator is driving the point A_2, the mechanism becomes

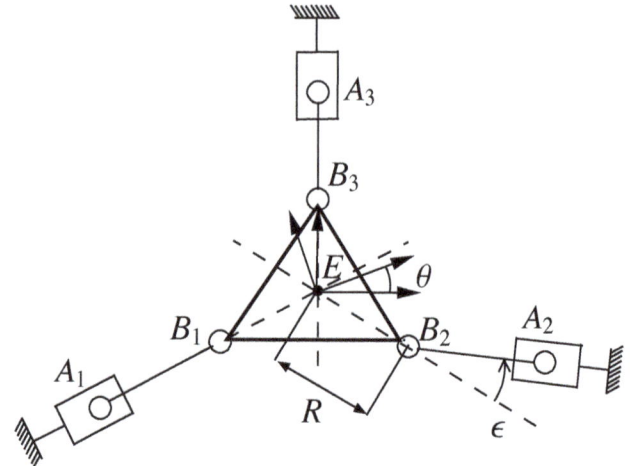

Figure 7. 3PRR parallel mechanism close to singularity.

equivalent in the represented configuration to a revolute joint, whose rotation amplitude is simply related to the actuated displacement δq of point A_2:

$$\delta\theta = \frac{1}{R\sin(\epsilon)}\delta q \tag{1}$$

with $R = \|EB_2\|$. We can easily set the amplitude of the rotation by modifying the value of ϵ, and obtain a high rotation/translation ratio. The obtained mechanical amplifier is also interesting because of its stiffness properties. In-plane stiffness is high due to its parallel nature, and out-of-plane stiffness is easily controlled by the width of the mechanism.

4.3   Design and experimentation

Selection of the device's geometrical parameters has been achieved using an iterative design process, analyzing the performances of the device with FEA. The prototype, represented in Fig. 8, exhibits displacements at the tip of the stabilizer shaft equal to 1.28 mm × 1.28 mm, which is consistent with the results of FEA and correspond to the design requirements.

5   What are the limits of the integration?

To deal with the interaction of the stabilizer with the trocar, RCM architecture has first been considered. Even if the device efficiency is satisfactory, its size still limits its ease of use. More importantly, placing the actuators outside the body simplifies the sterilization process but greatly increases the complexity of the control, because of the non-collocated sensor and actuator configuration. New requirements in the design of an active stabilizer appear: we need compliant architectures that allow the integration of actuators inside the stabilizer shaft, close to the stabilizer tip, and we need to

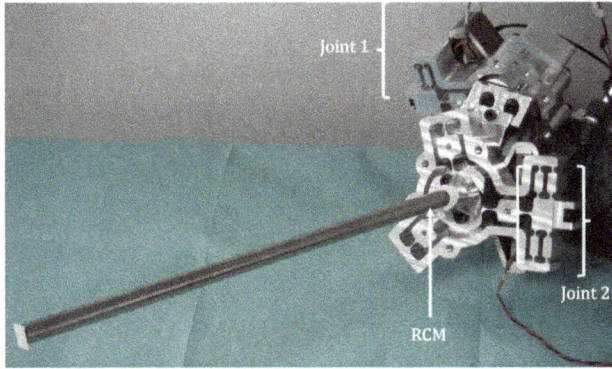

Figure 8. The Cardiolock 2 prototype.

identify methodologies for their design. In this paper, ongoing research on the design of a fully integrated device is presented with a proof-of-concept whose design is obtained using ant colony optimization associated with pseudo rigid body modeling.

5.1 Design methodology and architecture selection

5.1.1 Design methodology

As a first step, we consider the problem of the integration of a compliant mechanism and its piezoelectric actuator for a 1 DOF mechanism. Because of the strong size constraints, the actuator is considered to be positioned along the shaft axis, providing an axial displacement. A piezoelectric actuator that can be integrated in the shaft presents maximum displacements in the order of 50 microns. A compliant mechanism has therefore to be designed to first convert the displacement along the shaft axis into a displacement perpendicular to the shaft, and second to amplify this displacement. The stabilizer tip displacement amplitude should be in the order of 1 mm.

The synthesis of the mechanism can be made by a topology optimization or by selection and optimization of a predefined architecture. 1 DOF amplification mechanisms have been widely studied, and we thus elaborate a transformation mechanism from existing solutions before performing its optimization.

5.1.2 Mechanism selection

A review of the existing transformation mechanisms has been performed. Two mechanisms are of particular interest: the Scott-Russell mechanism (Tian et al., 2009) and the four-bar mechanism (Parkinson et al., 2001). The first one transforms a translation movement into another translation movement in the perpendicular direction and its oblong shape tends to be compatible with an integration inside the stabilizer shaft. The second one can amplify rotations (Sitti, 2003) as well as translations (Choi et al., 2010; Liaw and Shirinzadeh, 2008).

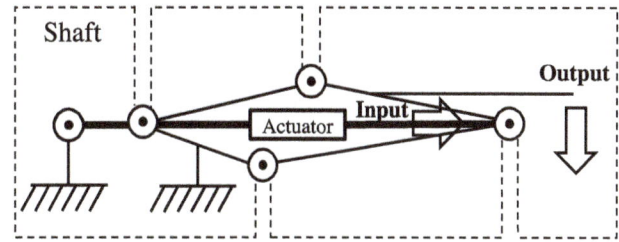

(a) Kinematic scheme of the proposed 4-bar mechanism. The dotted lines represent the contour of the bars in the structure.

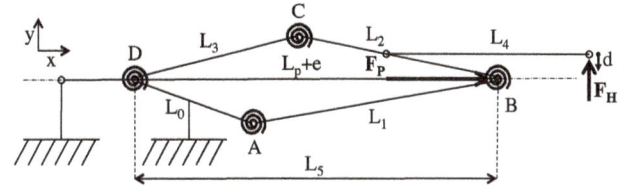

(b) Parameterization of the PRBM of the mechanism.

Figure 9. Description of the proposed active stabilizer with embedded actuation.

The compliant mechanism has to be manufactured from the stabilizer shaft, of cylindrical shape. In this context, the elements that compose the Scott-Russell mechanism have slender shapes and may lack stiffness. The four-bar mechanism is therefore chosen (Fig. 9a), with the integration of the piezoelectric actuator inside a tubular shaft. With this configuration, the force delivered by the actuator introduces tensile stresses in the mechanism bars, without any risk of buckling. To obtain the corresponding compliant mechanism, circular notch joints are preferred to leaf spring joints for their ease of manufacture and their better accuracy (Trease et al., 2004).

5.2 Optimization with Ant colony algorithm

5.2.1 Modeling

The use of a PRBM appeared in the design of Cardiolock 1 as an efficient way to describe the mechanism during its optimization. This approach is therefore still considered for this new device. The model is represented in Fig. 9b. Six geometrical parameters define the mechanism and its initial configuration (L_0, L_1, L_2, L_3, L_4 and L_5). Since the mechanism is integrated in a tube, its outer radius R and its thickness T are also needed to define the geometry. The compliant joints are described by their minimum thickness t, the radius r of their circular profile and their width, defined by their position with respect to the tube axis.

The relationship between the actuator displacement e and the output displacement d can be easily obtained by expressing loop closure equations. To obtain the achievable output displacement, a static model is derived, since the piezoelectric actuator maximum displacement depends on the stiffness of the mechanism acting against it. The compliant joint rotational stiffnesses are described using the model derived by Schotborgh et al. (2005), described as the most accurate and having the widest range of validity by Yong et al. (2008).

The stresses in the compliant joints can be expressed analytically. Stress concentration factors are included to improve the accuracy (Pilkey and Pilkey, 2008).

5.2.2 Ant Colony Optimization for compliant mechanisms

The compliant architecture is defined by sixteen geometrical parameters: six bar lengths, four joint thicknesses, four radii circular notches, one tube diameter and one tube thickness. During the optimization, we consider that the actuator is given, since we will use the actuator providing the maximum displacement that can be integrated in the tube, and the tube material properties are known. The objective of the optimization is then to maximize the output displacement d, while satisfying a set of constraints:

- The maximum stresses in the joint must be compatible with the material properties.

- The mechanism profile must remain inside the tube's outer shape.

- The angles between the mechanism bars must remain compatible with the configuration represented in Fig. 9b.

- The geometry of the joints must be compatible with the shape of the bars as represented in Fig. 9a.

- The model developed in Schotborgh et al. (2005) must remain valid, which includes the existence of a symmetry of the compliant joints with respect to their neutral axis.

- The actuator must fit in the tube.

This constitutes a set of seven linear constraints and fifteen non-linear constraints represented as a set of inequalities. For such a highly constrained problem, and a large number of parameters, gradient-based optimization algorithms usually have a low efficiency, particularly with the presence of many local minima in the optimization function. Metaheuristic optimizations, based on stochastic algorithms, can constitute interesting alternatives. Genetic algorithms and evolutionary algorithms are well known approaches. Hereafter, we propose to investigate the use of Ant Colony Optimization (ACO) in our context. ACO is essentially interesting for its ease of implementation because the method does not need a delicate tuning of many internal parameters to be efficient.

ACO mimics the behavior of ants that are able collectively to optimize the path between their nest and a source of food. ACO is therefore performed by randomly generating sets of

solutions, and iteratively restraining the generation of solutions around the most interesting solutions found in a previous step. Considered initially for combinatorial problems (Dorigo and Gambardella, 1997), it has been extended to continuous domain (Blum, 2005; Socha and Dorigo, 2008), which is our case. ACO has been proposed for structural topology optimization (Kaveh et al., 2008; Luh and Lin, 2009) but, to the authors knowledge, not yet considered for the optimization of a compliant mechanism.

Before introducing the adaptation of ACO to our context, we need to define a performance function to distinguish and rank possible solutions. The displacement d has to be maximized. For a given material, the objective is also to use in an optimal way its mechanical properties. In other words, it can be interesting to introduce into the performance function the closeness of the maximum stress to the admissible maximum stress for the material. The proposed performance function is expressed as:

$$\text{Performance } (\xi) = \frac{d}{1.01^{\text{abs}(\sigma_{\text{objective}} - \sigma_{\text{max}})}} \qquad (2)$$

with ξ the parameter set, $\sigma_{\text{objective}}$ the maximal admissible stress, σ_{max} the maximum stress in the mechanism and 1.01 a penalty factor chosen for the problem. Since ACO does not need any continuity or derivation property for the performance evaluation function, such a non-linear function can be considered and indeed will efficiently rank the solutions since the stress value is "locked" to the admissible value.

Adapted to our context, the algorithm is composed of four main steps (Fig. 10). In the first step, sets of geometrical parameters are generated randomly, using a uniform distribution law in all the parameter domain. To increase the chance of finding a viable mechanism, the constraints are divided into two sets of different importance. The most important constraints are initially considered before working in a parameter domain around the identified solutions to determine parameter sets that respects the whole set of constraints.

In the second step, for each parameter set that respects all the constraints, a new set of parameters is generated in a restrained domain surrounding the solution to improve the solution performance. The restrained domain corresponds to 20 % of the range of the parameters. To avoid a blockage of the solution evolution, the size of the restrained domain is increased by 4 % each time no better solution can be found. If the search space increase reaches 40 %, the second step is stopped. To rank the solutions, each mechanism is tested and associated to its performance value. Non-viable mechanisms are associated with a null value. It is important to note that the algorithm is quite robust with the thresholds used to restrain or widen the search space. Their values are easily set and do not strongly affect the success of the optimization.

In the third step, solution generation is performed by switching to a gaussian distribution law for the evolution of each parameter. The mean value and standard deviation are

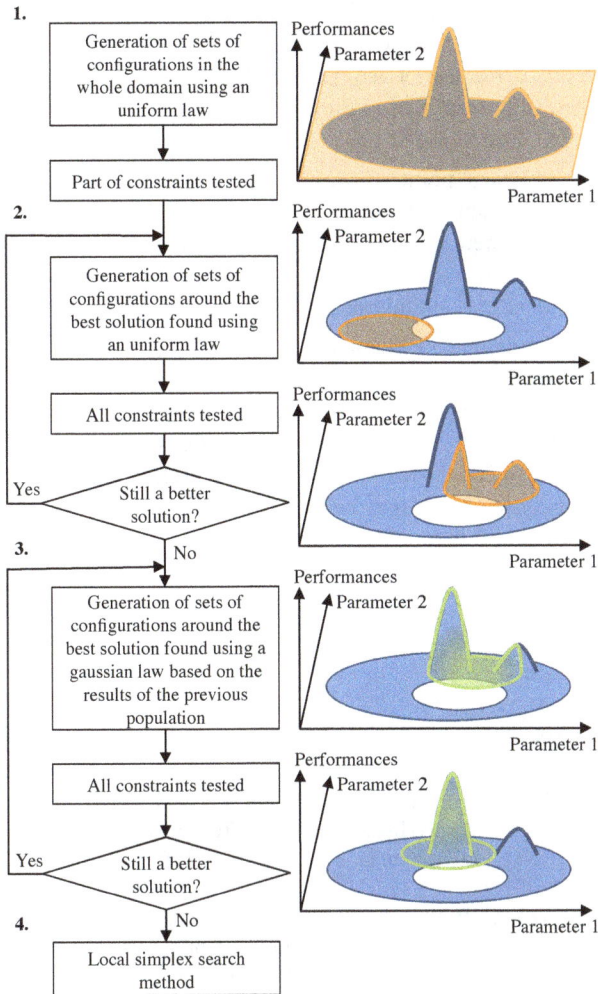

Figure 10. Algorithm structure with a simplified illustration of a two parameters problem for each step.

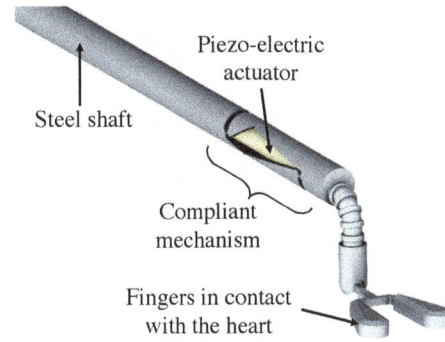

Figure 11. Cardiac stabilizer and its integrated compensation compliant mechanism.

computed using the sets of viable configurations determined in the second step.

The standard deviation of each parameter tends to zero during the third step. When all the values become small enough, local search using a simplex search method is performed to refine the solution.

5.3 Optimization of the active stabilizer

Among possible piezoelectric actuators, a device from PI, P-007.40, is considered. It has one of the largest available displacements (60 µm) with a diameter of 7 mm which allows its integration. Its length is equal to 50 mm, and the maximal force is 1150 N. A 12 mm diameter shaft in 35NCD16 steel is considered.

The ACO is implemented using the Matlab software. Even though such an implementation is not optimal, optimization for our problem was achieved in less than six hours.

With the optimal solution (Fig. 11), the stabilizer can exhibit a displacement of 0.9 mm, with maximal stress of 550 MPa. This latter value corresponds to the endurance limit of the material. It is important to note that the performance of the device is confirmed by comparison with FEA results, performed using PTC Pro/Mechanica. Relative errors in the estimation of the displacement and the stresses in the compliant joints do not exceed 5 %. The PRBM describes accurately the behavior of the device, and combining ACO with PRBM allows us to introduce an active stabilizer whose performance corresponds to the medical requirement.

5.4 Experimentation

Interesting results have been obtained in terms of mechanism optimization and design of the medical device. Experimentation has been carried out to confirm these results. For availability reasons, a slightly different piezoelectric actuator is chosen: the P-041.30 has a diameter of 12 mm, length of 80 mm and can provide a maximum displacement of 45 µm. A Z30C13 stainless steel tube of 18 mm diameter is chosen. The yield limit of the material is 500 MPa.

Optimization is performed using these new specifications. The results of the ACO are again in very good accordance with the FEA. The discrepancy in the estimation of the displacement d is below 4 % and below 7 % for the stresses. The device is designed to exhibit a maximal displacement of 0.5 mm.

The prototype is manufactured using CNC machining (Fig. 12). Its performance is evaluated using a visual marker located at the tip of the prototype and a high speed camera that allows the determination of the marker displacement. The maximum displacement is estimated as 0.5 mm (Fig. 13), the value obtained with the PRBM model as well as the FEA. Loads have been applied on the stabilizer tip to evaluate the consistency of PRBM model, FEA and experimental results. For a load up to 5.7 N, above the target value of 5 N, relative errors between the three displacement values are below 5 %. Finally an analysis of the first eigenfrequency

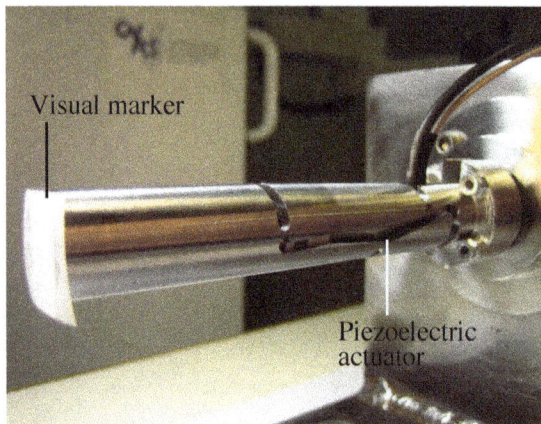

Figure 12. Prototype assembly and its machined compliant mechanism.

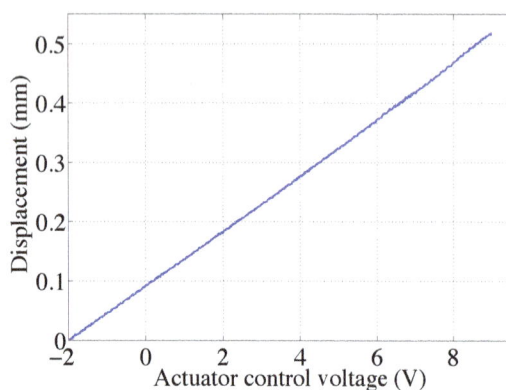

Figure 13. Experimental measure of the displacement.

has been performed. Its value is around 200 Hz, with 10 Hz of measurement uncertainty, whereas FEA simulation provides a value of 203 Hz.

Experimental results are very close to those obtained with the numerical simulation and the PRBM. This tends to show that the proposed modeling reliably describes the behavior of the device. As a consequence, the performance of the device with the 12 mm diameter tube, compatible with the medical application, should be experimentally achievable.

6   Conclusions

In this paper, the use of compliant mechanisms in the context of the design of a new surgical device has been presented. First, we have shown that combining compliant mechanisms with piezoelectric actuation can allow the development of a new surgical device. Then, in a second step, we have outlined that the surgical requirements can lead to the development of original compliant architectures. A remote center of motion architecture has been developed based on original mechanical amplifiers. The development of active stabilizers is now

related to the development of highly integrated compliant architectures, and their associated design methodologies. We have proposed the use of Ant Colony Optimization, which gives interesting results according to the first numerical and experimental evaluation of our proof-of-concept.

From a medical point of view, active stabilization is a promising approach, and current results on a fully integrated stabilizer open new perspectives. The application area of such a mechanism could actually be widened, for instance, to tremor compensation in microsurgery or micropositioning.

From a compliant mechanism design point of view, using Ant Colony Optimization for the optimization of PRBM of compliant mechanisms may constitute a new approach for the synthesis of compliant structures. Further work in the design of mechanisms such as parallel architecture for multi-DOF systems will now be carried out.

Edited by: A. Barari

References

Awtar, S., Trutna, T., Nielsen, J., Abani, R., and Geiger, J.: FlexDex: A Minimally Invasive Surgical Tool with Enhanced Dexterity and Intuitive Actuation, ASME Journal of Medical Devices, 4, 8 pp., 2010.

Bachta, W., Renaud, P., Laroche, E., Forgione, A., and Gangloff, J.: Design and control of a new active cardiac stabilizer, IEEE/RSJ Int. Conf. on Intelligent Robotos and Systems – IROS, 404–409, 2007.

Bachta, W., Renaud, P., Laroche, E., Forgione, A., and Gangloff, J.: Cardiolock: an active cardiac stabilizer, first in vivo experiments using a new robotized device, Comput. Aided Surg., 13, 243–254, 2008.

Bachta, W., Renaud, P., Laroche, E., and Gangloff, J.: Cardiolock2: Parallel singularities for the design of an active heart stabilizer, in: ICRA, 3839–3844, 2009.

Blum, C.: Ant colony optimization: Introduction and recent trends, Phys. Life Rev., 2, 353–373, 2005.

Cattin, P., Dave, H., Grunenfelder, J., Szekely, G., Turina, M., and Zund, G.: Trajectory of coronary motion and its significance in robotic motion cancellation, Eur. J. Cardio-Thorac., 25, 786–790, 2004.

Choi, K.-B., Lee, J. J., and Hata, S.: A piezo-driven compliant stage with double mechanical amplification mechanisms arranged in parallel, Sensor. Actuat. A-Phys., 161, 173–181, 2010.

Dorigo, M. and Gambardella, L. M.: Ant colonies for the travelling salesman problem, Biosystems, 43, 73–81, 1997.

Frecker, M. I., Powell, K. M., and Haluck, R.: Design of a Multifunctional Compliant Instrument for Minimally Invasive Surgery, J. Biomech. Eng., 127, 990–993, 2005a.

Frecker, M. I., Schadler, J., Haluck, R. S., Culkar, K., and Dziedzic, R.: Laparoscopic Multifunctional Instruments: Design and Testing of Initial Prototypes, JSLS, J. S. Laparoend., 9, 105–112, 2005b.

Gosselin, C. and Angeles, J.: The optimum kinematic design of a spherical three-degree-of-freedom parallel manipulator, J. Mech. Transm.-T., 111, 202–207, 1989.

Gosselin, C. and Angeles, J.: Singularity analysis of closed-loop kinematic chains, IEEE T. Robotic. Autom., 6, 281–290, 1990.

Gregorio, R. D.: Kinematics of the 3-RSR wrist, IEEE T. Robotic. Autom., 20, 750–754, 2004.

Howell, L.: Compliant mechanisms, Wiley-IEEE, 2001.

Kaveh, A., Hassani, B., Shojaee, S., and Tavakkoli, S.: Structural topology optimization using ant colony methodology, Eng. Struct., 30, 2559–2565, 2008.

Lemma, M., Mangini, A., Redaelli, A., and Acocella, F.: Do cardiac stabilizers really stabilize? Experimental quantitative analysis of mechanical stabilization, Interactive Cardiovascular and Thoracic Surgery, 4, 222–226, 2005.

Liaw, H. C. and Shirinzadeh, B.: Robust generalised impedance control of piezo-actuated flexure-based four-bar mechanisms for micro/nano manipulation, Sensor. Actuat. A-Phys., 148, 443–453, 2008.

Luh, G.-C. and Lin, C.-Y.: Structural topology optimization using ant colony optimization algorithm, Appl. Soft Comput., 9, 1343–1353, 2009.

Parkinson, M. B., Jensen, B. D., and Kurabayashi, K.: Design of compliant force and displacement amplification micromechanisms, Proceedings of ASME Design Engineering Technical Conferences, DET2001/DAC-21089, 2001.

Pham, H. and Chen, I.: Workspace and Static Analyses of 2-DOF Flexure Parallel Mechanism, in: Int. Conf. on Control, Automation, Robotics and Vision, 968–973, 2002.

Pilkey, W. D. and Pilkey, D. F.: Peterson's Stress Concentration Factors, 3rd Edn., John Wiley & Sons, 2008.

Ranganath, R., Nair, P. S., Mruthyunjaya, T. S., and Ghosal, A.: A force-torque sensor based on a Stewart Platform in a near-singular configuration, Mech. Mach. Theory, 39, 971–998, 2006.

Rebello, K.: Applications of MEMS in surgery, Proc. IEEE, 92, 43–55, 2004.

Schotborgh, W. O., Kokkeler, F. G. M., Tragter, H., and van Houten, F. J. A. M.: Dimensionless design graphs for flexure elements and a comparison between three flexure elements, Precis. Eng., 29, 41–47, 2005.

Sitti, M.: Piezoelectrically actuated four-bar mechanism with two flexible links for micromechanical flying insect thorax, IEEE-ASME T. Mech., 8, 26–36, 2003.

Socha, K. and Dorigo, M.: Ant colony optimization for continuous domains, Eur. J. Oper. Res., 185, 1155–1173, 2008.

Stoughton, R. and Arai, T.: Kinematic optimization of a chopsticks-type micromanipulator, in: Japan-USA Symp. on Flexible Autom., USA, 4472–4477, 1992.

Tian, Y., Shirinzadeh, B., Zhang, D., and Alici, G.: Development and dynamic modelling of a flexure-based Scott-Russell mechanism for nano-manipulation, Mech. Syst. Signal Pr., 23, 957–978, 2009.

Trease, B., Moon, Y.-M., and Kota, S.: Design of Large-Displacement Compliant Joints, J. Mech. Design, 127, 788–798, 2004.

Yong, Y. K., Lu, T.-F., and Handley, D. C.: Review of circular flexure hinge design equations and derivation of empirical formulations, Precis. Eng., 32, 63–70, 2008.

9

Dynamics of a gravity car race with application to the Pinewood Derby

B. P. Mann, M. M. Gibbs, and S. M. Sah

Mechanical Engineering and Material Science Department, Duke University, Durham, NC 27708, USA

Correspondence to: B. P. Mann (brian.mann@duke.edu)

Abstract. This paper investigates the underlying physics of a gravity car race. This work seeks to provide a sound theoretical basis to elucidate the design considerations that maximize performance while simultaneously dispelling false assertions that may arise from incomplete analyses. The governing equations are derived and solved analytically to predict race times; trend analyses are then performed along with a sensitivity analysis to ascertain the most important factors that influence performance. The inferences from a conservative energy balance are then compared with the predictions from the full set of differential equations, which include the dissipative terms associated with air resistance and friction.

1 Introduction

Gravity car races have provided their enthusiast numerous thrills over the years. While some of the longer-standing competitions include the Soap Box Derby and the Pinewood Derby, new events, such as the Extreme Gravity Racing Series and the Wile Street-luge Sliders of the X-Games, have also recently emerged. Although the race vehicles from these events can drastically vary in their size, shape, and complexity, they also share many common challenges. For example, they are all driven by the force of gravity and must minimize the forces that oppose the vehicle motions, such as wind resistance and friction.

The pinewood derby is one of the more distinctive events. It originated as a Cub Scout competition where elementary school children raced a car assembled from a kit consisting of a block of wood, four nails and wheels. A small industry has sprung up around the pinewood derby, with countless internet sites and books offering tips, tricks and even enhanced car parts to the estimated 43 million children that have built pinewood derby cars since its founding (Garguilo and Garguilo, 2011; Pedigo, 2002; Reinke, 2010). While there are many who offer advice, only a few scientific investigations have been published and it is rare to find accurate explanations on how certain modifications could result in faster race times (Coletta and Evans, 2008). In reference Coletta and

Evans (2008), Coletta and Evans used an algebraic function to obtain an analytical expression for the time and speed as function of the distance traveled along the track. Their analysis included the rotational energy of the wheels, rolling friction, and air resistance. In an early work, Cowley et al. (1989) obtained an approximation for the race time by considering the curved section of the track as two straight parts.

This paper seeks to provide a sound theoretical basis for making car modifications from the derivations herein. Trends in the peak velocity and race time are investigated from an energy balance and the governing equations. In addition, we compare the results of our analyses to a series of experimental results that verify the trends unveiled in our analyses also occur experimentally. As a part of our theoretical investigations, a sensitivity analysis was performed to ascertain the relative importance of five key parameters on reducing race times.

The work contained in this paper is organized as follows. The next section considers the conservative system and then performs an energy balance to derive an expression for the peak velocity. In Sect. 3, the equation of motions are derived for the two sections of the track, namely the straight and curved regions, taking into account the nonconservative forces. Analytical solutions are then derived for the governing equations. A sensitivity analysis is performed to determine the relative importance of altering the cross sectional

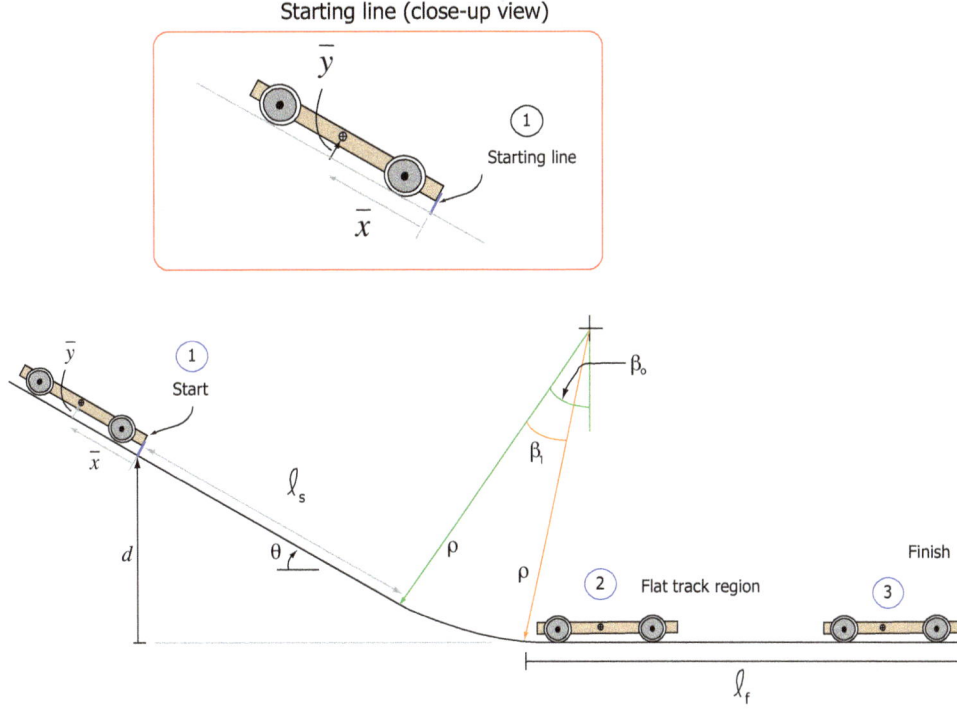

Figure 1. Illustration of a car at the starting line (upper left) with the location of the center of mass marked by $\bar{x}$ and $\bar{y}$. Bottom illustration shows the car at several locations along the track and the important track geometry.

area, car mass, friction coefficient, and wheel parameters to reduce the race times. The trends from the theoretical investigations are then compared with a series of experimental tests in Sect. 4. Finally, the last section provides a discussion of the combined theoretical and experimental investigations. The discussion also explains several additional opportunities for improving performance beyond the conclusions made from theoretical and experimental investigations.

2 Conservative system energy balance

An energy balance often provides a useful alternative to directly solving the governing differential equations and is used here to elucidate how design changes can influence the vehicle performance. The generic form of an energy balance is

$$\mathcal{T}_{1\rightarrow2} + \mathcal{U}_{1\rightarrow2} = \mathcal{W}_a - \mathcal{W}_d, \tag{1}$$

where $\mathcal{T}_{1\rightarrow2}$ is the change in kinetic energy, $\mathcal{U}_{1\rightarrow2}$ is the change in potential energy, $\mathcal{W}_a$ is the added work of external forces, and $\mathcal{W}_d$ is the work due to energy dissipation over the time interval from t_1 to t_2. Figure 1 shows the car at three locations along the track, i.e. the starting line, at the entry into the final horizontal section, and at the finish; this section will use these three locations to gain insight when applying an energy balance. Consider first the transition from the starting line (location one) to the beginning of the final

straight section (location two). When the system is treated as a rigid body, as opposed to the point mass assumption of Jobe (2004), both translational and rotational energy terms appear in the kinetic energy

$$\mathcal{T}_{1\rightarrow2} = \frac{1}{2}mv^2 + \frac{N}{2}\left(I_w\dot{\phi}^2\right), \tag{2}$$

where m is the total vehicle mass, v is the velocity of the center of mass, and N is the number of wheels, with a mass moment of intertia I_w, which rotate with an angular velocity $\dot{\phi}$. We next assume the wheels roll without slip, which allows the car's velocity to be written in terms of the angular rotations of the wheels $v = r_o\dot{\phi}$, where r_o is the wheel radius. After defining the wheel mass moment of inertia in terms of a radius of gyration, the change in kinetic energy can be written as

$$\mathcal{T}_{1\rightarrow2} = \frac{1}{2}m\left(1 + N\frac{m_o}{m}\left(\frac{k}{r_o}\right)^2\right)v^2, \tag{3}$$

where m_o is the mass and k the radius of gyration of a single wheel. The change in potential energy between the two locations is given by

$$\mathcal{U}_{1\rightarrow2} = mg\left(\bar{y}(1 - \cos\theta) - \bar{x}\sin\theta - d\right), \tag{4}$$

where d is the vertical distance to the starting position on the track, $\bar{y}$ is the vertical distance from the track to the center of mass, $\bar{x}$ is the horizontal distance from the front of the car

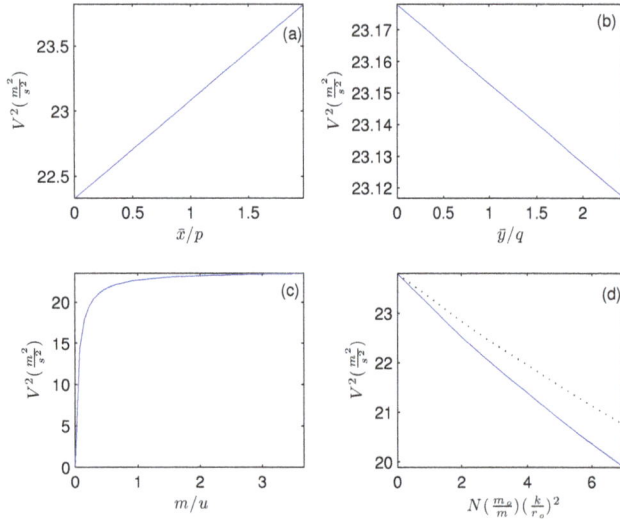

Figure 2. Energy balance trends showing the trends in v^2 while varying $\bar{x}$ **(a)**, $\bar{y}$ **(b)**, m **(c)**, and wheel parameters **(d)**. The dotted line in **(d)** represents a car with three wheels contacting the track or $N = 3$. The parameters listed in Table 1 and the normalization constants $p = 0.09$ m, $q = 0.013$ m, and $u = 0.1091$ Kg were used.

to the center of mass, and θ is angle of incline of the track (see Fig. 1 for geometry). Assuming no work is added or dissipated from the system, Eqs. (3) and (4) can be inserted into Eq. (1) to obtain an expression for the velocity

$$v^2 = \frac{2g\left(d + \bar{x}\sin\theta + \bar{y}(\cos\theta - 1)\right)}{1 + N\frac{m_o}{m}\left(\frac{k}{r_o}\right)^2}. \tag{5}$$

This energy balance solution alone can be used to provide much insight into how design changes will influence the vehicle velocity. More specifically, v^2 will increase linearly for linear changes in $\bar{x}$, but v^2 will decrease linearly for linear changes in $\bar{y}$; this suggests the center of mass should be located as far back and close to the track as possible. Focusing on the denominator of Eq. (5), the combined terms $N\frac{m_o}{m}\left(\frac{k}{r_o}\right)^2$ increase the denominator to be greater than one; thus the smaller this grouping of terms can be made, the greater the increase in v^2. Setting m_o equal to zero in this expression gives the same result as if the car was being modeled as a point mass. This assumption, which is commonly made in pinewood derby analyses, neglects the reduction in race time that can be achieved by manipulating the combined terms shown in Fig. 2d.

To illustrate the importance of $\bar{x}$ and $\bar{y}$, along with the terms that appear in the denominator, we have varied the different model parameters to ascertain their influence on v^2. For example, Fig. 2 shows the trends in the car's peak velocity for changes in the location of the center of mass and wheel parameters – at least in the absence of any dissipative forces. In these plots, we have normalized the horizontal axes

Table 1. Parameters used for energy balance trend analysis.

Parameter	Value
θ	26°
d	1.17 (m)
$\bar{x}$	98 (mm)
$\bar{y}$	13 (mm)
N	4
g	9.81 (m s^{-2})
m	0.191 (Kg)
m_o	2.65×10^{-3} (Kg)
r_o	1.51 (mm)
k	10.7 (mm)
I_w	3.06×10^{-7} (Kg m^{-2})

by the parameters of an out-of-the-box pinewood derby car and restricted the range to attainable limits. After applying the track geometry, parameters given in Table 1, we observe v^2 to increase and decrease linearly with changes in $\bar{x}$ and $\bar{y}$, respectively, with $\bar{x}$ having nearly four times the impact on v^2 than $\bar{y}$. However, the total mass of the car and the combined terms that appear in the denominator dominate the expression for v^2 – increasing the mass to double the out-of-box parameter has more than 15 times the impact of doing the same to $\bar{x}$. The v^2 trend in plot (d) highlights the benefit of reducing the m_o/m ratio, which can be accomplished by either removing mass from the wheels or by increasing the car mass relative to the wheel mass. This plot additionally shows that substantial increases could be achieved in the peak velocity provided it were possible to reduce the mass moment of inertia for the wheels, i.e. this is captured by a reduction in the radius of gyration k. Thus, we highlight that removing material from the outer portion of the wheel would reduce both m_o and k and should have a double effect to increase the peak velocity.

In summary, we have presented an energy balance that suggests $\bar{y}$ should be as small as possible and $\bar{x}$ and m should be as large as possible. However, it is important to recognize certain practical and physical restrictions that also constrain these values; for example, locating $\bar{x}$ behind the rear axle would cause the car to tip over, so the rear axle should be located as far back as possible and $\bar{x}$ should be just in front of the rear axle. Similarly, $\bar{y}$ must provide clearance between the car bottom and the raised center of the track. While Fig. 2 also suggests that lifting one wheel off the track will increase the car's velocity, additional energy losses occur if the fourth wheel is not low enough to contact the raised center of the track, which helps to maintain alignment. The energy balance neglects dissipative forces and therefore v^2 continues to grow as mass is added to the car; this trend does not reflect the fact that after a certain point velocity reduction due to friction will outweigh the benefit of adding additional mass. We therefore consider the results of this section as guidelines with certain limits and explore this notion even further in the upcoming sections. In the next section, the equations of motion are derived with the inclusion of the nonconservative forces.

3 Equations of motion and their analyses

This section derives the equation of motion for a prototypical pinewood derby car. The governing equations are later used to further explore the influence of design choices and parameter uncertainty on the race times of the vehicle. The derivation that follows has been split into two parts with separate derivations for the curved and straight sections of the track. To complete the analysis, it was assumed that the car wheels would roll without slip and that the length of the car is negligible compared to the length of the track.

3.1 Straight track sections

Consider the inclined portion of the track shown in Fig. 1. Applying notations from the previous sections, the kinetic energy becomes

$$T = \frac{1}{2}m\left(1 + N\frac{m_o}{m}\left(\frac{k}{r_o}\right)^2\right)\dot{s}^2,\tag{6}$$

where s is the distance the car's center of mass has traveled along a straight section of the track and $\dot{s} = v$ is the vehicle's velocity along the flat section of the track. The potential energy of the system is given by

$$\mathcal{U} = mg\left(d + \bar{x}\sin\theta + \bar{y}\cos\theta - s\sin\theta\right).\tag{7}$$

In addition to the conservative forces, the nonconservative forces of air resistance and sliding friction between the wheels and axles must be taken into account. Here, we note sliding friction causes a moment that opposes the wheel rotation. Applying Lagrange's equation, where the nonconservative forces from drag and friction are included, results in the following governing equation:

$$m\left[1 + N\frac{m_o}{m}\left(\frac{k}{r_o}\right)^2\right]\ddot{s} - mg\sin\theta = -\frac{1}{2}\rho_a AC_D\dot{s}|\dot{s}| - 2f_A\frac{r_i}{r_o} - 2f_B\frac{r_i}{r_o},\tag{8}$$

where f_A is the sliding friction force on a front wheel and f_B is the sliding friction force on a back wheel, ρ_a is the air density, C_D the drag coefficient, and A is the projected cross-sectional area. To derive the expression for the nonconservative forces, we note that a roll with slip condition was applied. To resolve the frictional forces f_A and f_B, a Coulomb friction law was applied to write $f_A = \mu_k N_A$ and $f_B = \mu_k N_B$, where μ_k is the kinetic coefficient of friction and N_A and N_B are the normal forces that act at the locations shown in Fig. 3; next, the moments were summed about the rear axle to obtain expression for the normal forces

$$N_A = (m - 4m_o)\frac{(h_o - \bar{y}_c)(\ddot{s} - g\sin\theta) + (\ell_B - \bar{x}_c)g\cos\theta}{\ell_B - \ell_A}\tag{9}$$

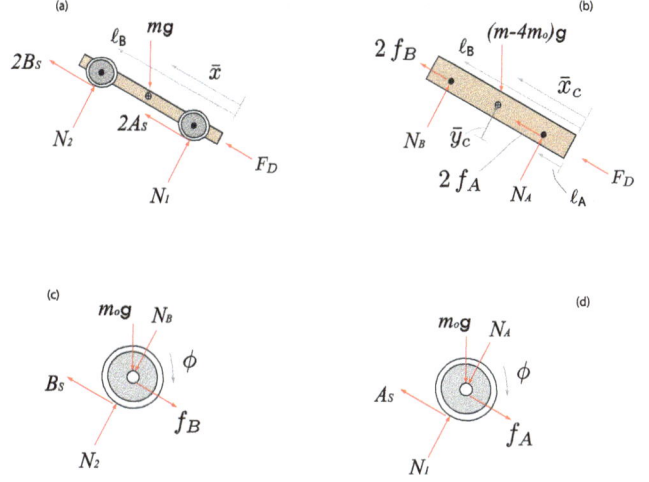

Figure 3. Free body diagram of the forces on (**a**) the assembled car, (**b**) only the car body, (**c**) the rear wheel, and (**d**) the front wheel.

summing the moments about an axis through A and solving for N_B in a similar manner gives

$$N_B = (m - 4m_o)\frac{(\bar{y}_c - h_o)(\ddot{s} - g\sin\theta) - (\ell_A - \bar{x}_c)g\cos\theta}{\ell_B - \ell_A}\tag{10}$$

where ℓ_A is the distance from the front of the car to the front axle, ℓ_B is the distance from the front of the car to the rear axle, and $\bar{x}_c$ is the center of mass location for the car body (see Fig. 1). After inserting the expressions for $f_A = \mu_k N_A$ and $f_b = \mu_k N_B$ into Eq. (8), the governing equation takes the form

$$\ddot{s} + \gamma\dot{s}|\dot{s}| = \eta\tag{11}$$

where the parameters γ and η are given by

$$\gamma = \frac{\rho_a AC_D}{2m\left(1 + N\frac{m_o}{m}\left(\frac{k}{r_o}\right)^2\right)},\tag{12a}$$

$$\eta = \frac{\sin\theta - 2\mu_k\frac{r_i}{r_o}\left(1 - 4\frac{m_o}{m}\right)\cos\theta}{1 + N\frac{m_o}{m}\left(\frac{k}{r_o}\right)^2}g.\tag{12b}$$

Before departing this section, we note the general form of Eq. (11) can be applied to either the horizontal ($\theta = 0$) or inclined ($\theta \neq 0$) sections of the track.

3.1.1 Analytical solution for the straight track sections

This section derives an exact analytical solution for the straight sections of the track. Along the straight track the sign of the car's velocity is always positive therefore $|\dot{s}| = \dot{s}$ in Eq. (11). The analysis starts by substituting $v = \dot{s}$ into Eq. (11) to obtain

$$\dot{v} + \gamma v^2 = \eta\tag{13}$$

where v is the vehicle's velocity. Here, we note the value of η in Eq. (13) can be either positive or negative depending on which part of the straight track the car is located. When the car is on the inclined section η is positive, but it takes on a negative value for the horizontal section of the track. Therefore the solution to Eq. (13) must consider the two parts of the straight track. In the case of the inclined track we begin by rearranging Eq. (13) to give the following differential relationship:

$$\frac{\mathrm{d}v}{\eta - \gamma v^2} = \mathrm{d}t. \tag{14}$$

After integrating this relationship and solving for v, the analytical expression for the vehicle's velocity becomes

$$v = \sqrt{\frac{\eta}{\gamma}} \tanh\left(\sqrt{\gamma\eta}\,t + \tanh^{-1}(\sqrt{\frac{\gamma}{\eta}}v_0) \right), \tag{15}$$

where $v_0 = v(0)$ is the initial velocity on the inclined track. To obtain the analytical expression for the car's position, Eq. (15) was integrated to obtain

$$s = s_o + \frac{1}{\gamma} \ln\left[\cosh\left(\sqrt{\gamma\eta}\,t + \tanh^{-1}(\sqrt{\frac{\gamma}{\eta}}v_0) \right) \right] - \frac{1}{\gamma}$$
$$\ln\left[\cosh\left(\tanh^{-1}(\sqrt{\frac{\gamma}{\eta}}v_0) \right) \right], \tag{16}$$

where s_o is the initial position of the vehicle. Assuming the car starts from rest, where $s_o = v_o = 0$, the analytical solutions for the velocity and position can be simplified to

$$v = \sqrt{\frac{\eta}{\gamma}} \tanh(\sqrt{\gamma\eta}\,t), \tag{17a}$$

$$s = \frac{1}{\gamma} \ln\left(\cosh(\sqrt{\gamma\eta}\,t) \right). \tag{17b}$$

We next consider the horizontal section of the track and presume the car enters this section at $t = t_1$. The solution to Eq. (13) is given by

$$v = -\sqrt{-\frac{\eta}{\gamma}} \tan\left(\sqrt{-\gamma\eta}(t - t_1) - \tan^{-1}(\sqrt{\frac{-\gamma}{\eta}}v_1) \right), \tag{18}$$

where $v_1 = v(t_1)$ is the initial velocity of the car as it enters the horizontal section of the track. The position of the car along the horizontal section of the track is obtained by integrating Eq. (18), which yields the following following expression:

$$u = -\frac{1}{\gamma} \ln\left[\sec\left(-\sqrt{-\gamma\eta}(t - t_1) + \tan^{-1}(\sqrt{\frac{-\gamma}{\eta}}v_1) \right) \right]$$
$$-\frac{1}{\gamma} \ln\left[\sec\left(\tan^{-1}(\sqrt{\frac{-\gamma}{\eta}}v_1) \right) \right], \tag{19}$$

where u is position along the horizontal section of the track. In deriving this result, it is important note that $u(t_1) = 0$ was applied to obtain Eq. (19).

3.2   Curved track equation of motion

This section derives the governing equation for the curved section of the track and is followed by the development of an approximate analytical solution. We first express the kinetic energy of the system using the roll without slip condition and a radius of gyration description for the wheel mass moment of inertia, as in the previous expression for the kinetic energy, to obtain the following:

$$\mathcal{T} = \frac{1}{2}m\left[(\rho - \bar{y})^2 + N\frac{m_o}{m}\left(\frac{k}{r_o}\right)^2 \rho^2 + \frac{I_G}{m} \right]\dot{\beta}^2, \tag{20}$$

where ρ is the track radius of curvature, I_G is the car's mass moment of inertia, and β is the angular position of the center of mass. The potential energy of the system is given by

$$\mathcal{U} = mg(\rho - \bar{y})(1 - \cos\beta). \tag{21}$$

Applying Lagrange's equation, where the nonconservative forces from drag and friction are included, results in the following governing equation:

$$m\left[(\rho - \bar{y})^2 + N\frac{m_o}{m}\left(\frac{k}{r_o}\right)^2 \rho^2 + \frac{I_G}{m} \right]\ddot{\beta} + mg(\rho - \bar{y})\sin\beta$$
$$= -\frac{\rho_a}{2}AC_D\bar{\rho}^3\dot{\beta}|\dot{\beta}| - 2\mu_k\rho\frac{r_i}{r_o}(N_A + N_B), \tag{22}$$

To resolve the normal forces N_A and N_B, the moments were summed about the front and rear axles, which yields

$$N_A = \frac{m - 4m_o}{\ell_B - \ell_A}\left[(\ell_B - \bar{x}_c)(g\cos\beta + (\rho - \bar{y})\dot{\beta}^2) \right.$$
$$\left. +(r_o - \bar{y}_c)(g\sin\beta + \ddot{\beta})(\rho - \bar{y})) \right] + \frac{I_g\ddot{\beta}}{\ell_B - \ell_A}, \tag{23}$$

$$N_B = \frac{m - 4m_o}{\ell_B - \ell_A}\left[-(r_o - \bar{y}_c)(g\sin\beta + (\rho - \bar{y})\ddot{\beta}) \right.$$
$$\left. +(\bar{x}_c - \ell_A)(g\cos\beta + (\rho - \bar{y})\dot{\beta}^2] - \frac{I_G\ddot{\beta}}{\ell_B - \ell_A}, \tag{24}$$

Equations (23)–(24) can then be combined to obtain the governing equation for the curved section of the track,

$$\ddot{\beta} + \mu\dot{\beta}|\dot{\beta}| + \alpha\cos\beta + \omega^2\sin\beta = 0, \tag{25}$$

where the parameters μ, ω^2, α and $\tilde{m}$ are given by

$$\mu = \frac{\left[\frac{\rho_a}{2}AC_D\bar{\rho}^2 + 2\mu_k\rho\frac{r_i}{r_o}(m - 4m_o) \right](\rho - \bar{y})}{\tilde{m}}, \tag{26a}$$

$$\omega^2 = \frac{mg(\rho - \bar{y})}{\tilde{m}}, \tag{26b}$$

$$\alpha = \frac{2\mu_k\rho\frac{r_i}{r_o}(m - 4m_w)g}{\tilde{m}}, \tag{26c}$$

$$\tilde{m} = m\left[(\rho - \bar{y})^2 + N\frac{m}{m_o}\left(\frac{k}{r_o}\right)^2 \rho^2 + \frac{I_G}{m} \right]. \tag{26d}$$

3.2.1 Approximate solution for the curved region

In order to find an approximate solution, we Taylor expand Eq. (25) and introduce a small parameter ϵ which gives

$$\ddot{\beta} + \epsilon\mu\dot{\beta}|\dot{\beta}| + w^2\beta + \epsilon\,c_2\,\beta^2 + c_1 = 0, \qquad (27)$$

where

$$c_1 = \alpha, \quad c_2 = \frac{-\alpha}{2}. \qquad (28)$$

The book-keeping parameter ϵ will serve as a perturbation parameter and will be set equal to unity at the end. In order to find an approximate solution to Eq. (27), we use the method of Krylov-Bogoliubov-Mitropolsky (KBM), see Mickens (1996) and Minorsky (1962). Equation (27) can be written as

$$\ddot{\beta} + w^2\beta + c_1 = \epsilon\,F(\beta,\dot{\beta}), \qquad (29)$$

where

$$F(\beta,\dot{\beta}) = -\mu\dot{\beta}|\dot{\beta}| - c_2\,\beta^2. \qquad (30)$$

We assume that the solution to Eq. (27) takes the following form

$$\beta = a\,\cos\psi - \frac{c_1}{w^2} + \epsilon\,u_1(a,\psi) + \epsilon^2\,u_2(a,\psi) + ..., \qquad (31)$$

where the $u_i(a,\psi)$ are periodic functions of ψ, with period 2π, and the quantities a and ψ are functions of time defined by the following equations:

$$\dot{a} = \epsilon\,A_1(a) + \epsilon^2\,A_2(a) + ..., \qquad (32)$$

$$\dot{\psi} = w + \epsilon\,B_1(a) + \epsilon^2\,B_2(a) + ..., \qquad (33)$$

The functions $u_i(a,\psi)$, $A_i(a)$ and $B_i(a)$ are to be chosen in such a way that Eq. (31), after replacing a and ψ by the functions defined in Eqs. (32)–(33), is a solution to Eq. (27). Taking into account Eqs. (31), (32) and (33), the first derivative of β takes the form

$$\dot{\beta} = -a\,w\,\sin\psi + \epsilon\left(A_1\cos\psi - aB_1\sin\psi + w\,\frac{\partial u_1}{\partial\psi}\right)$$

$$+\epsilon^2\left(A_2\cos\psi - aB_2\sin\psi + A_1\,\frac{\partial u_1}{\partial a} + B_1\,\frac{\partial u_1}{\partial\psi} + w\,\frac{\partial u_2}{\partial\psi}\right) + ..., \qquad (34)$$

On the other hand, the right-side of Eq. (29) can be rewritten to the form:

$$\epsilon F(\beta,\dot{\beta}) = \epsilon F(a\cos\psi - \frac{c_1}{w^2}, -aw\sin\psi) + \epsilon^2\left[u_1 F_\beta(a\cos\psi - \frac{c_1}{w^2}, -aw\sin\psi)\right.$$

$$\left.+\left(A_1\cos\psi - aB_1\sin\psi + w\,\frac{\partial u_1}{\partial\psi}\right)F_{\dot{\theta}}(a\cos\psi - \frac{c_1}{w^2}, -aw\sin\psi)\right]. \qquad (35)$$

Substituting Eqs. (31)–(35) into Eq. (29), collecting the terms with like powers of ϵ, and setting them to zero, gives

$$\frac{\partial^2 u_1}{\partial\psi^2} + u_1 = F_0(a,\psi) + 2A_1\,\sin\psi + 2aB_1\,\cos\psi, \qquad (36)$$

$$\frac{\partial^2 u_2}{\partial\psi^2} + u_2 = F_1(a,\psi) + 2A_2\,\sin\psi + 2aB_2\,\cos\psi. \qquad (37)$$

The functions $A_i(a)$, $B_i(a)$, and $u_i(a,\psi)$ can be found by first expanding $F_j(a,\psi)$ and $u_i(a,\psi)$ into a Fourier series:

$$F_j = g_{j,0}(a) + \sum_{n=1}^{\infty}[g_{j,n}\,\cos(n\psi) + h_{j,n}\,\sin(n\psi)], \qquad (38)$$

$$u_i(a,\psi) = p_{i,0} + \sum_{n=1}^{\infty}[p_{i,n}\,\cos(n\psi) + q_{i,n}\,\sin(n\psi)], \qquad (39)$$

where

$$g_{j,n} = \frac{1}{\pi}\int_0^{2\pi} F_j(a\cos\psi - \frac{c_1}{w^2}, -aw\sin\psi)\,\cos(n\psi)\,d\psi, \qquad (40)$$

$$h_{j,n} = \frac{1}{\pi}\int_0^{2\pi} F_j(a\cos\psi - \frac{c_1}{w^2}, -aw\sin\psi)\,\sin(n\psi)\,d\psi, \qquad (41)$$

and then equating the harmonics of the same order. It should be noted that the integration above is broken into two parts- one with the limits 0 and π and the other with the limits π and 2π. For example,

$$h_{0,1} = \frac{1}{\pi}\int_0^{2\pi} F_0(a\cos\psi - \frac{c_1}{w^2}, -aw\sin\psi)\,\sin(\psi)\,d\psi$$

$$= \frac{1}{\pi}\int_0^{2\pi} -\mu(-aw\sin\psi)|aw\sin\psi|\sin\psi - c_2(a\,\cos\psi - \frac{c_1}{w^2})^2\sin\psi\,d\psi$$

$$= \frac{a^2 w^2}{\pi}\left[\int_0^{\pi}\sin^3\psi\,d\psi - \int_{\pi}^{2\pi}\sin^3\psi\,d\psi\right] = \frac{8a^2 w^2}{3\pi}. \qquad (42)$$

After doing the above calculations we found

$$\beta = a\,\cos\psi - \frac{c_1}{w^2} + \epsilon\left[\frac{c_1 c_2 a}{w^4} - \frac{2c_1 c_2 a}{3w^4}\cos 2\psi - \frac{c_1 c_2}{4w^4}\right.$$

$$\left.\cos 3\psi - \frac{8\mu a^2}{9\pi}\sin 2\psi - \frac{\mu a^2}{3\pi}\sin 3\psi\right]$$

$$\dot{\beta} = -aw\,\sin\psi + \epsilon\left[-\frac{4w\mu a^2}{3\pi}\cos\psi + \frac{c_1 c_2 a}{w^3}\sin\psi\right.$$

$$+w\left(\frac{4c_1 c_2 a}{3w^4}\sin 2\psi + \frac{3c_1 c_2}{4w^4}\sin 3\psi - \frac{16\mu a^2}{9\pi}\right.$$

$$\left.\left.\cos 2\psi - \frac{\mu a^2}{\pi}\cos 3\psi\right)\right], \qquad (43)$$

where

$$\dot{a} = -\frac{4w\mu}{3\pi}a^2\,\epsilon + a^2\left[\frac{14\mu c_1 c_2}{15\pi w^3} - \frac{4\mu c_2}{9\pi w}a\right]\epsilon^2, \qquad (44)$$

$$\dot{\psi} = w - \frac{c_1 c_2}{w^3}\epsilon + \left[\frac{4\mu^2 w a^2}{5\pi^2} + c_1 c_2^2(\frac{2a}{3w^5} - \frac{c_1}{2w^7})\right]\epsilon^2. \qquad (45)$$

Figure 4 shows a comparison of the analytical solutions for the inclined, curved and the horizontal sections, Eqs. (15)–(19) and Eqs. (43)–(45), with the numerical results, Eqs. (13)

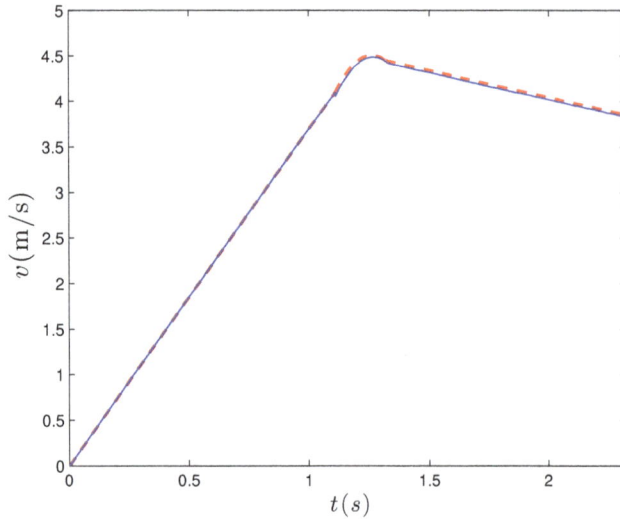

Figure 4. Comparison of the analytical solutions (solid line) for the inclined, curved, and horizontal sections of the track with the numerical simulation (dashed line).

Table 2. Parameters used for the analysis of a prototypical pinewood derby car.

Parameter	Value
θ	$26°$
d	1.17 (m)
$\bar{x}$	0.098 (m)
$\bar{y}$	0.013 (m)
N	4
g	9.81 (m s^{-2})
m	0.191 (Kg)
m_o	2.65×10^{-3} (Kg)
r_o	0.0151 (m)
r_i	2.54×10^{-3} (m)
k	0.0107 (m)
ℓ_f	4.21 (m)
ℓ_s	2.16 (m)
β_1	26
β_o	34.3
μ_k	0.167
C_D	1.1
A	0.00146
ρ_a	1.2041
I_w	3.06×10^{-7} (Kg m^{-2})
I_G	2.5205×10^{-4} (Kg m^{-2})

and (27), for the parameters given in Table 2 and for $\epsilon = 0.01$. From the figure it is clear that the analytical and numerical results are in agreement. In the next section, the derived theoretical results are used to study trends in the velocity and race time for hypothetical changes in the car's physical parameters.

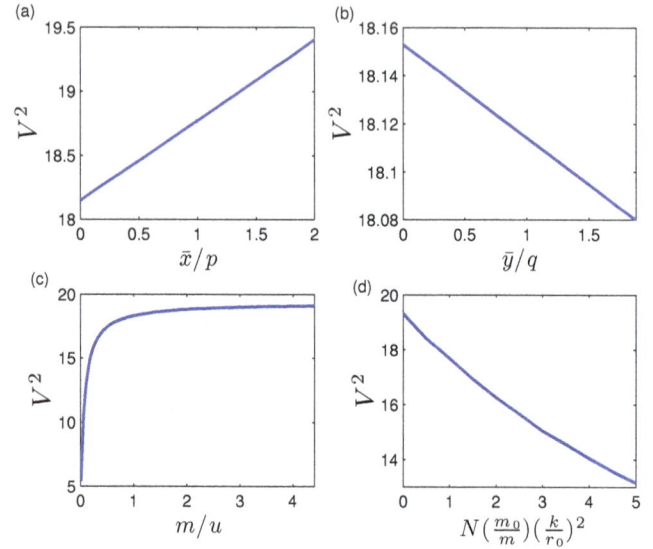

Figure 5. Theoretical trends in v^2 with the dissipative forces included; results show the effects of varying $\bar{x}$ (**a**), $\bar{y}$ (**b**), m (**c**), and the wheel parameters (**d**) with parameters given in Table 2 and the normalization constants $p = 0.09$ m, $q = 0.013$ m, and $u = 0.1091$ Kg.

3.3 Trend studies

A conservative energy balance was used in Sect. 2 to explore trends in the car's peak velocity as key parameters, or groups of parameters, were changed. In this section, we have included the dissipative forces that appear in the governing equations, see Eqs. (13) and (25), to generate trend studies and highlight some additional behavior of interest. While the studies shown in this section were obtained from numerical simulation, we have also validated their accuracy with the analytical solutions, as shown previously in Fig. 4.

The first series of results are shown in Fig. 5. It is interesting that the trends of Fig. 5 are very similar to those presented previously in Fig. 2, results that were obtained by ignoring the dissipation. For example, trends in v^2 due to changes in $\bar{x}$, $\bar{y}$, and the wheel parameters are nearly identical. Similarly, both figures show regions where v^2 can dramatically increase for changes in m and other regions where the peak velocity changes very little for increases in m. However, the results from Figs. 2 and 5 are not identical and certain important differences do appear. In particular, the dissipative forces are shown to significantly decrease the peak velocity.

Figure 6 further examines the effect of increasing mass with plots of the car displacement and velocity. It is shown that the car with highest mass is the fastest one, i.e. the first one to reach the end of the track, see Fig. 6a–b. While peak velocity has already been shown to increase with mass in Fig. 5, the additional insight from Fig. 6 is that the velocity is also larger on other sections of the track. It should be noted that we have chosen two substantially different mass values for the purposes of illustration, i.e. this causes more

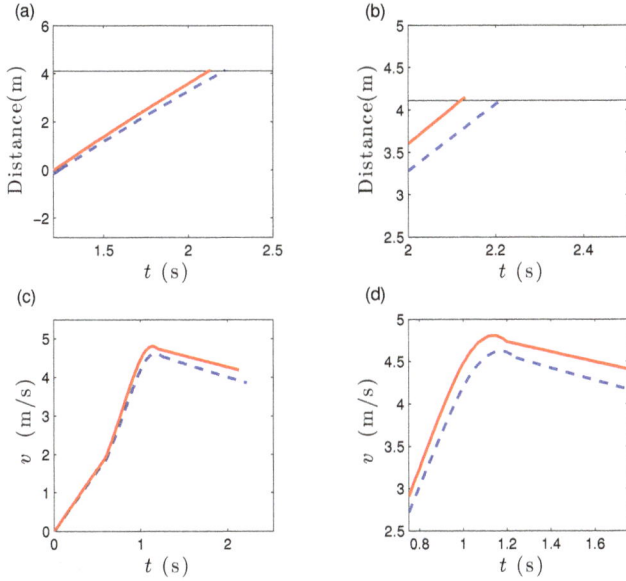

Figure 6. Displacement and velocity plots for two values of the car mass $m = 0.05$ Kg (dashed), $m = 0.35$ Kg (solid) and with the remaining parameters given in Table 2. The horizontal line in (**a**) and (**b**) represents the finish line. The car with highest mass is the first one to reach the end of the track.

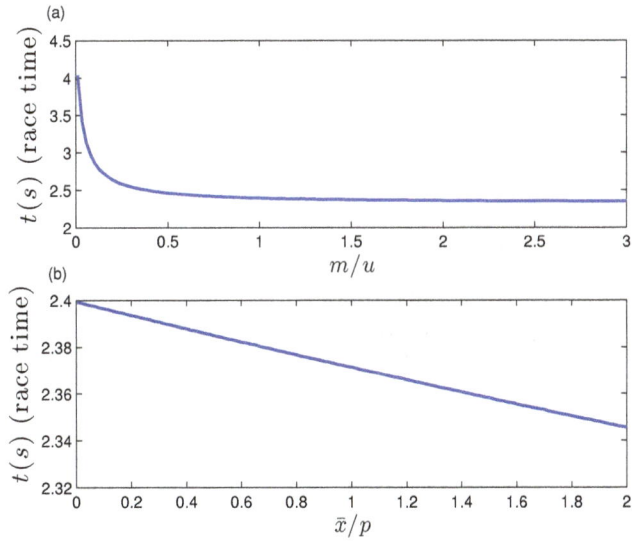

Figure 7. The race time of the car as a function of the car mass (**a**) and the x-direction location for the car center of mass (**b**) with parameters given in Table 2 and the normalization constants $p = 0.09$ m, and $u = 0.1091$ Kg. Beyond a certain mass value the race time barely changes whit increased mass (a).

noticeable differences in the distance and the velocity time histories shown in Fig. 6. Since the race time and not the displacement and velocity is the typical quantity of interest, Fig. 7 shows the trends in race time for variations in the car mass and $\bar{x}$. Focusing on the results of Fig. 7a, one can see regions where altering the mass can either make a large or

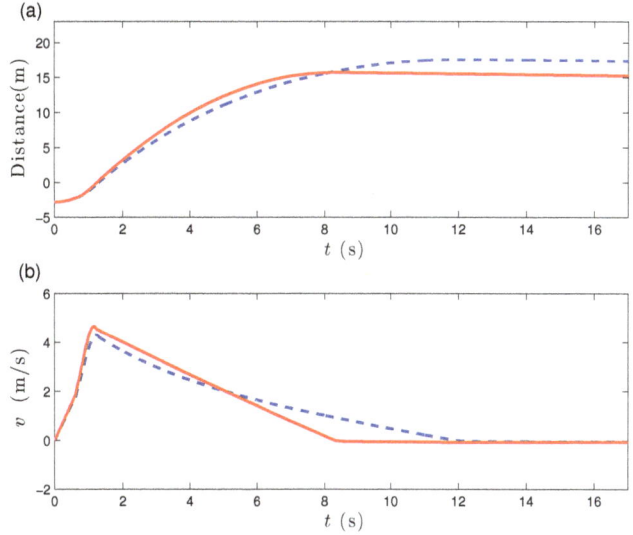

Figure 8. Displacement and velocity for two values of the car mass $m = 0.02$ Kg (dashed), $m = 0.2$ Kg (solid) and with parameters given in Table 2 and $\mu_k = 0.2$. The heavier (solid) car stops before the lighter car when the track is sufficiently long.

nearly insignificant difference on the race time. In contrast, the results of Fig. 7b show the race time only changes linearly with $\bar{x}$.

Given the evidence presented thus far, it might seem reasonable to conclude a car with a larger mass will always finish the first. However, this is not the case and one example where this is not the case is shown in Fig. 8. For this example, the heavier car actually stops before the lighter car when the track is sufficiently long. While this might seem like an obvious case, since the heavier car stopped, other cases also exist where neither car stops, as shown in Fig. 9a. These trends were further explored by generating the 3-D plot of the race time as function of both the mass and the friction coefficient, shown in Fig. 9b, and it can be seen that as μ_k becomes larger the race time increases with the mass, as opposed to decreasing as originally expected. This is significant because the conservative energy balance analyses completely misses this behavior. The disadvantages of adding mass to the car are only revealed when using the full set of differential equations.

3.4 Sensitivity analysis

For nearly any design exercise it is important to consider which factors or design parameters influence performance the most. To gain insight into this question, we performed a sensitivity analysis on the car race time as a function of several key parameters. In the results that follow, we show results from a normalized sensitivity analysis and plots of the UMFs (uncertainty magnification factors), which were determined analytically (Coleman and Steele, 1999; Frey and Patil; Hamby, 1994). The general form for the i-th UMF

(a)

(b)

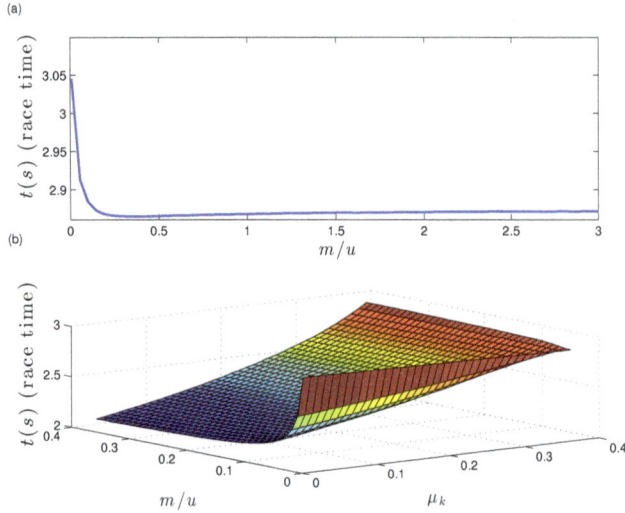

Figure 9. The race time of the car as a function of the car mass with parameters given in Table 2 and $\mu_k = 0.35$ **(a)**. The race time of the car as a function of both the car mass and the coefficient friction μ_k with parameters given in Table 2 and the normalization constants $u = 0.1091$ Kg **(b)**.

factor can be expressed as follows:

$$\text{UMF}_i = \frac{X_i}{t} \frac{\partial t}{\partial X_i} \tag{46}$$

where the sensitivity coefficient $\frac{\partial t}{\partial X_i}$ is the ratio of the change in output, in our case the race time, to the change in input while all other parameters are held fixed. The quotient $\frac{X_i}{t}$ is introduced to normalize the coefficient by removing the affects of units. Therefore a large value of UMF for a certain parameter means that the parameter has more influence on the race time than a parameter with a small UMF value. In Fig. 10, the UMF factor for different parameters is shown for the case of the inclined track, Eqs. (12a), (12b) and (17b). We consider five parameters, namely the projected area of the car A, the total car mass m, the kinetic coefficient of friction μ_k, the wheel radius r_o, and the inner radius r_i. In Fig. 10 the parameters X_i are divided by the parameters considered in our experiment and it can be seen that the dependence of UMF on the parameters A, μ_k and r_i is linear while it is nonlinear in the case of m and r_0. Also, it can be deduced from the figure that overall the radius of the wheel r_o has more influence on the race time compared to the other parameters, while the projected area of the car A has less influence. For the parameters considered in our experiment, the race time is sensitive in a decreasing order to $r_o, \mu_k, r_i, m,$ and A.

In the next section, experimental investigations are performed and compared to the analytical results.

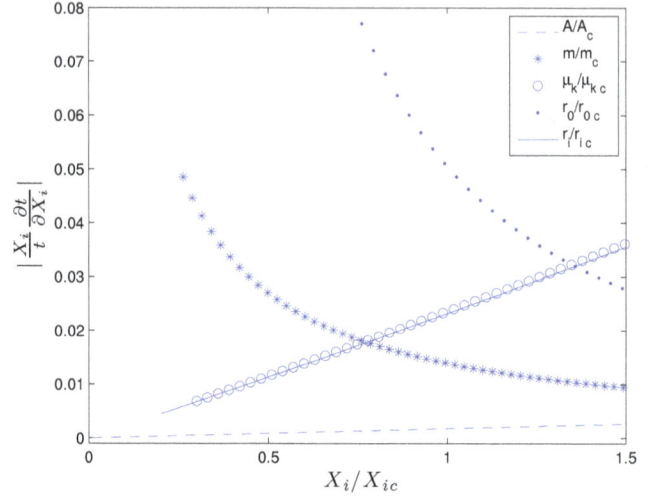

Figure 10. The uncertainty magnification factor as we change the normalized cross section area (dashed line), mass (stars), friction coefficient (circles), outer wheel radius (dotted) and inner wheel radius (solid line) with parameters given in Table 2.

4 Experimental investigations

This section describes the experimental investigations performed to validate the trends described in the earlier theoretical sections of the paper. We first describe the experimental track and vehicles used in the experimental trials then proceed to discussing the parameter identification methods. These sections are followed by a discussion of the trends observed in the experimental trials.

4.1 Track description

A picture of the experimental track is shown in Fig. 11. As in the theoretical studies, the geometry for the aluminum track can be divided into three distinct regions; the initial inclined region is connected to a relatively short curved section that is then followed by a final straight section that is horizontal. The gravity car begins the race at the starting position on the inclined section. The car is held in place by a spring-loaded peg that protrudes from the track center and is withdrawn to initiate the race. The car then travels down the inclined, curved, and horizontal sections of the track until it reaches the finish line where an automatic timer records the elapsed time. The timer is triggered at the start of the race by a microswitch attached to the start gate. The finish of the race is determined when the car breaks an infrared beam of light that is shone down from the timer onto the sensors in the track. While the car is traveling down the length of the track it also occasionally makes contact with a raised center portion which acts to recenter the vehicle.

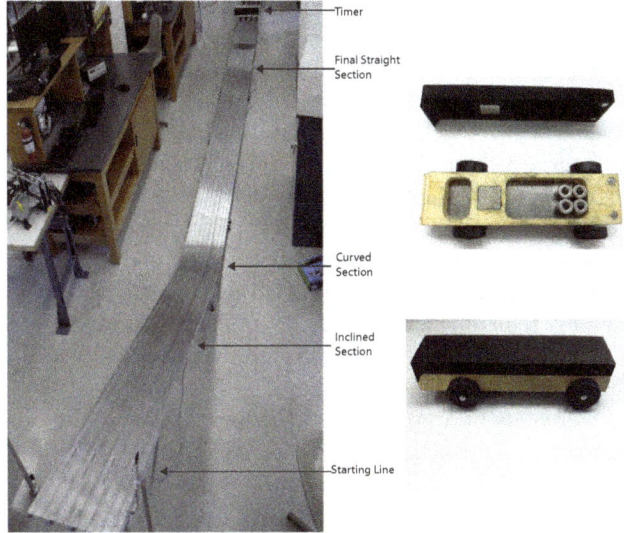

Figure 11. Experimental track and car body used during the experimental investigations. The outer shell of the car is removable to allow for trials with different masses and centers of mass.

4.2 Gravity car description

The gravity car is comprised of a base made from a standard pinewood derby kit that has been modified to be fitted magnetically with four plastic shells which differ in shape. The base is hollow and weights can be placed inside to vary the car's center of mass. Four nails act as axles which slide into grooves on the bottom face. A standard base with interchangeable shells was created to isolate the effects of different car bodies on race times by eliminating uncertainty due to variable wheel alignments, friction coefficients, etc. It is common in the pinewood derby to use graphite to reduce friction between the axles and the wheels, but we did not use graphite in our experiment because spills and variability of application would cause inconsistent results.

4.3 Identification of car parameters

Several parameters had to be experimentally identified during the course of our investigations. In particular, separate experiments were required to obtain the radius of gyration for the wheels, the mass moment of inertia for the car, and the friction coefficient between each wheel and axle. An instrumented pendulum apparatus was constructed to obtain the vehicle's mass moment of inertia. The procedure consisted of attaching the car to a pendulum platform, measuring the car-pendulum system oscillations, and then using the period between oscillations $T = \dfrac{2\pi}{\omega\sqrt{1-\zeta^2}}$, where ζ is the damping ratio obtained via logarithmic decrement and ω is the linear natural frequency of the car-pendulum system, to extract ω. Using the parallel axis theorem, the mass moment of inertia

was obtained from the expression

$$I_G = \frac{Mg\bar{d}}{\omega^2} - I_\mathrm{p} - md^2 \tag{47}$$

where M is the total mass of the car and pendulum platform, d is the perpendicular distance from the pivot point to the car center of mass, $\bar{d}$ is the perpendicular distance from the pivot point to the combined car-pendulum center of mass, and I_p is the mass moment of inertia for the pendulum platform. Instrumented wheel spin tests were used to estimate the radius of gyration for the wheels and the wheel-axle friction coefficient. In these tests, the average angular deceleration was determined from time between n complete rotations of the wheel; a laser tachometer was used to detect full tire rotations and monitor the time between successive rotations. If we let t_1 be a reference time, t_n be n revolutions into the future, and t_r denote r revolutions into the future, the following expression gives the angular deceleration

$$\alpha = \frac{4\pi}{t_r - t_n}\left(\frac{r}{t_r - t_1} - \frac{n}{t_n - t_1}\right), \tag{48}$$

which can be related to the friction coefficient and tire mass moment of inertia through a summation of forces and moments. More specifically, the following was obtained by summing moments about the axle

$$\alpha = -\frac{\mu_\mathrm{k} m_o g r_i}{I_o}. \tag{49}$$

To solve for μ_k and I_o independently, a special attachment was affixed to the wheels and the spin test was repeated; the second set of experiments that allowed the unknowns μ_k and I_o to be identified.

4.4 Experimental trends

Two tests are performed to validate the theoretical trends for car mass and horizontal center of mass discussed in earlier sections. The first holds all of the car parameters constant while varying the car mass. As mass was added between trials, a constant center of mass was maintained by changing the locations of magnetic weights in the hollowed out car body. These tests used the block shell shown at the bottom right of Fig. 11. The second series of tests changed the horizontal center of mass by moving a weight along the length of the car.

The experimental mass plot (a) verifies the theoretical advantage of increasing the car's mass seen in both the energy balance and the numerical simulation of the car's equation of motion, Eqs. (12) and (22). The plot also highlights the diminishing returns of adding mass as the overall mass of the car increases. This trend is exaggerated in the energy balance balance plot (Fig. 2) because it does not take into account the energy lost due to non-conservative forces. The energy balance plot shows the race time levels out at around half of the

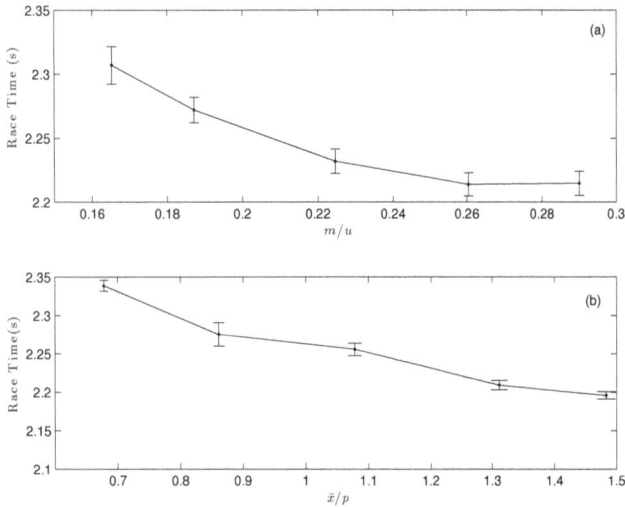

Figure 12. Experimental results of race times for cars with varying mass (**a**) and varying the horizontal center of mass (**b**). The mass is normalized by the out-of-the-box car mass of $u = 0.1091$ kg and $\bar{x}$ is normalized by the out-of-the-box center of mass $p = 0.09$ m.

mass of an out-of-the-box car. The plot of race times resulting from the numerical simulation of the car's equations of motion (Fig. 5) portrays a more accurate representation of the relationship between mass and race time. This plot shows the advantage of adding mass starting to level out at around two times the out-of-the-box car mass, which is consistent with the experimental data presented in Fig. 12. As mass is added to the car, it increases the normal force acting on all four wheels, as seen in Eqs. (11a) and (11b). Higher magnitude normal forces correspond to more friction between the axle and the wheel which slows the car down, a phenomena which is not reflected in the energy balance plot. The horizontal center of mass trends found experimentally also comply with both theoretical and simulated results. The simulated results and experimental plots indicate that changing $\bar{x}$ between 1 and 1.5 times the out of the box parameter lowers race times at a faster rate. Increasing $\bar{x}$ moves the center of mass farther from the starting line, which allows the car to continue accelerating along the straight section of the track for longer than a car with a center of mass closer to the front wheels. The advantage gained by a longer acceleration time corresponds to a faster race time. One downside to moving the center of mass to the extremities of the car, one that is not accounted for in the mathematical model, is a higher instance of wheel impacts with the center track partition. Moving the center of mass to either extreme – the front or the rear – causes the car to wobble along the track. It should be noted, however, that in the experimental results, the advantages gained by increasing the center of mass offset the unfavorable wobbling condition.

5　Discussion

Experiments showed that increasing the car's mass results in a faster race time, a result that is consistent with theoretical studies of the energy balance and the numerical simulation of the equation of motion. However, when the car's mass exceeds a certain limit, it actually leads to a reduction in race time because of the increased friction caused by a larger mass. This phenomena is not accounted for in the simple energy balance, but comes to light through theoretical investigations. Experimental findings also showed increasing the horizontal center of mass will decrease the race time, a result that validates theoretical findings.

In addition to making the modifications suggested by our theoretical and experimental results, the following techniques can be employed to further reduce race times. Polishing the axles with progression of coarse to fine sandpaper is an effective way to reduce friction and improve race times. Reducing the radius of gyration by removing mass from the outer portion of the wheels is another modification that can effectively reduces race times. If given the freedom to modify the wheel significantly, our sensitivity analysis showed that changing the outer radius of the wheels has one of the largest effects on reducing race time; one could alternatively decrease the inner radius of the wheel or implement a variety of techniques to reduced the friction between the wheels and their axle. Race times were the least sensitive to the cross sectional area of the car. This finding suggests that the significant amount of time and resources put into creating sleek car bodies would be better used to change parameters that the sensitivity analysis found to have a greater impact on race time. If the rules of a given pinewood derby race do not permit wheel alterations, the most efficient parameter to manipulate is the friction between the wheel and the axle.

Another modification that decreases race times involves attaching a long thin object to the front of the car. This technique that takes advantage of the starting mechanism of the racetrack. The car begins to move as soon as the extension loses contact with the retractable pin, which rotates out of the way at the start of the race, essentially giving the modified car a running start. In summary, our findings add to the existing body of knowledge about how to modify gravity cars to reduce race times and offers a ranked order of importance for these modification based on a sensitivity analysis.

Beyond the specific problem studied herein, we believe some of the research findings are generic and could provide insights in other application areas where the relationship between aerodynamics, gravity, and friction alter performance, e.g. skateboarding, rollerblading, and snowboarding. For example, the energy balance expression could also be used to predict trends in peak velocity for a skateboarder.

Acknowledgements. The authors thank Patrick McGuire for his help in building the experiment. Partial Support from US National Science Foundation is gratefully acknowledged.

Edited by: M. Teodorescu

References

Chapman, W. L., Bahill, A. T., and Wymore, A. W.: Engineering Modeling and Design, CRC Press, Boca Raton, 1992.

Coleman, H. W. and Steele, W. G.: Experimentation and Uncertainty for Engineers, 2nd Edn., New York, Wiley, 1999.

Coletta, V. P. and Evans, J.: Analysis of a model race car, Am. J. Phys., 76, 903–907, 2008.

Cowley, E. R.: Pinewood Derby Physics, Phys. Teach., 27, 610–612, 1989.

Frey, H. C. and Patil, S. R.: Identification and review of sensitivity analysis methods, Risk Anal., 22, 553–578, 2002.

Garguilo, J. and Garguilo, S.: Winning Pinewood Derby Secrets, Pinewood Pro, 2011.

Hamby, D. M.: A Review of the techniques for parameter sensitivity analysis of environmental models, Environ. Monit. Assess., 32, 135–154, 1994.

Jobe, J. D.: Physics of the Pinewood Derby With Engineering Applications, Missouri City, TX, Self-publishing, 2004.

Meade, D.: Pinewood Derby Speed Secrets, Fox Chapel Publishing Company, Inc., 2006.

Mickens, E. R.: Oscillations in planar dynamic systems, World Scientific, 1996.

Minorsky, N.: Nonlinear Oscillations, Van Nostrand, Princeton, N.J., 1962.

Pedigo, T. L.: How to Win a Pinewood Car Derby, Winning Edge Publications, Inc., 2002.

Reinke, P.: Pinewood Winning by the Rules, Eloquent Books, 2010.

Solzak, T. A. and Polycarpou, A. A.: Engineering outreach to cub scouts with hands-on activities pertaining to the pinewood derby car race, Int. J. Eng. Educ., 22, 1077–1096, 2006.

Torvi, D. A.: Using Pine Wood Derby Cars to Introduce Mechanical Engineering to Students in a First Year General Engineering Design Course, 4th Canadian Design Engineering Network (CDEN)/Canadian Congress on Engineering Education (C2E2) Conference, Winnipeg, MB, 22–24 July, 2007.

10

Study of design parameters for squeeze film air journal bearing – excitation frequency and amplitude

C. Wang and Y. H. J. Au

Advanced Manufacturing and Enterprise Engineering, School of Engineering and Design, Brunel Univ., UK

Abstract. The paper presents a design of squeeze film air journal bearing based on the design rules derived from CFX and FEA simulation study of an air film in between two flat plates, one of which was driven in a sinusoidal manner. The rules are that the oscillation frequency should be at least 15 kHz and that the oscillation amplitude be as large as possible to ensure a greater film thickness and to allow the bearing to reach its stable equilibrium quickly. The proposed journal bearing is made from AL2024-T3, of 20.02 mm outer diameter, 600 mm length and 2 mm thickness. Three 20-mm long fins are on the outer surface of the bearing tube and are spaced 120° apart; three longitudinal flats are milled equi-spaced between the fins and two piezoelectric actuators are mounted lengthwise on each flat. Such a design produces a modal shape on the bearing tube which resembles a triangle. When excited in this mode at the frequency of 16.37 kHz, and a voltage of 75 V AC with 75 V DC offset acting on the piezoelectric actuators, the air gap underneath of the bearing tube behaves as a squeeze air film with a response amplitude of 3.22 μm. The three design rules were validated by experiments.

1 Introduction

Precision engineering dictates that bearings used in machine tools must be capable of producing high precision motion with low friction and wear and generating very little heat in an oil-free condition (Stolarski and Chai, 2006b). Whilst aerostatic and aero-dynamic bearings can meet these requirements, they do come with bulky ancillary equipment, such as air compressors and hoses, and hence not very portable. A search for better bearings was an activity that has exercised the minds of many researchers.

To study the bearing behaviour of a thin air film between two surfaces, the Reynolds equation is used. Stolarski and Chai (2006a) identified three mechanisms from the equation that would show a pressure-generating phenomenon, which gives the bearing its load-carrying capability. The first refers to the "physical wedge" as is found in hydrodynamic bearings where the fluid flows through a wedge; the second requires the two surfaces to contract or expand in-plane in order to create a variable velocity on the bearing surfaces; the third requires that the two bearing surfaces move normal to each other with an oscillating velocity and is known as the "squeeze film" effect. Stolarski and Chai (2006a) asserted that the pressure generated by the hydrodynamic and squeeze film effects is of a similar order of magnitude and hence the justification for exploring the latter in the design of a new type of bearing. Squeeze film bearings have the significant advantage due to the fact that they do not require air compressors and connecting hoses; the equipment needed for generating the squeeze film action is far smaller and it can be miniaturised to the extent that it becomes a single package with the bearing.

In the design investigated by Stolarski and Woolliscroft (2007) and Yoshimoto et al. (1995), the squeeze film air bearings, made from Aluminium, used elastic hinges to ensure easy flexing of the bearing plates when driven at and around the fundamental frequency of a few KHz by stack piezoelectric actuators. The presence of the elastic hinges helps increase the dynamic response resulting in a thicker air film but because of the intricate machining required, the manufacturing cost increased. In addition, the driving frequency, being of a few kHz, is within the sensitive audible range, which can cause annoyance. Yoshimoto et al. (2006) and Ono et al. (2009) proposed another design of bearing that was driven at ultrasonic frequency with single-layer piezoelectric actuators to avoid problem of audible noise during operation. The use of the single-layer piezoelectric actuators can reduce the power consumed.

The purposes of work reported in this paper are:

1. To develop a model that affirms the existence of positive pressure developed in a squeeze-film air bearing;

2. To develop a finite-element model for a single layer piezoelectric actuator that uses realistic boundary conditions;

3. To design a journal squeeze film air bearing using the design rules derived from the modelling of the squeeze air film.

In the rest of this paper, Sect. 2 presents a theory of the squeeze air film for flat plates, proving the existence of asymmetrical pressure that produces a levitation force. Section 3 considers the design rules of a squeeze film bearing leading to the design of a journal bearing, which was subsequently tested to establish its dynamic characteristics. Section 4 deals with the discussion of the simulation and experimental results. Finally, Sect. 5 gives the conclusion.

2 Theory of squeeze air film for flat plates

2.1 Pressure profile

Consider an air film that is squeezed between two flat plates having relative sinusoidal motion of frequency ω at amplitude e, in the direction of the film thickness, as shown in Fig. 1a. The pressure at a point in the air film is governed by the Reynolds equation,

$$\frac{\partial}{\partial X}\left(H^3 p \frac{\partial p}{\partial X}\right) + \frac{\partial}{\partial Z}\left(H^3 p \frac{\partial p}{\partial Z}\right) = 2\sigma \frac{(\partial PH)}{\partial \tau} \tag{1}$$

Equation (1) is given in a non-dimensional form (Stolarski, 2009). X, Y and Z are the coordinates of a point in the air film expressed as a fraction of its length, width and thickness respectively; $P = \frac{p}{p_0}$ is the ratio of the instantaneous pressure to the initial pressure; H the ratio of the instantaneous to the initial film thickness; σ the squeeze number; and $\tau = \omega t$ the non-dimensional time obtained as the product of the angular frequency of oscillation ω and time t.

The squeeze number is defined as $\sigma = \frac{12\mu\omega L^2}{p_0 h_0^2}$ where μ is the dynamic viscosity and L the length of the air film.

The instantaneous film thickness $h = h_0 + e \sin(\tau)$ and hence the instantaneous film thickness ratio, being h/h_0 is given by $H = 1 + \varepsilon \sin(\tau)$.

2.2 Equation of motion of flat plates

If now the bottom plate is given a sinusoidal motion normal to its surface while the top plate is supported by the air film,

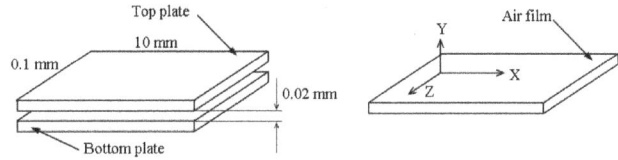

Figure 1. (a) Dimensions of the plates and the air film thickness; and (b) the air film in the rectangular coordinate system used.

the top plate will be caused to move in a manner governed by the Newton's Second Law, namely

$$m\frac{dv}{dt} = F_L - mg \tag{2}$$

In Eq. (2), m and v are the respective mass and velocity of the top plate, F_L is the levitation force exerted by the air film and mg is the weight force of the top plate.

Equation (1) computes the pressure profile of the air film between the two plates. From the pressure profile is obtained the force of levitation F_L. The corresponding value of velocity v can then be found from Eq. (2), from which the displacement of the top plate is calculated by integration. This displacement, together with the sinusoidal motion of the bottom plate, changes the air film thickness h and hence the thickness ratio H. This, in turn changes the pressure profile as computed from Eq. (1). By repeating the afore-mentioned calculation procedure over time, it is possible to determine the corresponding displacement response of the top plate.

2.3 Simple model of oscillating plates and boundary conditions

The simple model considered is the one as shown in Fig. 1a, with dimensions of the plates and the initial air film thickness as indicated. The bottom plate was given a sinusoidal motion whilst the top plate, of mass 6.24×10^{-3} kg, responded also with a sinusoidal motion. The air film, as shown in Fig. 1b, is assumed to have no leakage around the three sides of its edges, namely the left, front and back sides; but there is leakage from the right side. Such assumptions are justified by the following considerations:

1. The left side is on the plane of symmetry of the complete air film. In other words, what is shown in Fig. 1b is only the right half and hence there is no sideway flow of air across the symmetry plane.

2. The front and back sides do not have air flow across them because of the symmetrical arrangement of the three pairs of bearing plates in the proposed journal bearing design – see Fig. 13 – and the synchronised driving of them.

3. The right side, however, is exposed to the atmosphere and leakage is expected.

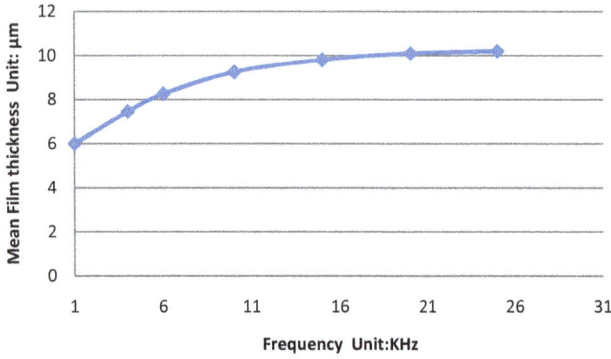

Figure 2. Mean air film thicknesses versus excitation frequency of bottom plate.

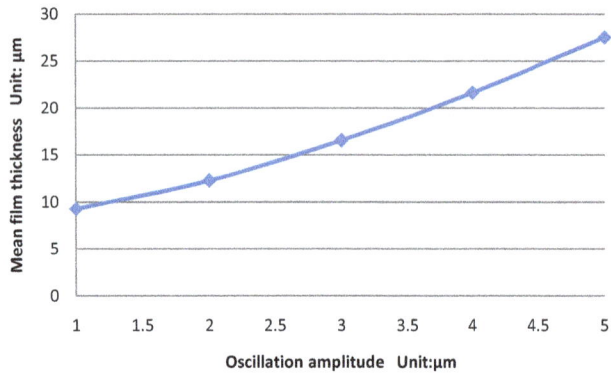

Figure 3. Mean film thickness versus oscillation amplitude of bottom plate (oscillation frequency at 10 kHz).

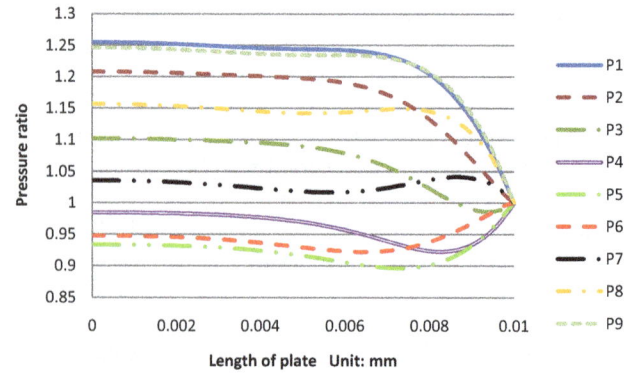

Figure 4. Steady-state pressure distribution in the x-direction (Fig. 2b) of air film in a period of oscillation of bottom plate in frequency of 10 KHz.

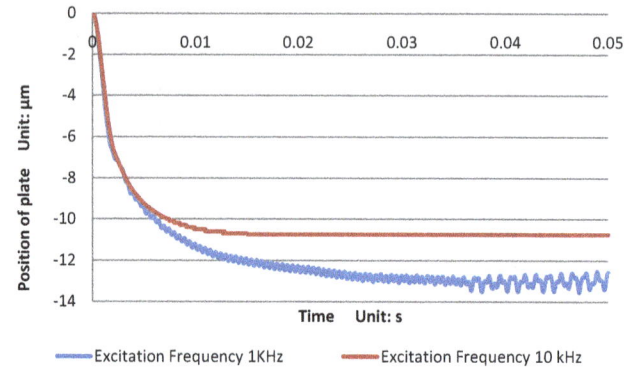

Figure 5. Transition to equilibrium position of top plate for two excitation frequencies of bottom plate at 1 kHz and 10 kHz.

The computation was performed with CFX (Guo et al., 2005) instead of home-built programming codes (Stolarski and Woolliscroft, 2007) to save time.

2.4 Calculation results

Figure 2 shows the relationship between the air film thickness and the oscillation frequency of the bottom plate. It is observed that the mean air film thickness at first increases with oscillation frequency but it reaches a constant value beyond around 15 kHz.

Figure 3 shows the relationship between the mean air film thickness and the oscillation amplitude with the oscillation frequency of the bottom plate kept at 10 KHz. It is noted that as the oscillation amplitude increases the mean air film thickness increases in an exponential fashion.

Figure 4 shows the steady-state pressure distribution of the air film over a period of oscillation along the x-axis (Fig. 2b) from the left edge (x = 0 mm) to the right edge (x = 10 mm) where the air film interfaces with the atmosphere whose pressure ratio P is 1. There are 9 pressure profile curves shown and they represent the pressure at different time instants in the cycle of oscillation such that the time interval between successive points, for example P1 and P2, is constant, being 12.5 µs. It is noted that the mean pressure ratio in the film at any distance is above unity; thus an up-thrust is created to levitate the top plate. The same conclusion was drawn by the authors in their paper using the theory of ideal gas law (Wang and Au, 2011).

Figure 5 shows the transition to the final equilibrium position of the top plate from the initial film thickness of 20 µm at the two oscillation frequencies of the bottom plate, namely 1 kHz and 10 kHz. The observation from Fig. 3 that the mean film thickness increases with oscillation frequency below 15 kHz is seen also to hold true here. In addition, at higher oscillation frequency of the bottom plate the response of the top plate shows greater stability, with no residual oscillation, achieved at around 0.037 s.

3 Design of the proposed squeeze film air bearing

Figures 2, 3 and 5 highlight some rules for the design of squeeze air film bearings. Specifically:

1. According to Fig. 2, to ensure a greater film thickness, the oscillation frequency imposed on the air film should be high, preferably above 15 kHz because the end leakage becomes insignificant.

2. Figure 3 points to the fact that the greater the oscillation amplitude of the air film, the greater is its mean thickness.

3. Figure 5 suggests that with a greater oscillation frequency of the air film, the bearing reaches its stable equilibrium position much more quickly.

Using these three design rules, a design of the squeeze film air journal bearing (Ha et al., 2005; Zhao and Wallaschek, 2009), as shown in Fig. 13, is proposed. This bearing is in the shape of a hollow round tube with three longitudinal flats milled equi-spaced around the circumference. Two piezoelectric actuators are mounted length-wise on each flat and they are driven simultaneously by an AC voltage with a DC offset. The material and geometry of the tube are such chosen that at least one modal frequency exists which is above 15 kHz and has a desirable modal shape.

With such a design, the x-axis (Fig. 1b) of the bearing plate is aligned with the longitudinal axis of the bearing tube, the y-axis with the radial axis, and the z-axis with the tangential axis to the circumference. Since the film thickness is very small in relation to the width or length of the bearing, the bearing plate can be assumed to be flat. Leakage is only significant in the longitudinal directions (both positive and negative directions of the x-axis) but otherwise virtually non-existent in the radial (y-axis) or tangential (z-axis) direction. To first approximation the model presented in Sect. 2 holds.

The modal shape of choice should be one that produces purely a radial deformation of the tube wall without the tube experiencing any torsion. Therefore the only possible modal shape for the design as shown in Fig. 13 has to be a triangle similar to Fig. 14. To encourage the tube to distort into a triangle, three external fins are added, which in effect partitions the tube into three 120° sectors. The fins do not cover the whole length of the tube but are foreshortened. This is to make sure that both ends of the tube do not deform or at least not as much as the inner sections of the tube; consequently the end leakage in the longitudinal direction can be further minimised.

At the desired mode as described above, the three fins can be imagined to have only radial motion and the tube wall between any consecutive pair of fins flexes about the fins as its end supports. Its static and dynamic behaviour is modelled in the next section.

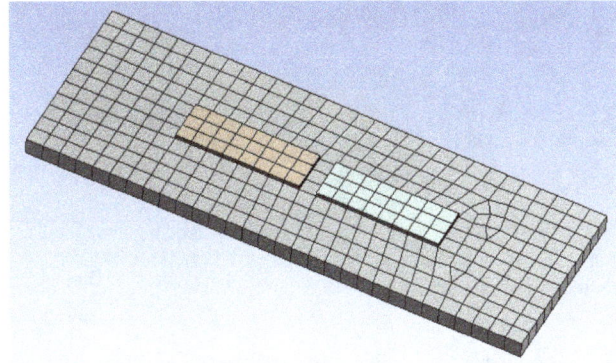

Figure 6. Model of rectangular plate with two single-layer piezoelectric actuators mounted on top surface.

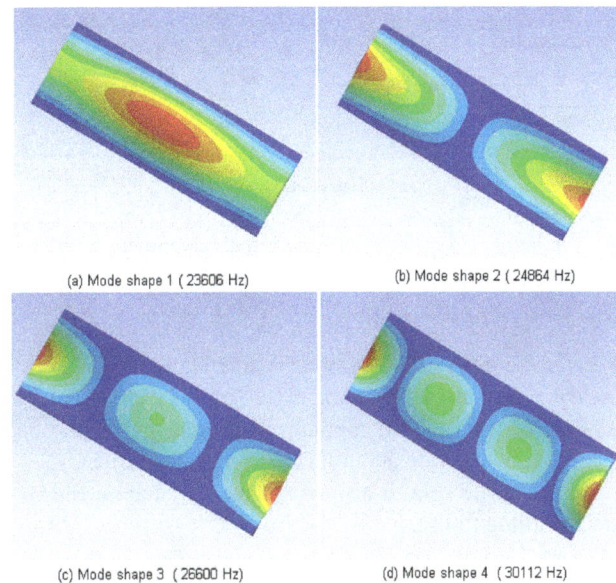

(a) Mode shape 1 (23606 Hz) (b) Mode shape 2 (24864 Hz)

(c) Mode shape 3 (26600 Hz) (d) Mode shape 4 (30112 Hz)

Figure 7. Mode shapes and natural frequencies of rectangular plate; colour red indicates maximum deformation and colour blue, no deformation.

3.1 Modal analysis

Figure 6 shows a flat plate with a width of 21 mm (equal to the circumferential width of the tube between two consecutive fins), length of 60 mm (the length of the tube) and thickness of 2 mm (the thickness of the tube). The plate has a built-in support along the two 60-mm edges. Two single-layer piezoelectric actuators are mounted on the top surface of the plate as a driving unit; the dimensions of the single-layer piezoelectric actuator are 15 mm × 5 mm × 0.5 mm.

Using ANSYS Workbench, a modal analysis was performed on the flat rectangular plate. The first four natural frequencies and the corresponding mode shapes are as shown in Fig. 7. Mode shape 1, obtained by exciting the two single-layer piezoelectric actuators at 23 606 Hz (Yoshimoto et al.,

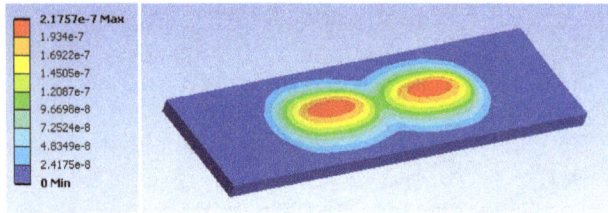

Figure 8. Static deformation of rectangular plate when a 150 V DC was applied to the two actuators.

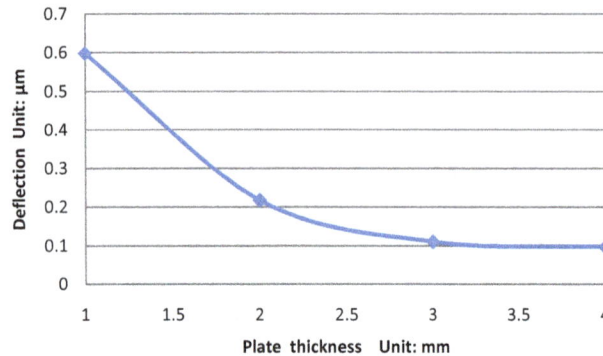

Figure 9. Relationship between maximum static deformation and plate thickness.

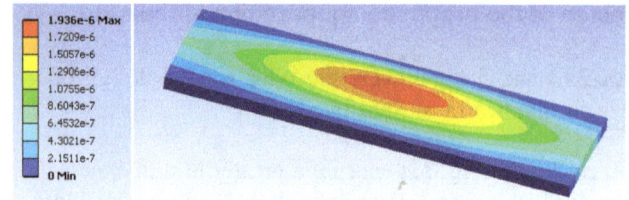

Figure 10. Dynamic response of plate when driven at the fundamental natural frequency of 23 606 Hz.

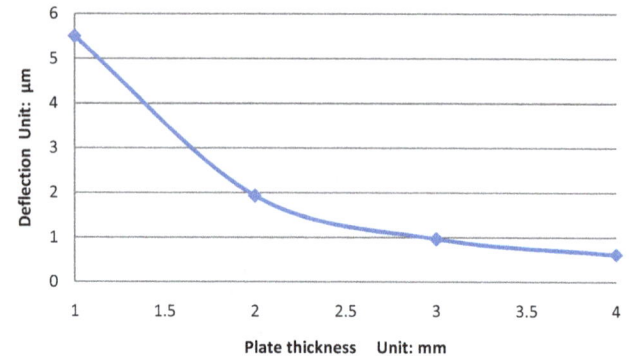

Figure 11. Dynamic deformation versus plate thickness when excited at natural frequency of plate.

2006), gives the most desirable mode shape feature. This is because the maximum deformation region occurs right at the centre of the plate, resulting in minimum leakage at the two opposite 21-mm edges. In addition, the excitation frequency, being higher than 15 kHz, is also conducive to reducing end leakage according to Fig. 2.

3.2 Static analysis

When a DC voltage is applied to two single-layer piezoelectric actuators mounted on the rectangular plate, a static deformation results. A larger deformation is preferred to a smaller one because of the thicker squeeze air film that it creates. Evidently the deformation is a function of the plate thickness.

Figure 8 shows the static deformation that appears on the rectangular plate when a 150 V DC was applied to the piezoelectric actuators with 0 V and 150 V on the respective bottom and top surfaces. A maximum deformation of 0.21757 μm is seen to occur at two regions, as shown in red in Fig. 8, symmetrically disposed from the centre line parallel to the short sides of the plate.

The relationship between the thickness of the plate and the deformation is shown in the Fig. 9. It is observed that the plate deflection increases disproportionately as the plate thickness decreases. The static deformation of a 1-mm thick plate is about 3 times as big as that of 2-mm thickness; however, as thickness increases from 3 to 4 mm, the static deformation hardly changes, being around 0.1 μm.

3.3 Dynamic analysis

Dynamic analysis is used to determine the dynamic response of a structure under a sinusoidal excitation force. The excitation force in this case is created from the alternate expansion and compression of the piezoelectric actuators when they are driven by an AC voltage at 75 V on top of a DC offset also at 75 V. To take advantage of the mechanical gain at resonance, the excitation frequency was chosen to coincide with the Mode 1 frequency of 23 606 Hz, as identified in Sect. 3.1, in order to achieve maximum dynamic response, thus giving a better floating performance as suggested by Fig. 3.

Figure 10 shows the dynamic response of the rectangular plate when excited at 23 606 Hz with the AC 75 V and the DC offset 75 V as occurring at the central region (shown in red) where the amplitude is 1.936 μm (Yoshimoto et al., 2006) . As expected the response of the plate is of the same shape as mode shape 1. By varying the beam thickness from 1 to 4 mm, the maximum amplitude of response is seen to decrease in an exponential manner as shown in Fig. 11, similar to that observed in the static analysis (Fig. 9). A comparison between the two analyses is presented in Fig.12: the difference in magnitude is striking; for a 2-mm thick plate, the dynamic deformation is about 9 times as large as the static deformation.

If maximum dynamic deformation was the only design criterion, then the thinner the plate the better. But the modal frequency drops as the plate thick decreases. Thus, for example,

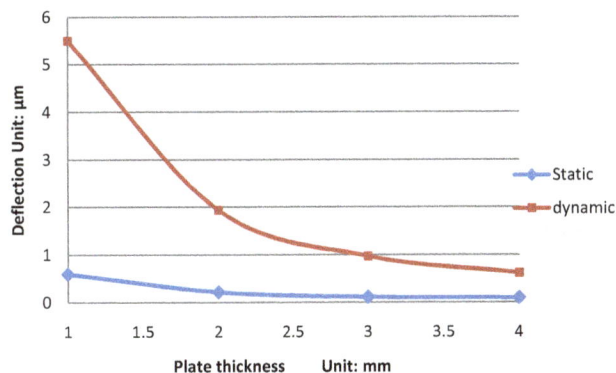

Figure 12. Comparison between static and dynamic plate deformations.

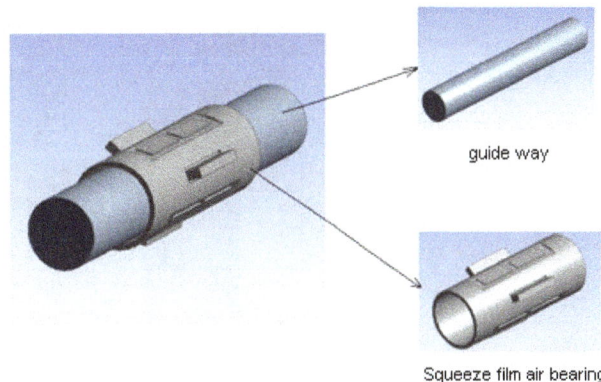

Figure 13. Squeeze film air journal bearing with six single-layer piezoelectric actuators mounted on three milled flat surfaces, shown with the guide way in the shape of a round bar.

the 1-mm thick plate has a mode 1 frequency of 11.835 kHz, which is well below the threshold frequency of 15 kHz, thus creating substantial end leakage (Fig. 2) and a long transition time to equilibrium (Fig. 5); furthermore the resulting strain in the driving piezoelectric actuators may be too high to cause them to fracture.

3.4 Experimental results for the designed squeeze film air journal bearing

Based on the results of the static and dynamic analyses and of the three design rules formulated in Sect. 3, the final design is created, which is as shown in Fig. 13. The journal bearing is made from the material AL2024-T3, and has a diameter of 20.02 mm, a length of 60 mm and a thickness of 2 mm. Three fins, each 20 mm long, are positioned 120° apart on the outer circumference of the bearing tube; they are designed to provide a desirable modal shape of a triangular cross-section when excited by the actuators. This enables the air gap underneath the actuators to behave effectively as a squeeze air film. The round bar has a diameter of 19.99 mm and the surface was produced by cylindrical grinding.

From the FEA modal modelling, the Mode 13 was identified to have the desired deformed geometry of a triangle and it has the modal frequency of 16.37 kHz, which is above the 15 kHz threshold. The corresponding mode shape is as shown in Fig. 14 where the red end of the colour spectrum denotes greater deformation. It can be observed that the outer edges of the round sleeve do not appear to deform much while the middle section deforms noticeably.

3.4.1 Experimental set up

Figure 15 shows the set up of equipment for the dynamic response experiment. The following items of equipment were used:

1. A signal generator – 0 to 5 V and 0 to 100 kHz (S J Electronics)

Figure 14. Mode shape 13 of squeeze film air journal bearing (not to scale).

2. An actuator driver – ENP-1-1U (Echo Electronics)

3. An actuator driver monitor – ENP-50U (Echo Electronics)

4. A capacitance displacement sensor and a gauging module – MicroSense 5810; measurement bandwidth up to 100 kHz; measurement ranges ±100 µm; accuracy ±49.69 nm and resolution 22.3 nm (Ixthus)

5. A data acquisition card – PXI 6110 (National Instruments)

The signal generator created a sinusoidal wave which was amplified by the actuator driver and shaped by the actuator monitor to provide an excitation signal, with a 75 V DC offset and a 75 V peak-to-zero AC sinusoid. This excitation signal was used to drive the single layer piezoelectric actuators. The vibration response of the structure caused by the actuators was measured by the capacitance displacement sensor, whose output was sampled into a PC via the data acquisition card driven by a LabVIEW program.

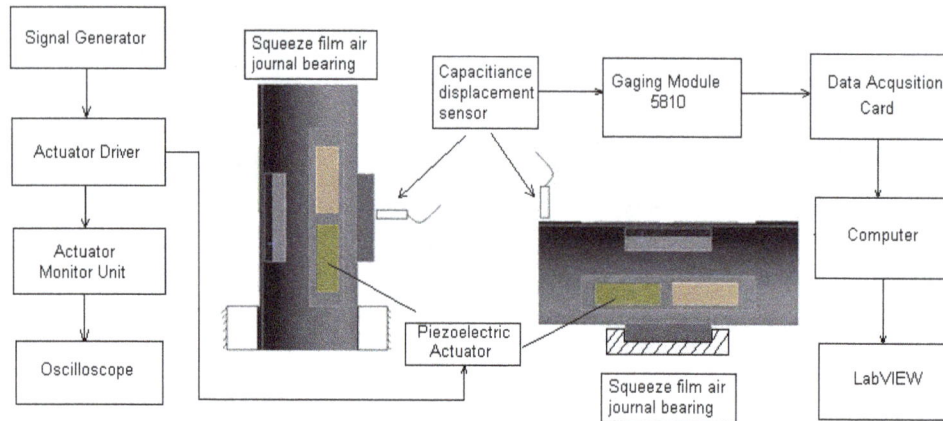

Figure 15. Schematic diagram of the experimental set-up.

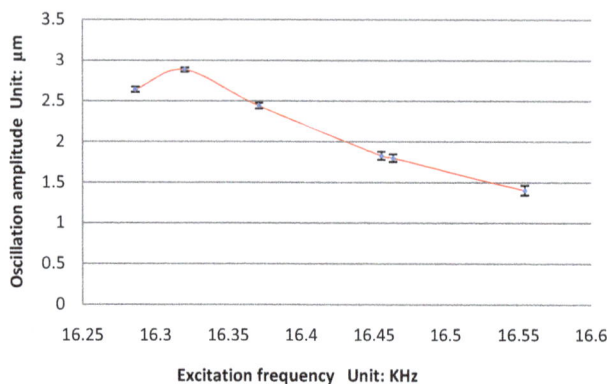

Figure 16. Oscillation amplitude at the centre of the bearing shell at Mode 13 versus excitation frequency at 75 V AC input with 75 V DC coupling; error bars represent ±2 standard errors.

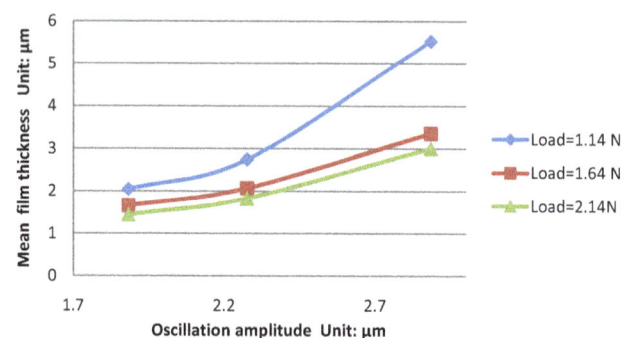

Figure 17. Mean film thickness versus oscillation amplitude at three load levels.

3.4.2 Oscillation amplitude on vibrating shell of journal bearing and excitation frequency

The maximum amplitude of oscillation at the vibrating shell of the bearing was measured within a range of frequencies, 16.28 kHz to 16.55 kHz, in the vicinity of the predicted Mode 13 frequency of 16.37 Hz. The results are as shown in Fig. 16. The measurements were made 10 times and it is the average that is shown on the graph; the corresponding error bar represents ±2 standard errors. The narrow extent of the error bars suggests good measurement repeatability and high precision of the displacement amplitude obtained. From Fig. 16, it is observed that the natural frequency for Mode 13 was 16.32 kHz; and that the amplitude of oscillation was 2.88 μm, compared to the simulated result of 3.22 μm from ANSYS Workbench modelling.

3.4.3 Mean film thickness and oscillation amplitude

According to the second design rule the mean squeeze film thickness increases with the amplitude of oscillation of the shell. Experiments were conducted to validate this assertion. Figure 17 shows the relationship between the mean film thickness and the oscillation amplitude at three load levels. The loading was implemented by attaching a weight to the journal bearing and three loads were studied, namely 1.14 N, 1.64 N and 2.14 N. These loads were hung on the bottom fin – see Fig. 14 – such that there was a squeeze air film at the top and another pair symmetrically disposed at 120∘ on either side. In Fig. 17, it can be seen that the mean film thickness increases in an exponential fashion with increasing oscillation amplitude for all three different loads.

3.4.4 Comparison between experimental and theoretical mean film thicknesses

With the journal bearing loaded as described in Sect. 3.4.3 corresponding to the orientation of the three squeeze films as shown in Fig. 14, an approximate simulation CFX model

Table 1. Comparison between flat plate and curved shell in respect of dynamic response.

	Flat Rectangular Plate 2 mm thick	Tubular Bearing with Curved Shell 2 mm thick
Natural frequency	23 606 Hz	16 368 Hz
Dynamic deformation	1.99 μm	3.22 μm

was formulated and analysed. Two simplifying assumptions were made:

1. As the surrounding sleeve is loaded, the squeeze film at the top has a thickness which is much smaller than that at the other two squeeze film situated towards the bottom (Fig. 14), such that the levitation force created is solely due to the top squeeze film;

2. The top squeeze film is flat and there is no leakage when in operation; and

3. The upper plate of the top squeeze film translates bodily up and down with no deformation.

The CFX simulation modelled an air film which was 30 mm long, 0.1 mm wide and 0.03 μm thick using the same set of boundary conditions as that described in Sect. 2.3. Since the bearing shell of the top squeeze film did not oscillate as a rigid body but rather flexed itself at the Mode 13 natural frequency, the average amplitude of oscillation of 1.31 μm was used in the simulation. This average was obtained by measuring the modal shape of the bearing shell at the top squeeze film with the bearing driven at the Mode 13 natural frequency of 16.32 kHz and then taking an arithmetical average from these measurements.

Figure 18 shows the results from the simulation as compared to those obtained from the experiment for different loadings. There is broad agreement between the two in terms of the trend and of the values of the mean film thickness, with better fit towards the higher loading. It is surmised that the better fit could be due to the fact that the first assumption becomes more correct as the loading increases; further work needs to be done in this area. The agreement between the theoretical and experimental results serves, in some way, to validate the simulation results in Sect. 2, from which the three design rules were derived.

4 Discussions

In this paper the flat rectangular plate was first studied in order to identify important design rules and parameters that would better inform subsequent design activity, leading to the design of a tubular bearing with fins. A comparison between the dynamic simulation results obtained from the flat plate and the curved shell is informative; Table 1 summarises the effort.

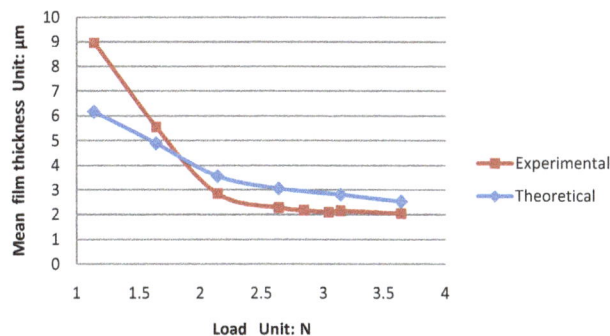

Figure 18. Comparison of theoretical and experimental for mean film thickness.

The flat plate is seen to have a higher natural frequency than the curved counterpart. This is due to two reasons: (1) the flat plate was subjected to a more severe end-fixing condition, namely built-in fixing along the full length, compared to the partial fixing via the foreshortened fins, which themselves are free to move radially, on the curved shell; (2) the curved shell has a milled flat for the actuators, which reduces the shell's stiffness in bending locally.

The foreshortened fins can be responsible for the larger dynamic response at 3.22 μm compared to the 1.99 μm on the severely edge-constrained plate.

5 Conclusions

Three design rules for squeeze air film bearings were produced and verified from the research:

1. To ensure a greater film thickness and to reduce end leakage, the oscillation frequency imposed on the air film should be high, preferably above 15 kHz (Fig. 2).

2. The greater the oscillation amplitude of the air film, the greater is its mean thickness (Fig. 3).

3. With a greater oscillation frequency of the air film, the bearing reaches its stable equilibrium position much more quickly (Fig. 5).

Based on the design rules, a journal squeeze air film bearing was designed (Figs. 13 and 14) and analysed. The bearing was designed to be driven at its 13th mode at the frequency of 16.37 kHz at which the amplitude response was 3.22 μm and the modal shape produced a squeezing action on the air film between the journal shell and the bearing shaft.

These three design rules were validated from experiments conducted on the journal squeeze air film bearing, as demonstrated in Figs. 17 and 18.

Edited by: A. Barari

References

Guo, Z. L., Hirano, T., and Kirk, R. G.: Application of CFD analysis for rotating machinery – Part 1: Hydrodynamic, hydrostatic bearings and squeeze film damper, J. Eng. Gas Turb. Power, 127, 445–451, 2005.

Ha, D. N., Stolarski, T. A., and Yoshimoto, S.: An aerodynamic bearing with adjustable geometry and self-lifting capacity. Part 1: self-lift capacity by squeeze film, International Journal of Engineering Tribology, 219, 33–39, 2005.

Ono, Y., Yoshimoto, S., and Miyatake, M.: Impulse-Load dynamics of squeeze film gas bearings for a linear motion guide, J. Tribol., 131, 1–6, 2009.

Stolarski, T. A.: Numerical modelling and experimental verification of compressible squeeze film pressure, Journal of Tribology International, 43, 356–360, 2009.

Stolarski, T. A. and Chai, W.: Load-carrying capacity generation in squeeze film action, Int. J. Mech. Sci., 48, 736–741, 2006a.

Stolarski, T. A. and Chai, W.: Self-levitation sliding air contact, Int. J. Mech. Sci., 48, 601–620, 2006b.

Stolarski, T. A. and Woolliscroft, S. P.: Performance of a self-lifting linear air contact, J. Mech. Eng. Sci., 221, 1103–1115, 2007.

Yoshimoto, S., Anno, Y., Sato, Y., and Hamanaka, K.: Floating characteristics of squeeze-film gas bearing with elastic hinges for linear motion guide, International Journal of JSME, 60, 2109–2115, 1995.

Yoshimoto, S., Kobayashi, H., and Miyatake, M.: Floating characteristics of a squeeze – film bearing for a linear motion guide using ultrasonic vibration, Journal of Tribology International, 40, 503–511, 2006.

Wang, C. and Au, Y. H. J.: Levitation Characteristics of a Squeeze Film Air Journal Bearing at its Normal Modes, Journal of Advanced Manufacturing Technology, under revision, 2011.

Zhao, S. and Wallaschek, J.: Design and modelling of a novel squeeze film journal bearing, in: Proceedings of the 2009 IEEE International Conference on Mechatronics and Automation, Changchun, China, 9–12 August 2009, 1054–1059, 2009.

11

Experimental comparison of five friction models on the same test-bed of the micro stick-slip motion system

Y. F. Liu[1], J. Li[1], Z. M. Zhang[2], X. H. Hu[2], and W. J. Zhang[1,2]

[1]Complex and Intelligent System Laboratory, School of Mechanical and Power Engineering,
East China University of Science and Technology, Shanghai, China
[2]Department of Mechanical Engineering, University of Saskatchewan, Saskatoon, Canada

Correspondence to: W. J. Zhang (chris.zhang@usask.ca)

Abstract. The micro stick-slip motion systems, such as piezoelectric stick-slip actuators (PE-SSAs), can provide high resolution motions yet with a long motion range. In these systems, friction force plays an active role. Although numerous friction models have been developed for the control of micro motion systems, behaviors of these models in micro stick-slip motion systems are not well understood. This study (1) gives a survey of the basic friction models and (2) tests and compares 5 friction models in the literature, including Coulomb friction model, Stribeck friction model, Dahl model, LuGre model, and the elastoplastic friction model on the same test-bed (i.e. the PE-SSA system). The experiments and simulations were done and the reasons for the difference in the performance of these models were investigated. The study concluded that for the micro stick-slip motion system, (1) Stribeck model, Dahl model and LuGre model all work, but LuGre model has the best accuracy and (2) Coulomb friction model and the elastoplastic model does not work. The study provides contributions to motion control systems with friction, especially for micro stick-slip or step motion systems as well as general micro-motion systems.

1 Introduction

Micro stick-slip motion systems can provide high resolution motions yet with a long motion range. The piezoelectric stick-slip actuator (PE-SSA), which is a hybridization of the piezoelectric actuator (PEA) and the stick slip actuator (SSA), is a typical example in these systems. By hybridization, it is meant that PE and SS are complementary to each other, according to the hybridization design principle proposed by Zhang et al. (2010).

The working process of the PE-SSA is demonstrated in Fig. 1. At position (1), a voltage is applied to the PEA and leads to an (relatively slow) expansion of the PEA, which pushes the stage moving to the right. The friction between the stage and the end effectorbrings the end effector to the position (2) (stick motion). When the applied voltage is shut down quickly, the PEA contracts quickly and the slip between the end effector and stage takes place due to the inertia of the end effector, which overcomes the friction resistance. The relative displacement S, with respect to its initial posi-

tion (1), is thus generated at position (3) (slip motion). If the aforementioned process is repeated periodically, the end effector will keep moving to the right as long as the physical system allows. The back forward motion of the end effector can also be obtained by reversing the actuation potential signal applied on the PEA.

It can be found from the aforementioned working process of the PE-SSA that the friction force between the end effector and stage plays both an active role in the forward stroke motion and a passive role in the backward stroke in the stick-slip motion system. The modeling of the friction for prediction and control of such a system is extremely important as well as difficult (Makkar et al., 2007; Li et al., 2008a). The difficulty lies in the possibility that the motion direction on the two contact surfaces may change the frictional effect (Zhang et al., 2011). Although numerous friction models have been studied in the context of macro, micro motion, such as Coulomb friction model, viscous friction model, Stribeck friction model, Dahl model, LuGre model,

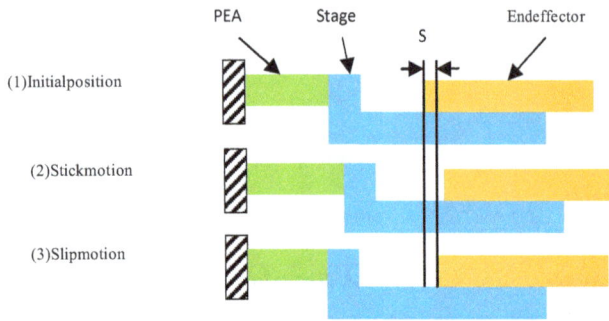

Figure 1. The working process of the PE-SSA.

Table 1. Friction characteristics captured by different models.

	Viscous	Stribeckeffect	Pre-sliding	Hysteresis
Coulomb	No	No	No	No
Viscous	Yes	No	No	No
Stribeck	Yes	Yes	No	No
Dahl	No	No	Yes	Yes
LuGre	Yes	Yes	Yes	Yes
Elastoplastic	Yes	Yes	Yes	–
Leuven	Yes	Yes	Yes	Yes
GMS	Yes	Yes	Yes	Yes

and elastoplastic friction model, the performances of these models in micro stick-slip motion systems are not well understood. This paper aims to investigate the performance of different friction models inmicrostick-slip motion systems. A survey of basic friction models is presented first and five of the mare selected for comparison. A common micro stick-slip motion system- the PE-SSA system which was introduced in our previous study (Li et al., 2008a) is used as a test-bed. Parameters in each friction model will be determined using the system identification technique. Performances of these models are then compared. Leuven and GMS models are out of the scope of the study because they are not commonly used in the step motion system such as the stick-slip motion system.

The remainder of the paper is organized as follows. Section 2 gives an overview and history of the five friction models. The mathematical descriptions of these friction models are presented in Sect. 3. The experimental setup is described in Sect. 4. Section 5 presents the analysis methodology. Section 6 presents the experimental results along with discussions, followed by conclusions in Sect. 7.

2 Overview of the friction models

The first friction model is Coulomb friction model (or called Amontons–Coulomb friction model), referring to the work done by Guillaume Amontons and Charles-Augustin de Coulomb in 1699 and 1785, respectively. In the Coulomb friction model, the friction force is the function of load and direction of the velocity. Morin (1833) found that the static friction (i.e. friction at zero sliding speed) is larger than the Coulomb friction. With respect to the static friction, the Coulomb friction is also called dynamic friction. Viscous friction was later introduced in relation to lubricants by Reynolds and it is often combined with Coulomb friction model. Stribeck (1902) experimentally observed that friction force decreases with the increase of the sliding speed from the static friction to Coulomb friction. The phenomenon is thus called Stribeck effect. The integration of the Coulomb friction, viscous friction, and Stribeck effect is often an idea

to obtain a more accurate friction model, which is called Stribeck model in literature.

Dahl (1968) first modeled friction as a function of the relative displacement of two contact surfaces, and it is thus called Dahl model. The model is based on the fact that friction force is dependent on the "micro motion" in ball bearings. The "micro motion", later called pre-sliding behavior, is that when the external force is not large enough to overcome the static friction, the asperities on two contact surfaces will experience deformation that results in the pre-sliding motion. The asperities form a kind of spring-damping system. When the external force is sufficiently large, the spring is broken, leading to a relative sliding between two contact surfaces. The Dahlmodel successfully describes the so-called breakaway phenomenon.

Canudas et al. (1995) developed a friction model called LuGre model, named after the two universities, namely Lund and Grenoble. LuGre model incorporates the viscous friction and Stribeck effect into Dahl model. The problem of incorporating the pre-sliding behavior in the friction model is that both Dahl and LuGre models experience "drift" when there is an arbitrarily small bias force or vibration. The reason for this drift is that both Dahl and LuGre models only include a "plastic" component in their model when they describe the pre-sliding phenomenon.

To overcome this drift, Dupont et al. (2000) proposed a friction model based on LuGre model, in which the pre-sliding was defined as the elastoplastic deformation of asperities; i.e. the relative displacement is elastic (reversible) first and then it transits to the plastic (irreversible) stage. The model of Dupont et al. (2000) is thus called elastoplastic friction model.

Leuven friction model and generalized Maxwell-slip (GMS) model were proposed by Swevers et al. (2000) and Lampaert et al. (2002), respectively. The two models were developed based on the experimental findings that the friction force in the pre-sliding regime has a hysteresis characteristic with respect to the position. It is reported that the two models can improve the hysteresis behavior of the friction predicted with LuGre model.

The friction characteristics that are captured by the aforementioned models are listed in Table 1. To control a dynamic system for high accuracy, a common sense seems to go with the friction model that can capture more friction characteristics. However, such may not always be the case. Generally speaking, for a complex dynamic system such as frictional systems, a (complete) model may be viewed as an integration of a couple of sub-models, each of which captures one or more characteristics. While being integrated, each of them may produce "side effects" to the modeling of other characteristics (Li et al., 2008b, 2009), because these characteristics are coupled, changing with time and perhaps, the physical structure of the frictional system changes as well.

3 The mathematical description of friction models

3.1 Coulomb model

Coulomb friction model is represented using the following equation:

$$F = \begin{cases} F_c \cdot Sgn(\dot{x}) & \text{if } \dot{x} \neq 0 \\ F_{app} & \text{if } \dot{x} = 0 \text{ and } F_{app} < F_c \end{cases} \tag{1}$$

where F is friction force, $\dot{x}$ is sliding speed, F_{app} is applied force, F_C is the Coulomb friction force, which is defined as

$$F_c = \mu F_N \tag{2}$$

where μ is the Coulomb friction coefficient (or called the dynamic friction coefficient), and F_N is a normal load between two contact surfaces.

The Coulomb model is illustrated in Fig. 2a. When $F_{app} < F_c$, there is no sliding (i.e. $\dot{x} = 0$ in the "macro" sense) between two contact surfaces, and the Coulomb friction can take any value from zero up to F_c. If $\dot{x} \neq 0$, Coulomb friction only takes F_c or $-F_c$, depending on the direction of the sliding.

Coulomb friction model is commonly used in the applications such as the prediction of temperature distribution in bearing design and calculation of cutting force in machine tools due to its simplicity. However, it is often troublesome to use the Coulomb friction model in micro motion systems because of the "undefined" friction force at $\dot{x} = 0$.

3.2 Viscous friction model

Viscous friction model is given by

$$F = k_v \dot{x} \tag{3}$$

where F is the friction force, k_v the viscous coefficient, and $\dot{x}$ the sliding speed.

The viscous friction model is illustrated in Fig. 2b. In the viscous model, the friction force is a linear equation of the sliding speed. The application of the viscous model is limited because it has a poor representation in regions where there is no lubricant (Andersson et al., 2007).

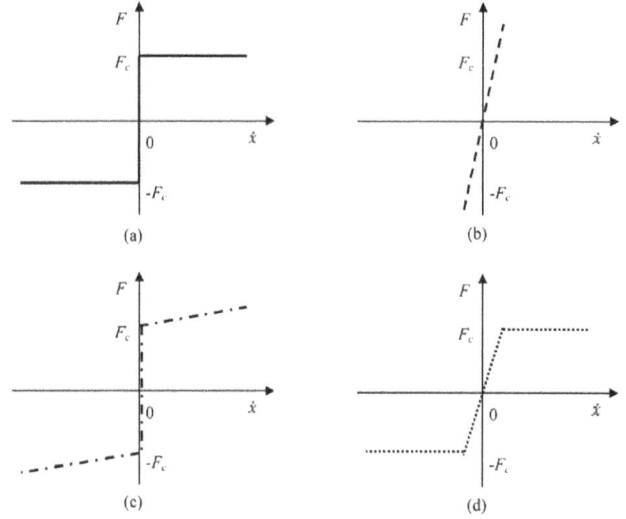

Figure 2. Friction force vs. sliding speed.

3.3 Integrated Coulomb and viscous model

There are two ways to combine viscous friction model and Coulomb model, leading to two different integrated Coulomb and viscous models. One is described by

$$F = \begin{cases} F_c \cdot Sgn(\dot{x}) + k_v \dot{x} & \text{if } \dot{x} \neq 0 \\ F_{app} & \text{if } \dot{x} = 0 \text{ and } F_{app} < F_c \end{cases} \tag{4}$$

This model (Eq. 4) is illustrated in Fig. 2c. The problem with this model is that the friction force at is still "undefined". To overcome this problem, the idea is to integrate the Coulomb model and the viscous model near $\dot{x} = 0$. This comes with the second model given by (Andersson et al., 2007)

$$F = \begin{cases} \min (F_c, k_v \dot{x}) & \text{if } \dot{x} \geq 0 \\ \max (-F_c, k_v \dot{x}) & \text{if } \dot{x} < 0 \end{cases} \tag{5}$$

This model (Eq. 5) is illustrated in Fig. 2d. The viscous coefficient determines the speed of the friction force transition from $-$ to $+$.

3.4 Stribeck friction model

Stribeck friction model is described by

$$F = \left(F_c + (F_s - F_c) e^{-\left(\left|\frac{\dot{x}}{v_s}\right|\right)^i} \right) Sgn(\dot{x}) + k_v \dot{x} \tag{6}$$

where F is friction force, $\dot{x}$ Sliding speed, F_c the Coulomb friction force, F_s the static friction force, v_s the Stribeck velocity, k_v the viscous friction coefficient, and i an exponent. It is clear from Eq. (6) that the Stribeck friction force takes F_s as the upper limit and F_c as the lower limit. The relation of friction versus sliding speed in Stribeck model is illustrated in Fig. 3.

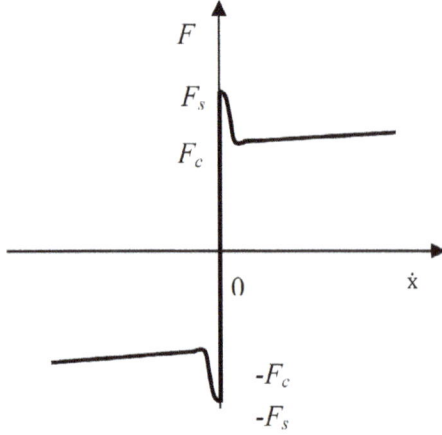

Figure 3. Stribeck friction model (Armstrong-Helouvry, 1993).

3.5 Dahl model

In Stribeck model, friction force is a function of the sliding speed. However, according to this model, friction force is "undefined" when it is less than F_s. Dahl developed a model to describe friction at this pre-sliding stage, which is given by

$$\frac{dF(x)}{dt} = \frac{dF(x)}{dx}\frac{dx}{dt} \tag{7}$$

with

$$\frac{dF}{dx} = \sigma_0 \left| - \frac{F}{F_c}sgn(\dot{x})\right|^i sgn\left(1 - \frac{F}{F_c}sgn(\dot{x})\right) \tag{8}$$

where F is the friction force, σ_0 the stiffness coefficient, and i the exponent which determines the shape of the hysteresis. In literature, Dahl model is often simplified with the exponent $i = 1$ and given by

$$\frac{dF}{dx} = \sigma_0\left(1 - \frac{F}{F_c}sgn(\dot{x})\right). \tag{9}$$

Dahl model does not describe the Stribeck effect (Canudas et al., 1995).

3.6 LuGre model

LuGre model has the following form (Canudas et al., 1995)

$$\begin{cases} F = \sigma_0 z + \sigma_1 \dot{z} + \sigma_2 \dot{x} \\ \dot{z} = \dot{x} - \sigma_0 \frac{\dot{x}}{g(\dot{x})} z \\ g(\dot{x}) = F_c + (F_s - F_c)e^{-\left(\left|\frac{\dot{x}}{v_s}\right|\right)^j} \end{cases} \tag{10}$$

where F is the friction force, σ_0 the contact stiffness, z the average deflection of the contacting asperities, σ_1 the damping coefficient of the bristle, σ_2 the viscous friction coefficient, x the relative displacement, F_c the Coulomb friction force, F_s the static friction force, $\dot{x}$ the sliding velocity, $g(\dot{x})$

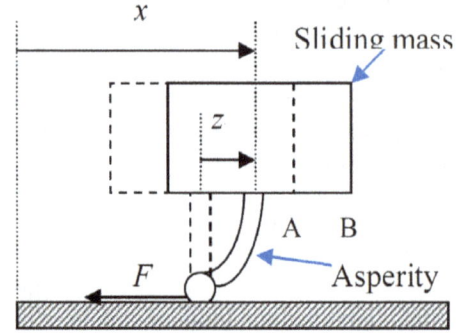

Figure 4. Illustration of deformation of asperity on the frictional surface.

the Stribeck effect, v_s the Stribeck velocity, j the Stribeck shape factor ($j = 2$ is often used in the literature). LuGre model integrates pre-sliding friction ($\sigma_0 z$), viscous friction ($\sigma_2 \dot{x}$), and Stribeck effect ($g(\dot{x})$) into one single model.

3.7 Elastoplastic friction model

The elastoplastic friction model is described by

$$\begin{cases} F = \sigma_0 z + \sigma_1 \dot{z} + \sigma_2 \dot{x} \\ \dot{z} = \dot{x}\left(1 - \sigma(z,\dot{x})\frac{z}{z_{ss}(\dot{x})}\right) \end{cases} \tag{11}$$

with

$$z_{ss}(\dot{x}) = \frac{g(\dot{x})}{\sigma_0} \tag{12}$$

$$g(\dot{x}) = F_c + (F_s - F_c)e^{-\left(\left|\frac{\dot{x}}{v_s}\right|\right)^j}. \tag{13}$$

In Eq. (11), $\sigma(z, \dot{x})$ is used to define the zones of the elastic and plastic deformation of asperities and given by

$$\begin{cases} \alpha(z,\dot{x}) = \begin{cases} 0, & |z| \leq z_{ba} \\ \alpha_m(*) & z_{ba} < |z| < z_{ss} \\ 1, & |z| \geq z_{ss} \end{cases} & Sgn(\dot{x}) \neq Sgn(z) \\ \alpha(z,x) = 0, & Sgn(\dot{x}) = Sgn(z) \end{cases} \tag{14}$$

where $Sgn(\dot{x}) = Sgn(z)$ represents that the sliding mass moves from position A to position B, as shown in Fig. 4, and $Sgn(\dot{x}) \neq Sgn(z)$ represents that the sliding mass moves from position B to A.

According to Equation (14), when $Sgn(\dot{x}) \neq Sgn(z)$, $\alpha(z, \dot{x}) = 0$. This represents that no slip occurs; the "sliding" mass is in an elastic deformation region, or two contact objects are in a stick phase. When $Sgn(\dot{x}) \neq Sgn(z)$, if $|z| \leq z_{ba}, \alpha(z, \dot{x}) = 0$. This represents that no slip occurs; the "sliding" mass is in an elastic deformation region, or two contact objects are in a stick phase. When $Sgn(\dot{x}) \neq Sgn(z)$, if $z_{ba} < |z| < z_{ss}, \alpha(z, \dot{x}) = \alpha_m(*)$. This represents that elastic deformation of the asperities starts to transit to the plastic deformation, i.e. transition phase. When $Sgn(\dot{x}) \neq Sgn(z)$, if $|z| \geq z_{ss}, \alpha(z, \dot{x}) = 1$. This represents that slip occurs; the

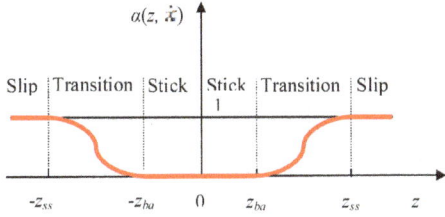

Figure 5. Dependency of $\alpha(z, \dot{x})$ on deformation of the asperity.

"sliding" mass is in a plastic deformation, or two contact objects are in a slip phase. z_{ba} and z_{ss} are given by

$$0 < z_{ba} \le z_{ss} \text{ and } z_{ss} = \text{Max}(z_{ss}(\dot{x})) = \frac{F_s}{\sigma_0}, \ \dot{x} \in R. \quad (15)$$

$\alpha_m(*)$ is typically in the following form:

$$\alpha_m(*) = \frac{1}{2}\sin\left(\pi\frac{z - \left(\frac{z_{ss}+z_{ba}}{2}\right)}{z_{ss} - z_{ba}}\right) + \frac{1}{2}. \quad (16)$$

The case when $Sgn(\dot{x}) = Sgn(z)$ in Eq. (14) are illustrated in Fig. 5.

From Fig. 5 it can be seen that in the elastoplastic friction model, $|z| \le z_{ba}$ is defined as the stiction zone. The stiction zone is to overcome the drift problem of Dahl and LuGre models. The transitions between stick and slip is given by $\alpha_m(*)$, and the slip zone is represented by $|z| \ge z_{ss}$ when it returns to LuGre model.

4 Experimental setup

The PE-SSA prototype is shown in Fig. 6. This system is composed of 1: frames, 2: friction plates, 3: temperature sensor, 4: weights, 5: end effector, 6: displacement sensor, 7: stage, 8: wheel, 9: PEA, and 10: vibration-isolated test bed. The PEA (Model: AE0505D16) purchased from NEC/TOKIN Corp. Is connected to the frame at one end, and its other end is connected to the stage. Friction plates were placed between the end effector and stage, and they were connected with the stage by screws. The weights were used to adjust the pressure between the end effector and stage. The wheel was used to support the stage. The temperature sensors are installed inside of the stage to measure temperature change in the system, which is not used in this work. The control system for the PE-SSA was designed as an open-loop system and implemented with dSPACE and Matlab/Simulink. The system was placed on the vibration isolated test bed in order to reduce disturbance to the system. More details about this prototype can be found in our previous work (Li et al., 2008a). During the experiments, the applied voltage to the PEA was a repeating saw tooth wave with amplitude of 30 V and frequency of 5 Hz. The displacements of the end effector were then measured with a KAMAN instrument (SMU 9000, Kaman) based on the eddy-current inductive principle and with a resolution of 0.01 μm.

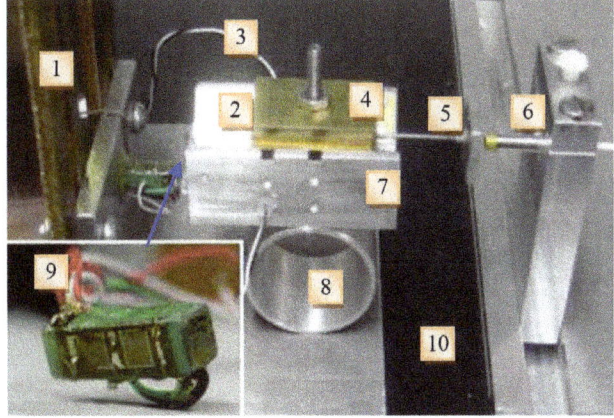

Figure 6. Experimental system of the PE-SSA (Li et al., 2008a).

5 The method for analysis

The method used to compare the performance of different friction models consists of the following steps.

– Step 1: get experimental data, i.e. the displacement of the PE-SSA with respect to time under a certain driven voltage and frequency.

– Step 2: model the PEA and stage and determine the parameters in the model (see Sect. 5.1).

– Step 3: model the PE-SSA by integrating the friction of the stick-slip motion into the model of the PEA and stage (see Sect. 5.2).

– Step 4: identify friction model parameters (see Sect. 5.3), including (a) use different friction models to represent friction in the PE-SSA model and (b) determine the parameters for each friction model.

– Step 5: compare the performance of the friction models in terms of their prediction of displacement, friction force, and sliding speed (i.e. relative speed between end effector and stage). The details are discussed in Sect. 6.

5.1 Modeling of the PEA and stage

Adriaens et al. (2000) showed that the PEA and stage can be modeled as a spring-mass-damper system which is shown in Fig. 7.

The governing equations of this spring-mass-damper system are given in Eqs. (17)–(19) as follows:

$$m\ddot{x}_p + c\dot{x}_p + kx_p = F_p \quad (17)$$

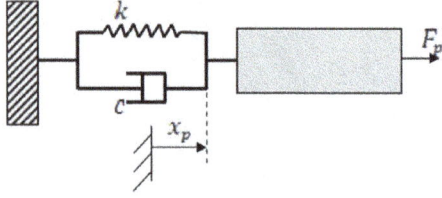

Figure 7. Physical model of the PE-SSA system (without friction) (Adriaens et al., 2000).

$$\begin{cases} m = \frac{4m_p}{\pi^2} + m_s \\ c = c_p + c_s \\ k = k_p + k_s \end{cases} \tag{18}$$

$$F_p = T_{em} u_p \tag{19}$$

where x_p is the displacement of the PEA, m_p the mass of PEA, c_p the damping coefficient of PEA, k_p the stiffness of PEA, m_s the mass of stage, c_s the damping coefficient of the stage, k_s the stiffness of stage, F_p the transducer force from the electrical side, T_{em} the electromechanical transducer ratio, and u_p the applied voltage on the PEA. In this study, Adriaens' model is taken due to its simplicity, which was validated by our previous study (Li et al., 2008b; Kang, 2007).

Equation (17) can be further written as

$$\ddot{x}_p + 2\xi\omega_n\dot{x}_p + \omega_n^2 x_p = K\omega_n^2 u_p \tag{20}$$

with

$$\begin{cases} \frac{c}{m} = 2\xi\omega_n \\ \frac{k}{m} = \omega_n^2 \\ \frac{T_{em}}{m} = K\omega_n^2 \end{cases} \tag{21}$$

where ξ is the damping ratio, ω_n the natural frequency, K and the amplified coefficient. The transfer function of the spring-mass-damper system can be written as,

$$G(s) = \frac{x_p(s)}{U(s)} = \frac{K\omega_n^2}{s^2 + 2\xi\omega_n s + \omega_n^2}. \tag{22}$$

In this study, the parameters ξ and are ω_n calculated from the following equations,

$$\begin{cases} \xi = \frac{\ln(os \times 100)}{\sqrt{\pi^2 + \ln^2(os \times 100)}} \\ \omega_n = \frac{\pi}{T_p\sqrt{1-\xi^2}} \end{cases} \tag{23}$$

where os is the overshoot of a step response and T_p is the peak time of a step response. The os and T_p are determined from the step response of the system. K is the ratio of output and input in the steady state of the step response. In our PEA system, we have $\xi = 0.2488$, $\omega_n = 6685.5$ rad s^{-1}, and $K = 0.096 \times 10^{-6}$ m v^{-1}, and they were obtained from the measured step responses of the system.

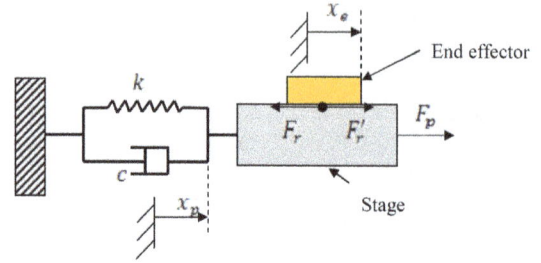

Figure 8. Physical model of the PE-SSA system (with friction) (Adriaens et al., 2000).

5.2 Modeling of the PE-SSA

In the PE-SSA, the friction force F_r between the end effector and stage is applied on the stage, as shown in Fig. 8.

Taking into account F_r, the following equation can be obtained from Eq. (17),

$$m\ddot{x}_p + c\dot{x}_p + kx_p + F_r = F_p. \tag{24}$$

On the end effector, friction force, denoted by that pushes the end effector move to the right, is given by

$$\begin{cases} F_r' = m_e\ddot{x}_e \\ F_r' = -F_r \\ x_e = x_p + x_{pe} \end{cases} \tag{25}$$

where x_e is the displacement of the end effector, and x_{pe} is the relative displacement between the end effector and stage. The stick-slip induced friction force F_r can be represented by any one of the aforementioned friction models.

5.3 Identification of parameters

The applied voltage to the PEA was a repeating sawtooth wave with an amplitude of 30 V and frequency of 5 Hz, and the displacements of the end effector were then measured by the displacement sensor. To identify parameters of in a particular friction model, the stick-slip induced friction force F_r in Eq. (25) is substituted by this friction model described in Sect. 3, and based on the experimental data the parameters in this friction model was identified using *Simulink* and *nonlinear fitting* (functions in Matlab software). The details for parameter identification technique were discussed in our previous study (Li et al., 2008b). The parameters determined for each friction model are listed in Table 2.

6 Results with discussion

The friction force and sliding speed predicted by Coulomb friction model are shown in Figs. 9 and 10, respectively. From Fig. 9 it can be seen that friction is either F_c or $-F_c$. From Figs. 9 and 10 it can be seen that the sign of the friction force is determined by the sign of the sliding speed. The

Table 2. Parameters determined for each friction model.

Parameter	Coulomb	Stribeck	Dahl	LuGre	Elastoplastic
μ	0.0587	N/A	N/A	N/A	N/A
k_v (N/mm)	N/A	40 101	N/A	N/A	N/A
F_c (N)	N/A	4.477	3.387	2.5	2.5
F_s (N)	N/A	5	N/A	3	3
v_s (m s^{-1})	N/A	1.6×10^{-8}	N/A	1.6×10^{-8}	1.6×10^{-8}
α_0 (N/mm)	N/A	N/A	1674	1670	1670
α_1 (Ns mm^{-1})	N/A	N/A	N/A	6	6
α_2 (Ns mm^{-1})	N/A	N/A	N/A	26	26
Z_{ba} (mm)	N/A	N/A	N/A	N/A	1×10^{-7}

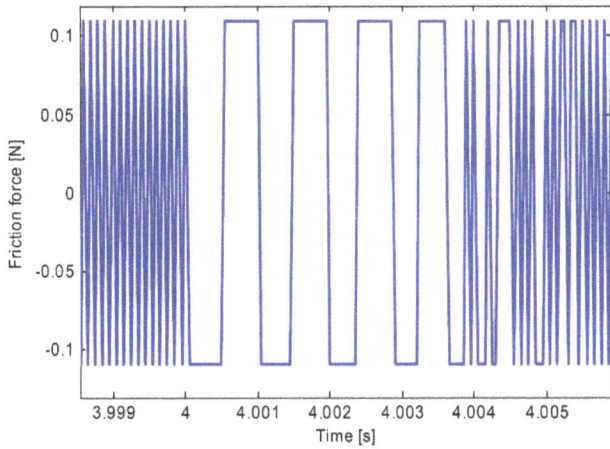

Figure 9. Friction predicted using Coulomb friction model (at around 4th second).

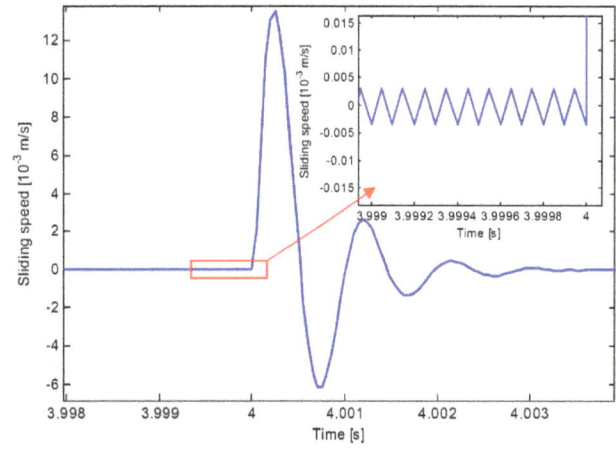

Figure 10. Sliding speed predicted using Coulomb friction model (at around the 4th second).

magnitude of F_c or $-F_c$ is dependent on the Coulomb friction coefficient.

It was found that the displacement predicted by the model fits either stick motion or slip motion but fails to fit both of them no matter what friction coefficient is chosen. Figure 11 shows a typical example of the displacement predicted by the model when $\mu = 0.0587$. It can be seen from this figure that the displacement in the stick motion period is well predicted, but the displacement predicted for the slip motion period cannot fit the experimental data and leads to huge errors after only a few cycles. The reason for this is analyzed as follows.

As previously discussed, the displacement is related to the friction force using the Newton's second law (see Eq. 25). The friction force in Coulomb friction model is a function of (1) the sign of sliding speed and (2) friction coefficient. The sliding speed (shown in Fig. 10) consists of two parts, i.e. sliding speed in stick motion and sliding speed in slip motion. Notice that the friction coefficient is the only effective variable in the Coulomb friction model. Once the friction coefficient required in stick motion is not the one required in slip motion, the Coulomb friction model will not work for the stick-slip motion system. This problem could be solved

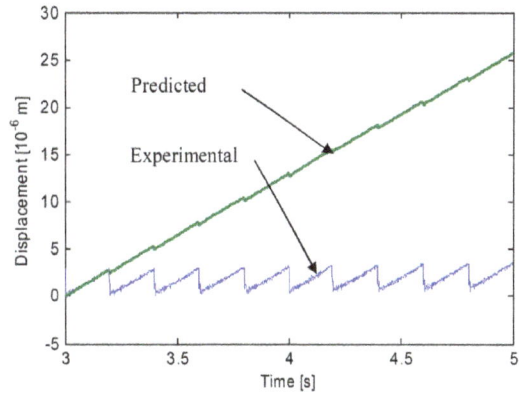

Figure 11. Displacement predicted by Coulomb model ($\mu = 0.0587$).

by considering an ideal situation, as reported in reported in Chang and Li (1999), where in the stick motion period, there is no sliding; correspondingly, in the stick motion period, the displacement of the end effector is determined by the displacement of the stage. In the slip motion period, the displacement of the end effector is determined by the Coulomb

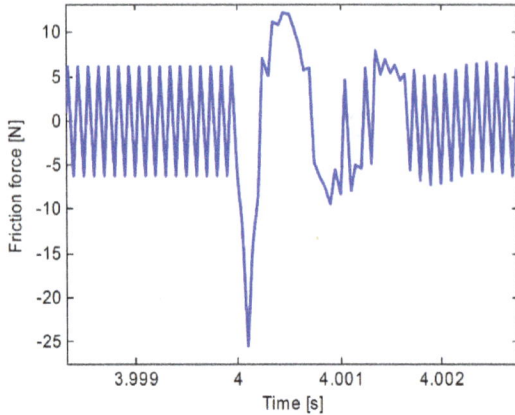

Figure 12. Friction force predicted using Stribeck model (at around the 4th second).

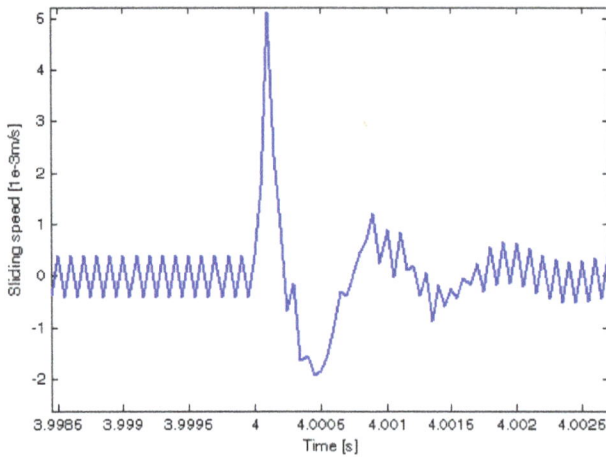

Figure 13. Sliding speed predicted using Stribeck model (at around the 4th second).

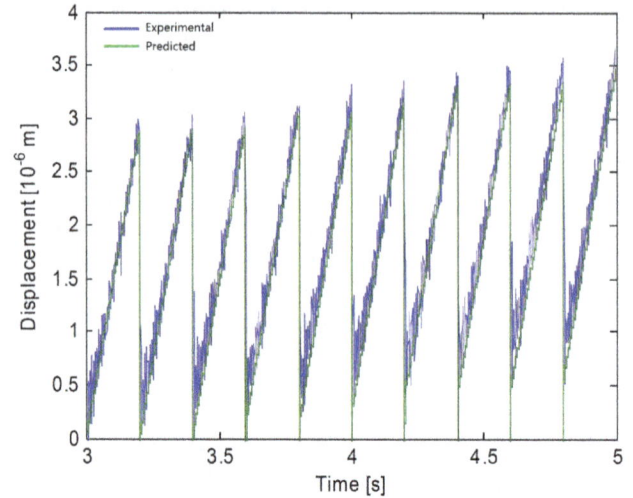

Figure 14. Experimental data and displacements predicted using Stribeck model.

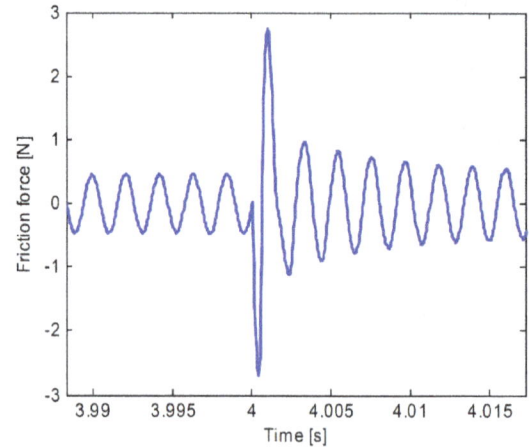

Figure 15. Friction predicted using Dahl model.

friction model. In other words, Coulomb friction model only models friction force in the slip motion period, instead of the entire stick-slip motion cycle. Another problem with the Coulomb friction model is the fluctuation in the sliding speed in stick motion shown in Fig. 10., which is supposed to be zero in theory. The fluctuation is caused by the use of *Sgn* function in the model (see Eq. 1).

6.1 Results of Stribeck friction model

The friction force and sliding speed predicted using the Stribeck friction model are shown in Figs. 12 and 13, respectively. The displacement predicted using the Stribeck friction model is shown in Fig. 14 from which it can be seen that Stribeck friction model can well predict friction both in stick and slip motion. Compared to the Coulomb friction model, Stribeck friction model can be used to model the friction force in the stick-slip motion. The reason for this might be that in the Stribeck model friction force is not only depen-

dent on the sign of sliding, but also on the sliding speed (see Figs. 12 and 13). From Fig. 12 it can be seen that the friction force predicted by the model fluctuates around zero in the stick motion, which further causes the sliding speed fluctuation (see Fig. 13) and displacement fluctuation (see Fig. 14). Such fluctuation results from the *Sgn* function in the Stribeck model.

6.2 Results of Dahl model

The friction force and sliding speed predicted using the Dahl model are shown in Figs. 15 and 16, respectively. From the two figures it can be seen that the friction force is not only dependent on the sign of sliding but also on the sliding speed. The displacement predicted using the Dahl model is shown in Fig. 17 from which it can be seen that Dahl model can well predict friction both in stick and slip motion. To observe the details, a scaled up picture from Fig. 17 (at around

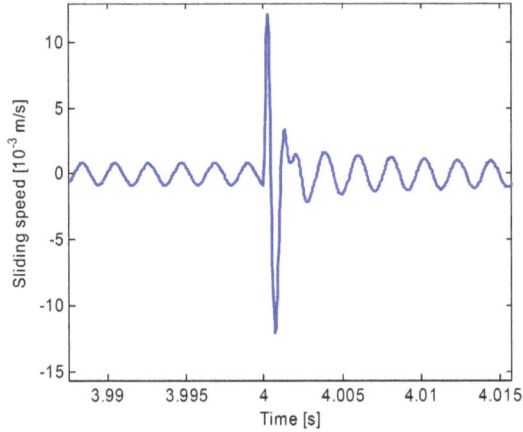

Figure 16. Sliding speed predicted using Dahl model.

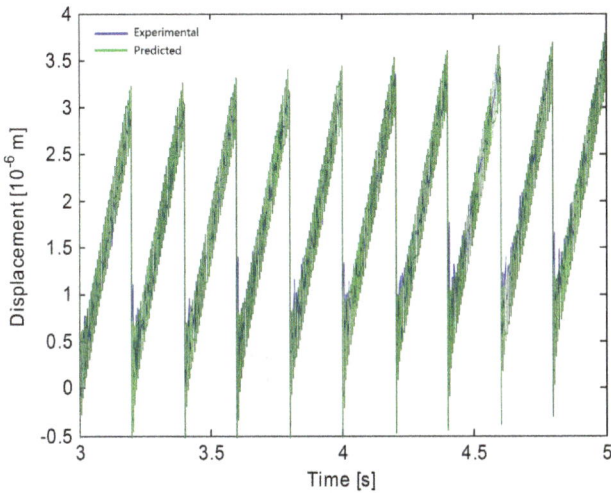

Figure 17. Experimental displacement and displacement predicted using Dahl model.

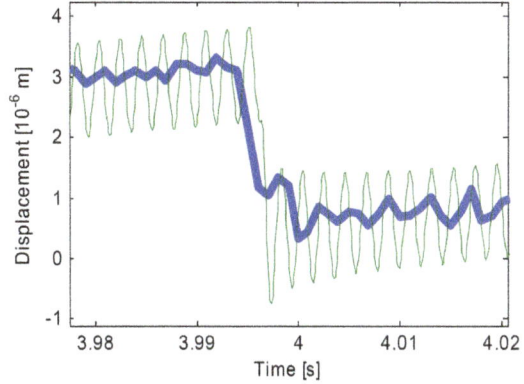

Figure 18. Scaled up picture from Fig. 16 (at around the 4th second).

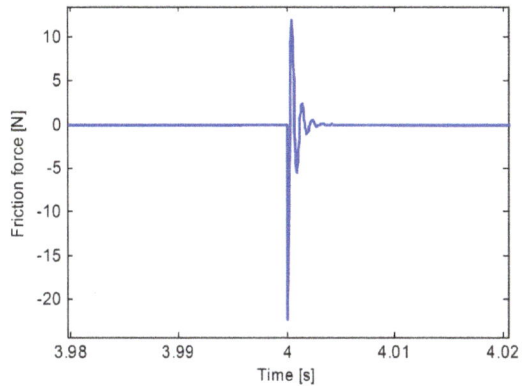

Figure 19. Friction predicted using LuGre model.

the 4th second) is shown in Fig. 18. From Fig. 18 it can be seen that the displacement predicted by the Dahl model is fluctuated around the experimental data. This is due to the sliding speed fluctuation, as shown in Fig. 16. The fluctuation (both in friction and sliding speed) results from the function of *Sgn* in the model. Compared to Stribeck model, the fluctuation (both in friction and sliding speed) in the Dahl model seems to be smoother than that in Stribeck model. The reason for this is that the friction force in the Dahl model is the integral of the *Sgn* function; see Eq. (9).

6.3 Results of LuGre model

The friction force and sliding speed predicted using the LuGre model are shown in Figs. 19 and 20, respectively. The displacement predicted using LuGre model is shown in Fig. 21 and its scaled up picture is shown in Fig. 22. From Figs. 19 and 20, it can be seen that the friction force is not

only dependent on the sign of sliding but also on the sliding speed, which is constant with the mathematical model. From Figs. 21 and 22 it can be seen that LuGre friction model can well predict friction both in stick and slip motion. Compared to Coulomb model, Stribeck model, and Dahl model (see Figs. 9, 10, 12, 13, 15 and 16), the friction force and sliding speed predicted by the LuGre model for stick motion are zero (see Figs. 19 and 20) instead of fluctuation around zero. From Figs. 21 and 22 it can be seen that the displacement predicted by the model during slip motion does not have any fluctuation, compared to Stribeck and Dahl models which are shown in Figs. 14 and 18, respectively. As we discussed before, the unwanted fluctuations in prediction of the Coulomb model, Stribeck model, and Dahl model are due to the *Sgn* function in the mathematical model. While this problem seems to be completely solved by the LuGre friction model, in which there is no *Sgn* function.

6.4 Results of the elastoplastic friction model

The displacement predicted by the elastoplastic model cannot fit the slip motion period no matter what initial parameters are chosen in order to determine the parameters in the

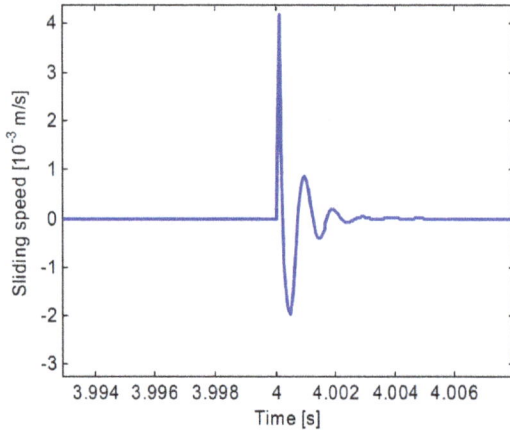

Figure 20. Sliding speed predicted using LuGre model.

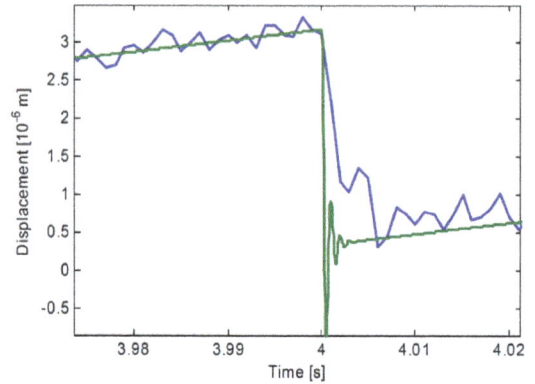

Figure 22. Scaled up picture from Fig. 21 (at around the 4th second).

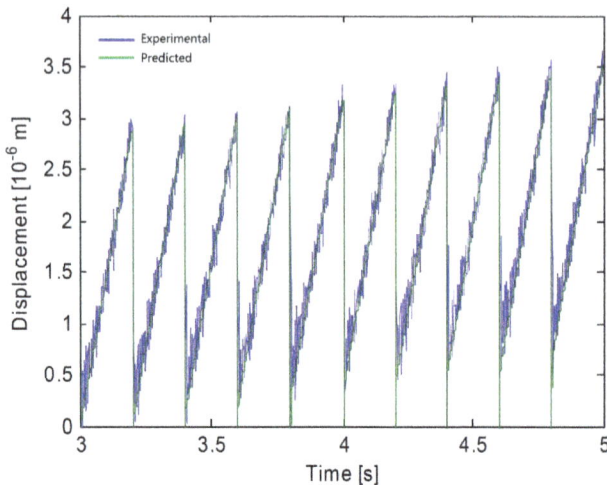

Figure 21. Displacement predicted using LuGre model.

remaining parameters were assumed (not determined by identification) as the same as those (which were already determined in Sect. 6.4) in LuGre model in order to find out the role of Z_{ba}. The identification results (not shown in this paper) show that the displacement predicted by the model still cannot fit the experimental data regardless of what Z_{ba} was chosen.

– Measure (4): comparing the elastoplastic model with LuGre model, it can be seen that if Z_{ss} and Z_{ba} approach to zero, the eleastoplastic model should turn back to LuGre model. In order to test the elastoplastic model, the parameters Z_{ss} and Z_{ba} were set as $Z_{ss} = 0.4$ nm and $Z_{ba} = 0.1$ nm, and the remaining parameters were determined as the same as those in LuGre model (which were already determined in Sect. 6.4 and listed in Table 2).

The friction force, sliding speed, and displacement predicted by the model identified in the Measure (4) are shown in Figs. 23–25, respectively. From Figs. 23 and 24 it can be seen that the friction force and sliding speed predicted by the elastoplastic model have the same characteristic with those predicted by LuGre model (see Figs. 19 and 20), i.e. no fluctuations in stick motion. However, the displacement predicted by the elastoplastic model cannot fit the slip motion period, as shown in Fig. 25. According to identification Measure (3) and Measure (4), it seems that Z_{ba} makes the elastoplastic model can not turn back to LuGre model, which is thus the cause that the elastoplastic model fails to model friction force in the micro stick-slip motion system. The plausible reason for this is discussed as follows.

The elastoplastic model expects that during slip motion, $sgn(\dot{x})$ should always be equal to $sgn(z)$ according to Eq. (14); in other words, when $sgn(\dot{x}) \neq sgn(z)$, the model predicts that stick motion occurs. The relation between $sgn(\dot{x})$ and $sgn(z)$ are shown in Figs. 26 and 27 (which is the amplified figure of Fig. 26) in which 0 represents $sgn(\dot{x}) = sgn(z)$ and 2 represents $sgn(\dot{x}) \neq sgn(z)$. It can be seen that during the stick motion period, $sgn(\dot{x})$ is

model using the system identification technique. The details are as follows.

– Measure (1): the parameters reported in Dupont et al. (2000) for the elastoplastic model were used as initial values in identification. However, the identification results (not shown in this paper) show that the displacement predicted by the model cannot fit the experimental data.

– Measure (2): the elastoplastic model was developed based on LuGre model, and it only has one more parameter Z_{ba} than LuGre model to be identified. Z_{ba} was initially set as 33 % of Z_{ss}, and the remaining parameters were initially set as the same as those in LuGre model. The identification results (not shown in this paper) show that the displacement predicted by the model still cannot fit the experimental data.

– Measure (3): similar with Step (2), Z_{ba} was initially set as 33 % of Z_{ss} and determined by identification, but the

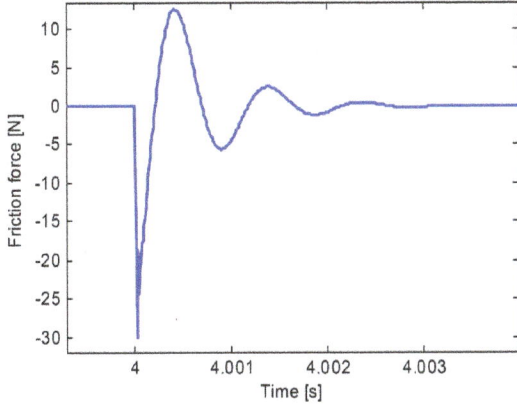

Figure 23. Friction predicted using the elastoplastic model.

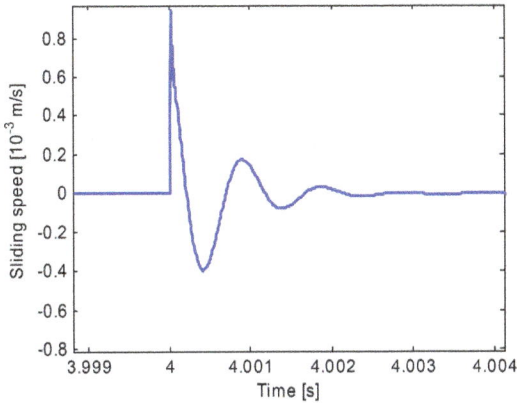

Figure 25. Experimental data and displacement predicted using the elastoplastic model.

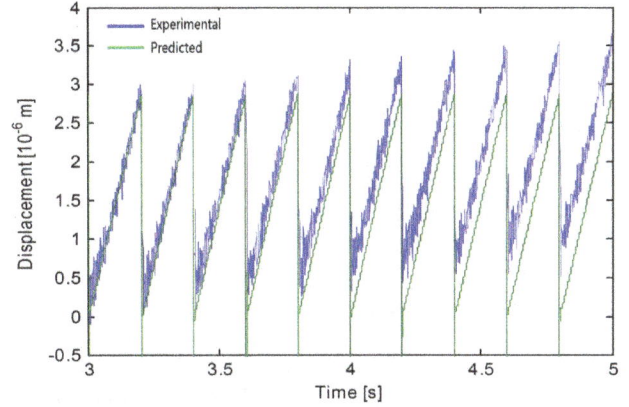

Figure 24. Sliding speed predicted using the elastoplastic model.

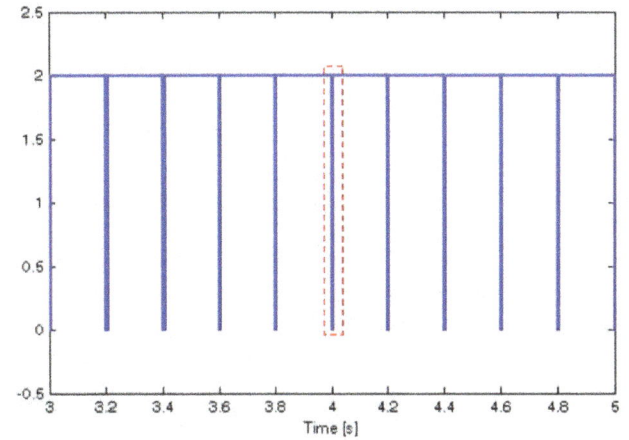

Figure 26. Relation of $sgn(\dot{x})$ and $sgn(z)$.

not equal to $sgn(z)$ over the whole period; as such, the model predicts that the stage and end-effector stick "together". During the slip motion period, however, it can be seen that $sgn(\dot{x})$ is not always equal to $sgn(z)$; in other words, the model predicts more stick motion than it really has. Therefore, the elastoplastic friction model leads to a larger static friction force and results in no relative motion in the slip motion period, as shown in Fig. 25. The reason that $sgn(\dot{x})$ is never always equal to $sgn(z)$ during the slip motion period is discussed as follows.

In stick-slip motion systems such as PE-SSA systems, the stage periodically extends and contract which makes $sgn(\dot{x})$ reverse when sliding occurs. However, according to the elastoplastic friction model, when $sgn(\dot{x})$ reverses, i.e. the end effector begins to move from B to A (Fig. 4), the deformation of contact as perities starts to decrease but still remain its initial direction until it reaches the original position, which leads to $sgn(\dot{x}) \neq sgn(z)$; correspondingly, the elastaplastic friction model predicts that no sliding occurs according to Eq. (14). This is the reason that the displacement predicted using the elastoplastic model (see Fig. 25) seems not to move at all in the slip motion period, which is not con-

sistent with the observed actual stick-slip motion phenomena.

6.5 Comparison of Stribeck, Dahl, and LuGre models

To compare Stribeck, Dahl, and LuGre models in terms of accuracy, an error index representing the deviation between the displacement predicted and experimental data is given by

$$e = \frac{1}{n}\sum_{k=1}^{n}|e_k| \tag{26}$$

where $|e_k|$ is the absolute value of the error between experimental data and the displacement predicted, n the number of displacement data, and $k = 1, 2, \ldots, n$.

The error index for each friction model is listed in Table 3.

It is shown that LuGre model has the least error index, and Stribeck model and Dahl model have larger error indexes. The reason is that the displacements predicted by Stribeck model and Dahl model fluctuate around the experimental data (see Figs. 14 and 17 or Fig. 18), which is caused

Table 3. Error index for friction models.

Models	Stribeck	Dahl	LuGre
Error index	$0.176\,\mu m$	$0.232\,\mu m$	$0.135\,\mu m$

by the *Sgn* function in the two models. LuGre model integrates, pre-sliding friction ($\sigma_0 z$) that is captured by Dahl model and viscous friction ($\upsilon_2 \dot{x}$) and Stribeck effect ($g(\dot{x})$) that are captured by Stribeck friction model, into one single model. Moreover, LuGre model does not have *Sgn* function and it has no fluctuation in displacement prediction (see Fig. 22). Thus LuGre model results in a better accuracy of friction force prediction in the micro stick-slip motion system.

7 Conclusions with further discussion

This paper first reviewed the friction models, i.e. Coulomb, viscous, combined Coulomb and viscous model, Stribeck, Dahl, LuGre, and the elastoplastic friction models. Five of them were applied to model friction force in the micro stick-slip motion. Parameters involved in each model were determined using the system identification technique. The performances of these models were compared. The plausible reasons for the difference in performance among these models applied in micro stick-slip motion were discussed.

This study concludes that Coulomb friction model is not adequate for describing the friction in the micro stick-slip motion. Stribeck model, Dahl model and LuGre model can all predict friction force in the micro stick-slip motion, and LuGre model has the best accuracy among the three. The elastoplastic friction model fails to model friction in the micro stick-slip motion because it introduces a larger static friction during the slip motion period that makes two contact surfaces seem to stick together. The failure of the elastoplastic model for the micro stick-slip motion system demonstrates our observation of the so-called "side effect" for complex dynamic systems. In the case of the elastoplastic model, it has more parameters to capture the friction phenomena, especially capturing the so-called "drift" phenomenon when LuGre model is used for the application of positioning system, but the model which resolves the drift problem also brings the side effect in the sense that the same modeling elements, i.e. elastic and plastic deformation zones, causes trouble to model friction force in the micro stick-slip motion.

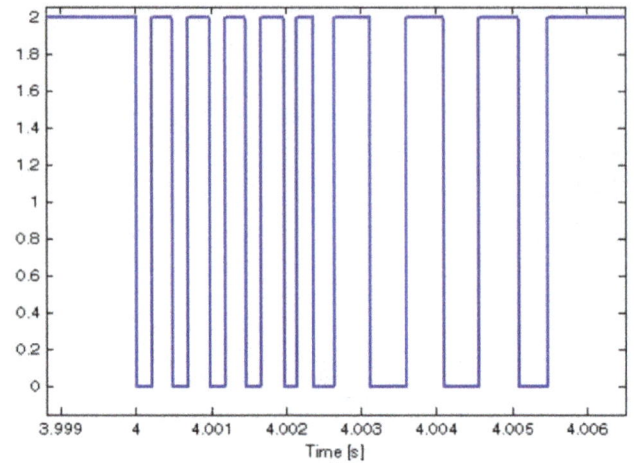

Figure 27. Scaled up picture from Fig. 26.

This study may also demonstrate an idea of using displacement measurement of micro motion system and system identification technique to investigate the dynamic friction in micro motion level. Further validation of this idea calls for some future study.

Appendix A

Table A1. Nomenclature.

c_p	the damping coefficient of PEA, $\mathrm{N s\,m^{-1}}$
c_s	the damping coefficient of the stage, $\mathrm{N s\,m^{-1}}$
F	the friction force, N
F_{app}	the applied force, N
F_c	the Coulomb sliding friction force, N
F_N	the normal load, N
F_r	friction force on the stage, N
F_r'	friction force on the end effector, N
F_s	the maximum static friction force, N
F_p	the transduced force from the electrical side, N
$g(\dot{x})$	the Stribeck curve
$G(s)$	the transfer function
i	an exponent
j	the Stribeck shape factor
K	the amplified coefficient, $\mathrm{m\,v^{-1}}$
k_p	the stiffness of PEA, $\mathrm{N\,m^{-1}}$
k_s	the stiffness of stage, $\mathrm{N\,m^{-1}}$
k_v	the viscous coefficient, $\mathrm{N s\,m^{-1}}$
m_p	the mass of PEA, kg
m_s	the mass of stage, kg
os	the overshoot of a step response
T_{em}	the electromechanical transducer ratio, $\mathrm{N\,v^{-1}}$
T_p	the peak time of a step response, s
u_p	the applied voltage on the PEA, V
v_s	the Stribeck velocity, $\mathrm{m\,s^{-1}}$
x	the relative displacement, m
$\dot{x}$	the sliding speed, $\mathrm{m\,s^{-1}}$
x_e	the displacement of the end effector, m
x_p	the displacement of the PEA, m
x_{pe}	the relative displacement between the end effector and stage, m
z	the average bristle deflection, m
z_{ba}	the stiction zone, m
z_{ss}	the parameter for transition from sticking to slipping, m
$\alpha(z, \dot{x})$	function of elastic and plastic deformation
$\alpha_m(*)$	the function of transition from sticking to slipping
μ	Coulomb friction coefficient
ξ	the damping ratio
ω_n	the natural frequency, $\mathrm{rad\,s^{-1}}$
σ_0	the stiffness of the bristles, $\mathrm{N\,mm^{-1}}$
σ_1	the damping coefficient of the bristles, $\mathrm{N s\,mm^{-1}}$
σ_2	the viscous friction coefficient, $\mathrm{N s\,mm^{-1}}$

Acknowledgements. The authors want to thank J. W. Li, Q. S. Zhang for their performing the experiments. This work has been supported by The Natural Sciences and Engineering Research Council of Canada (NSERC), National Natural Science Foundation of China (NSFC) (grant number: 51375166) and China Scholarship Council (CSC), and these organizations are thanked for the support.

Edited by: A. Barari

References

Adriaens, H. J. M. T. A., De Koning, W. L., and Banning, R.: Modeling piezoelectric actuators, IEEE/ASME T. Mechatron., 5, 331–341, 2000.

Andersson, S., Söderberg, A., and Björklund, S.: Friction models for sliding dry, boundary and mixed lubricated contacts, Tribol. Int., 40, 580–587, 2007.

Armstrong-Helouvry, B.: Stick slip and control in low-speed motion. IEEE T. Automat. Contr., 38, 1483–1496, 1993.

Chang, S. H. and Li, S. S.: A high resolution long travel friction-driven micropositioner with programmable step size, Rev. Scient. Instrum., 70, 2776–2782, 1999.

Canudas de Wit, C., Olsson, H., Åström, K. J., and Lischinsky, P.: A New Model for Control of Systems with Friction, IEEE T. Automat. Contr., 40, 419–425, 1995.

Dahl, P. R.: A solid friction model, Technical Report Tor-0158(3107-18)-1, The Aerospace Corporation, EI Segundo, CA, 1968.

Dupont, P., Armstrong, B., and Hayward, V.: Elasto-plastic friction model: contact compliance and stiction, Proc. Am. Control Conf., Chiacago, IL, 1072–1077, 2000.

Kang, D.: Modeling of the Piezoelectric-Driven Stick-Slip Actuators, Thesis of Master of Science, University of Saskatchewan, Saskatchewan, Canada, 2007.

Lampaert, V., Swevers, J., and Al-Bender, F.: Modification of the Leuven integrated friction model structure, IEEE T. Automat. Contr., 47, 683–687, 2002.

Li, J. W., Yang, G. S., Zhang, W. J., Tu, S. D., and Chen, X. B.: Thermal effect on piezoelectric stick-slip actuator systems, Rev. Scient. Instrum., 79, 046108, doi:10.1063/1.2908162, 2008a.

Li, J. W., Chen, X. B., An, Q., Tu, S. D., and Zhang, W. J.: Friction models incorporating thermal effects in highly precision actuators, Rev. Scient. Instrum., 80, 045104, doi:10.1063/1.3115208, 2008b.

Li, J. W., Zhang, W. J., Yang, G. S., Tu, S. D., and Chen, X. B.: Thermal-error modeling for complex physical systems: the-state-of-arts review, Int. J. Adv. Manufact. Technol., 42, 168–179, 2009.

Makkar, C., Hu, G., Sawyer, W. G., and Dixon, W. E.: Lyapunov-Based Tracking Control in the Presence of Uncertain Nonlinear Parameterizable Friction, IEEE T. Automat. Contr., 52, 1988–1994, 2007.

Morin, A.: New friction experiments carried out at Metz in 1831–1833, Proc. French Roy. Acad. Sci., 4, 1–128, 1833.

Stribeck, R.: Die wesentlichen Eigenschaften der Gleit- und Rollenlager, Zeitschrift des Vereins Deutscher Ingenieure, Duesseldorf, Germany, 36 Band 46, 1341–1348, 1432–1438, 1463–1470, 1902.

Swevers, J., Al-Bender, F., Ganseman, C. G., and Prajogo, T.: An integrated friction model structure with improved presliding behaviour for accurate friction compensation, IEEE T. Automat. Contr., 45, 675–686, 2000.

Zhang, W. J., Ouyang, P. R., and Sun, Z.: A novel hybridization design principle for intelligent mechatronics systems, Proceedings of International Conference on Advanced Mechatronics (ICAM2010), Osaka University Convention Toyonaka, Japan, 4–6, 2010.

Zhang, Z. M., An, Q., Zhang, W. J., Yang, Q., Tang, Y. J., and Chen, X. B.: Modeling of directional friction on a fully lubricated surface with regular anisotropic asperities, Meccanica Springer Netherlands, 2011.

12

Vector model of the timing diagram of automatic machine

A. Jomartov

Institute Mechanics and Mechanical Engineering, Almaty, Kazakhstan

Correspondence to: A. Jomartov (legsert@mail.ru)

Abstract. In this paper a vector model of timing diagram of automatic machine is developed, which allows us to solve a variety dynamic tasks by changing the parameters of timing diagram of its mechanisms. The connection between the parameters of the timing diagram of automatic machine and equations of motion mechanisms through functions of position and transfer functions of mechanisms is established. The vector model of timing diagram can be used to optimize the timing diagrams of looms and polygraphic machines.

1 Introduction

Modeling of the timing diagram is one of the main parts for design of automatic machines. A detailed analysis of the works on the theory of the timing diagram performed prior to 1965 is given in Petrokas (1970). Timing diagram is a sequence of machine operations performed by mechanisms depending on the angular displacement of the main shaft (Browne, 1965; Youssef and El-Hofy, 2008; Homer, 1998; Sandler, 1999; Natale C, 2003; Singh and Bhattacharya, 2006; Levner and Kats, 2007; Niir Board, 2009; Norton, 2009; Topalbekiroglu and Celik, 2009). Timing diagram allows determining of the position of each of the executive body at any position of the main shaft (see Fig. 1).

Timing diagram automatic machine is modeled by a directed graph (see Fig. 2) (Novgorodtsev, 1982). The disadvantages of this model are the lack of consideration for connections executive bodies displacements mechanisms and accounting precision of manufacturing.

Analysis of the methods of synthesis and analysis of timing diagram automatic machine showed the need for further development of optimization methods of timing diagram, taking into account the accuracy of manufacture and the dynamics of automatic machine.

2 Vector model of the timing diagram of automatic machine

The timing diagram of automatic machines can be represented as the vector polygons (Jomartov, 2010, 2011) (see Fig. 3). Let us replace the segments linear timing diagram by the vectors ℓ_{ij}. The vectors ℓ_{ij} is directed sequentially from one position to another position of mechanism, where i is the number of mechanisms, j is the number of position of i-mechanism, m_j is the number of positions of i-mechanism, n is the number of mechanisms.

The projection of vectors ℓ_{ij} on the x axis is α_{ij} – phase angles of actuation of mechanisms. The projection ℓ_{ij} on the y axis is the displacement δ_{ij} of j-position of i-mechanism.

$$\delta_{ij} = \frac{S_{ij}}{S_{\max}}, \quad S_{\max} = \max S_{ij}, \quad i = 1, \ldots, n; \ j = 1, \ldots, m_i,$$

where S_{ij} is the displacement of j-position of i-mechanism.

To explain the parameters α_{ij} and S_{ij} in Fig. 4 shows a diagram of the displacement of mechanism, in the figure denote: α_o – phase angles of actuation of mechanism in the position of open, α_d is the phase angles of actuation of mechanism in the position of dwell, α_c is the phase angles of actuation of mechanism in the position of close, S_o is the displacement of mechanism in the position of open, S_c is the displacement of mechanism in the position of close.

Let us introduce the vector P connecting the point of beginning and end of the cycle. The projection of the vector P on the x axis is 2π on the y axis is zero. The interaction of mechanisms with each other will reflect in the form of the vectors of connection c_{ik}, where $k = 1, \ldots, r_i$, r_i is the number of vectors of connection of i-mechanism. The projection of the vectors of connection to the x axis is the delay of actuation mechanism, and the projection on the y axis is the difference between the displacements of mechanisms.

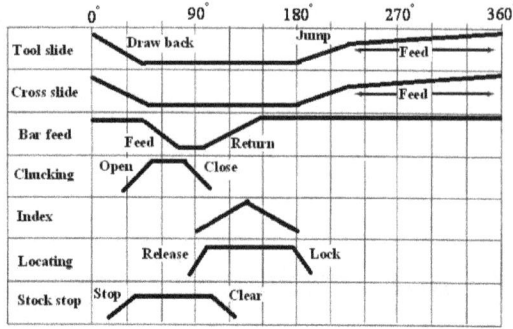

Figure 1. Linear timing diagram of working and auxiliary cams of a four-spindle bar automatic.

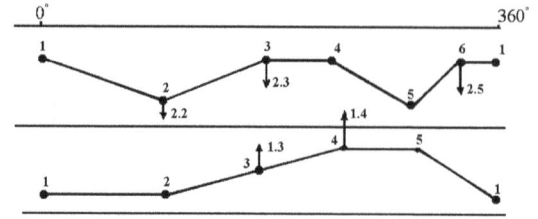

Figure 2. Presentation of timing diagram automatic machine as a directed graph.

Let us impose the timing diagram of mechanisms at each other using zero vectors (see Fig. 3) connecting the boundary points of timing diagram mechanisms in the y axis.

Let us compose the system of vector equations describing the works of mechanisms automatic machine (see Fig. 3).

$$\left. \begin{aligned} \sum_{j=1}^{m_i} \boldsymbol{\ell}_{ij} &= \boldsymbol{P}, \ i = 1, \ldots, n, \\ \boldsymbol{c}_{ik} &= \sum_{i=1}^{n} \sum_{j=1}^{m_i} b_{ij} \cdot \boldsymbol{\ell}_{ij} \end{aligned} \right\}, \tag{1}$$

where $b_{ij} \in \{0, \ \pm 1\}$.

Let us projected the vector Eq. (1) on the axis x and y.

$$\left. \begin{aligned} \sum_{j=1}^{m_i} \alpha_{ij} &= 2\pi, \ \sum_{j=1}^{m_i} \delta_{ij} = 0, \\ c_{ik}^x &= \sum_{i=1}^{n} \sum_{j=1}^{m_i} b_{ij}\alpha_{ij}, \ c_{ik}^y = \sum_{i=1}^{n} \sum_{j=1}^{m_i} b_{ij}\delta_{ij} \end{aligned} \right\} \tag{2}$$

On the phase angles of actuation of mechanisms α_{ij}, and displacements of mechanisms δ_{ij} impose constraints

$$\alpha_{ij} \geq \alpha_{ij}^{\mathrm{m}}, \ \delta_{ij}^{\ell} \geq \delta_{ij} \geq \delta_{ij}^{\mathcal{H}}, \tag{3}$$

where α_{ij}^{m} is the minimum allowable phase angles of actuation of mechanisms, $\delta_{ij}^{\ell}, \delta_{ij}^{\mathcal{H}}$ is the upper and lower limits assigned by the designer.

On the projection vectors of connection impose constraints

$$c_{ik}^{x\ell} \geq c_{ik}^x \geq c_{ik}^{x\ell}, \ c_{ik}^{y\ell} \geq c_{ik}^y \geq c_{ik}^{y\mathcal{H}} \tag{4}$$

where $c_{ik}^{x\mathcal{H}} = e_{ik}^x + \Delta c_{ik}^x, \ c_{ik}^{y\mathcal{H}} = e_{ik}^y + \Delta c_{ik}^y$ where e_{ik}^x, e_{ik}^y are the minimum permissible projection vectors of connection, Δc_{ik}^x, Δc_{ik}^y are the errors of the projections of vectors of connection, $c_{ik}^{x\ell}, c_{ik}^{y\ell}$ are the upper limits imposed by the designer.

Equation (2) and constraints (Eqs. 3 and 4) describe the collaboration works of mechanisms (timing diagram) of automatic machine.

3 A mathematical model of automatic machine based on the timing diagram of mechanisms

Lets define the connection between the differential equations of motion the automatic machine and the equations describing its timing diagram. In Fig. 5 shows the dynamic model of the machine, where c_i is the elasticity coefficients, β_i is the coefficients of resistance, J_i, I_i are the moments of inertia, $M_{\partial B}$ is the motor torque, M_i is moment of resistance, Π_i, $i = 1, \ldots, n$ is the function of position of mechanisms.

To compile the equations of motion mechanisms automatic machine (see Fig. 5), let us use Lagrange equations II (Wolfson, 1976):

$$\left. \begin{aligned} \frac{\mathrm{d}}{\mathrm{d}t}\left(\frac{\partial T}{\partial \dot\varphi_j}\right) - \frac{\partial T}{\partial \varphi_j} + \frac{\partial V}{\partial \varphi_j} &= Q_j + \sum_{i=1}^{m} \lambda_i h_{ij} \\ \sum_{j=1}^{m+n} h_{ij}\dot\varphi_j + h_i &= 0, \end{aligned} \right\} \tag{5}$$

where $\varphi_1, \varphi_2, \ldots, \varphi_n$ are the generalized coordinates, λ_i is Lagrange multipliers, h_{ij}, h_i are some functions, T is the kinetic energy of a holonomic system, V is the potential energy of the system, Q_j is the generalized force.

To establish the connection between the equations of timing diagram (2–4) and the dynamic Eq. (5), let us write the functions of position, the transfer functions of mechanisms of automatic machine in the following form:

$$\left. \begin{aligned} \Pi_i &= \Pi_{i1} \cdot [1 - L(\phi_i - \alpha_{i1})] + \sum_{j=2}^{m} \Pi_{ij}\left[1 - L\left(\phi_i - \sum_{r=1}^{j}\alpha_{ir}\right)\right] \cdot L\left(\phi_i - \sum_{r=1}^{j-1}\alpha_{ir}\right) \\ \Pi_i' &= \Pi_{i1}' [1 - L(\phi_i - \alpha_{i1})] + \sum_{j=2}^{m} \Pi_{ij}'\left[1 - L\left(\phi_i - \sum_{r=1}^{j}\alpha_{ir}\right)\right] \cdot L\left(\phi_i - \sum_{r=1}^{j-1}\alpha_{ir}\right) \\ \Pi_i'' &= \Pi_{i1}'' [1 - L(\phi_i - \alpha_{i1})] + \sum_{j=2}^{m} \Pi_{ij}''\left[1 - L\left(\phi_i - \sum_{r=1}^{j}\alpha_{ir}\right)\right] \cdot L\left(\phi_i - \sum_{r=1}^{j-1}\alpha_{ir}\right) \end{aligned} \right\} \tag{6}$$

where $i = 1, \ldots, n$, $L(x)$ is a step function of the form

$$L(x) = \begin{cases} 0, & x < 0, \\ 1, & x \geq 0 \end{cases}.$$

$\Pi_{ij}, \Pi_{ij}', \Pi_{ij}''$ are the functions of position, the first transfer function, the second transfer function on parts of phase angles of actuation α_{ij} of mechanisms.

Equation (6) establish a connection between the Eq. (5) describe the dynamics of the automatic machine and the Eqs. (2)–(4) timing diagram of the machine-automaton. This

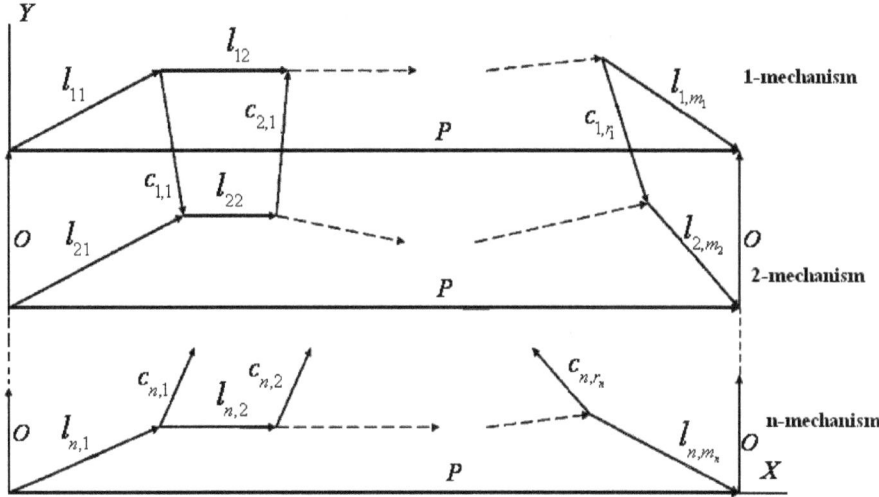

Figure 3. Vector model of the timing diagram of automatic machine.

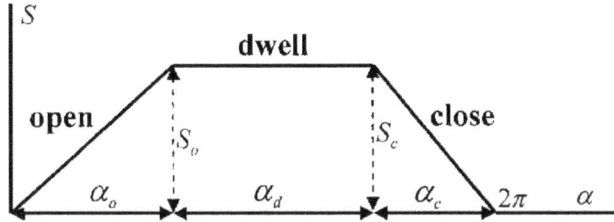

Figure 4. Diagram of the displacement of mechanism.

method allows to solve various optimization tasks, where the variable parameters are the phase angles α_{ij} and of the displacements δ_{ij} of timing diagram automatic machine.

4 An example

Let us show in more detail the connection between timing diagram of automatic machine and dynamics of mechanisms on the example of automatic machine with two cam mechanisms. The dynamic model is shown in Fig. 6, where φ_0, φ_1, φ_2 are generalized coordinates, M_D is the motor torque, M_1, M_2 are the moments of resistance, I_1, J_0, J_1, J_2 are the moments of inertia of mechanisms, c_0, c_1 are the coefficients of elasticity of shafts, $\Pi_i(\phi_i)$ is the function of position of cam mechanisms, $\Pi'_i(\phi_i)$ is the first transfer functions of the cams, $\Pi''_i(\phi_i)$ is the second transfer functions of the cams.

This dynamic model is described by the following equations

$$\left. \begin{array}{l} J_0\ddot{\phi}_0 + c_1(\phi_0 - \phi_1) = M_D, \\ \left(J_1 + I_1\Pi_1^2(\phi_1)\right)\ddot{\phi}_1 + I_1\Pi'_1(\phi_1)\Pi''_1(\phi_1)\dot{\phi}_1^2 + c_0(\phi_1 - \phi_0) + c_1(\phi_1 - phi_2) = -M_1\Pi'_1(\phi_1), \\ \left(J_2 + I_2\Pi_2^2(\phi_2)\right)\ddot{\phi}_2 + I_2\Pi'_2(\phi_2)\Pi''_2(\phi_2)\dot{\phi}_2^2 + c_2(\phi_2 - \phi_2) = -M_2\Pi_{2'}(\phi_2) \end{array} \right\} (7)$$

where

$$\Pi'_i(\phi_i) = \frac{d\Pi_i(\phi_i)}{d\phi_i}; \; \Pi''_i(\phi_i) = \frac{d^2\Pi_i(\phi_i)}{d\phi_i^2}; \; i = 1, 2.$$

Figure 7 shows vector timing diagram of automatic machine, which is described by the following equations:

$$\left. \begin{array}{l} l_{11} + l_{12} = P \\ l_{21} + l_{22} = P \\ c_{21} = l_{21} - l_{11} \end{array} \right\}. \tag{8}$$

Let us projected Eq. (8) on the x, y respectively

$$\left. \begin{array}{l} \alpha_{11} + \alpha_{12} = 2\pi \\ \alpha_{21} + \alpha_{22} = 2\pi \\ c_{11}^x = \alpha_{21} - \alpha_{11} \end{array} \right\} \tag{9}$$

$$\left. \begin{array}{l} \delta_{11} - \delta_{12} = 0 \\ \delta_{21} - \delta_{22} = 0 \\ c_{11}^y = \delta_{21} - \delta_{11} \end{array} \right\}. \tag{10}$$

Impose the constraints on the phase angles, displacements of mechanisms, and projections of vectors of connection

$$\left. \begin{array}{l} \alpha_{ij} \geq \alpha_{ij}^{min} \\ \delta_{ij}^{max} \geq \delta_{ij} \geq \delta_{ij}^{min} \\ c_{11}^{xmax} \geq c_{11}^x \geq c_{11}^{xmin} \\ c_{11}^{ymax} \geq c_{11}^y \geq c_{11}^{ymin} \end{array} \right\}. \tag{11}$$

The expressions (Eqs. 9–11) allow you to vary the phase angles and displacements of mechanisms of automatic machine, without disrupting their normal work.

To establish the connection between the equations describing of joint work of the mechanisms of automatic machine (Eqs. 9–11) and the dynamic Eq. (7), let us write the functions of position and the transfer functions of mechanisms of automatic machine (see Fig. 8) as follows:

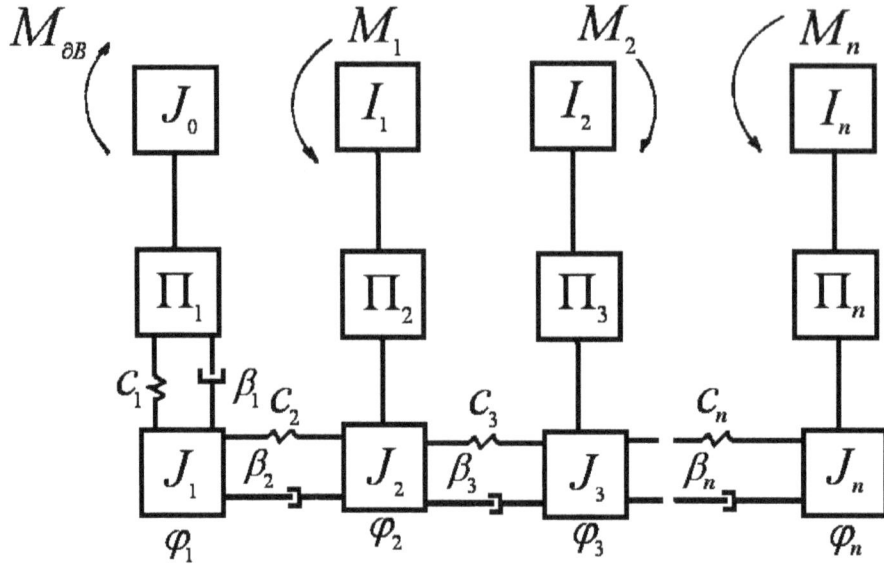

Figure 5. A dynamic model of automatic machine.

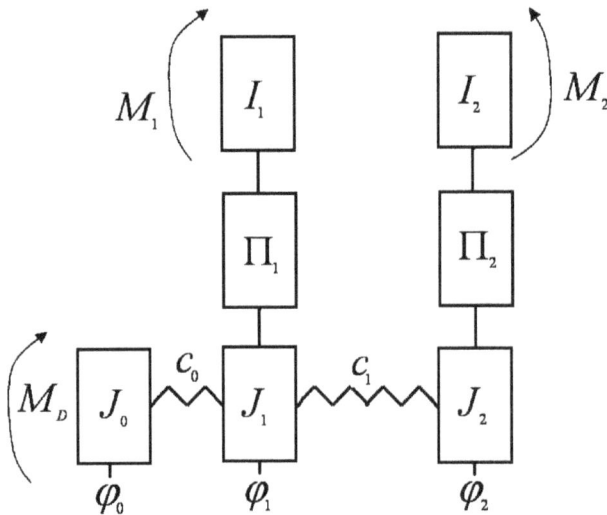

Figure 6. Dynamic model of automatic machine with two cams mechanisms.

$$\left.\begin{array}{l} \Pi_i = \Pi_{i1} \cdot [1 - L(\phi_i - \alpha_{i1})] + \Pi_{i2}[1 - L(\phi_i - (\alpha_{i1} + \alpha_{i2}))] \cdot L(\phi_i - \alpha_{i1}) \\ \Pi'_i = \Pi'_{i1} \cdot [1 - L(\phi_i - \alpha_{i1})] + \Pi'_{i2}[1 - L(\phi_i - (\alpha_{i1} + \alpha_{i2}))] \cdot L(\phi_i - \alpha_{i1}) \\ \Pi''_i = \Pi''_{i1} \cdot [1 - L(\phi_i - \alpha_{i1})] + \Pi''_{i2}[1 - L(\phi_i - (\alpha_{i1} + \alpha_{i2}))] \cdot L(\phi_i - \alpha_{i1}) \\ i = 1, 2 \end{array}\right\} \quad (12)$$

where

$$L(x) = \left\{ \begin{array}{l} 0, \ x < 0, \\ 1, \ x \geq 0. \end{array} \right. \cdot$$

Represent the generalized coordinates φ_1, φ_2 through the dimensionless coefficients k_1, k_2 where $\varphi_1 = 2\pi k_1$; $\varphi_2 = 2\pi k_2$; $k_1, k_2 \in [0, 1]$

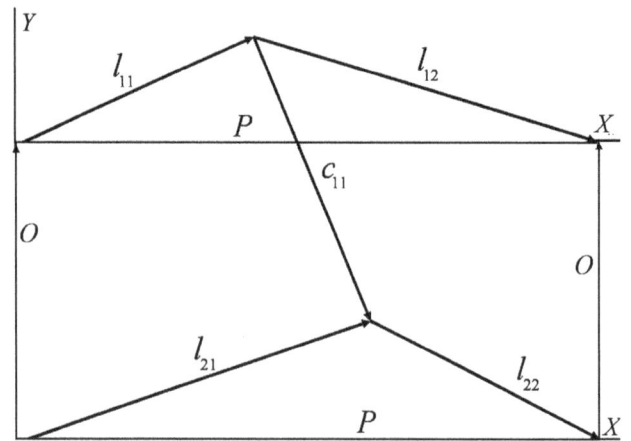

Figure 7. Vector timing diagram of automatic machine with two cams mechanisms.

$$\left.\begin{array}{l} \Pi_{ij} = a_{ij}(k_i) \, \delta_{ij} \\ \Pi'_{ij} = b_{ij}(k_i) \, \dfrac{\delta_{ij}}{\alpha_{ij}} \\ \Pi''_{ij} = d_{ij}(k_i) \, \dfrac{\delta_{ij}}{\alpha_{ij}^2} \end{array}\right\} \cdot \qquad (13)$$

$$i = 1, 2; \ j = 1, 2$$

$a_{ij}(k_i), b_{ij}(k_i), d_{ij}(k_i); \ i = 1, 2; \ j = 1, 2;$ are coefficients of the displacement, the velocity, the acceleration of mechanism in j-position.

Equations (12) and (13) establish a connection between the phase angles of actuation of mechanisms α_{ij} and magnitude of displacement of mechanisms δ_{ij} and of their functions of position and transfer functions that are explicitly included in the equations of motion of the automatic machine (Eq. 7). Now, depending on the chosen optimization criterion, by

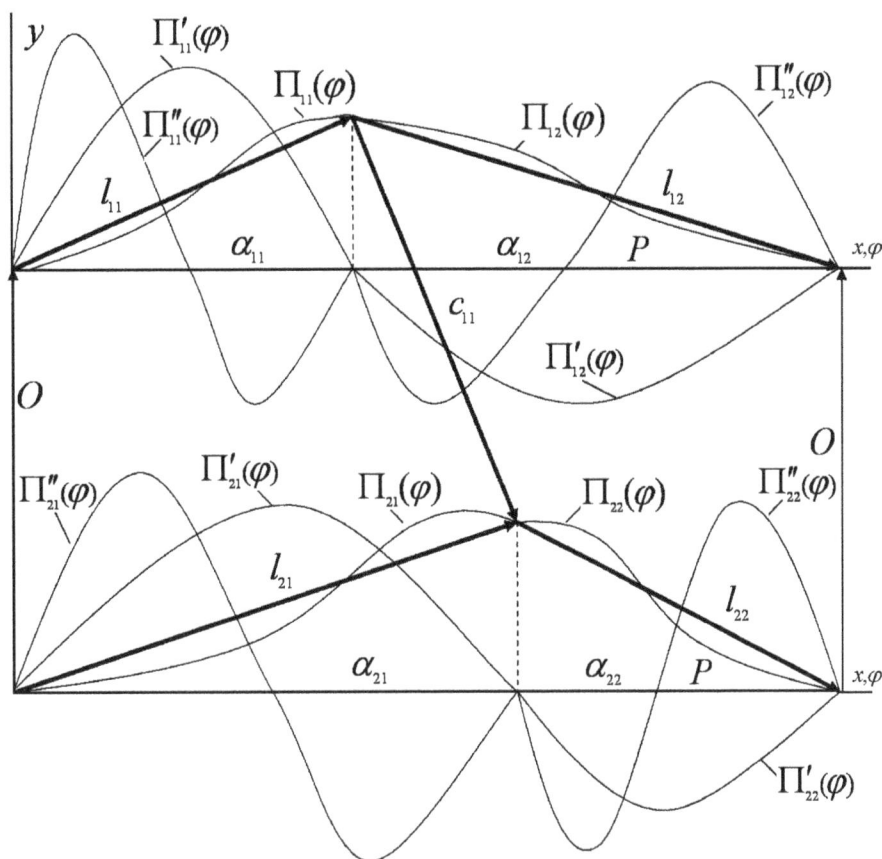

Figure 8. The functions of position and the transfer functions of mechanisms of automatic machine.

varying the parameters of timing diagram α_{ij} and δ_{ij}, can improve the dynamics of the automatic machine. As an optimization criterion can use the following expression:

$$\max_{\varphi_i} \left(\Pi_i'(\varphi_i) \, \Pi_i''(\varphi_i) \right).$$

5 Conclusions

The vector model of timing diagram is developed on the basis of representation timing diagram of the automatic machine as vector polygons, which allows solving various dynamic tasks at the expense of change of parameters of timing diagram.

The mathematical model of the automatic machine with elastic links on the basis of the timing diagram of its mechanisms is received.

The equations of connection between parameters timing diagram of the automatic machine and equations of dynamics through functions of position and transfer functions of mechanisms were received.

The model of timing diagram of the automatic machine is the only method which allows to solve a variety of dynamic tasks, by optimizing its timing diagrams, at this time.

Edited by: A. Barari

References

Browne, J. W.: The Theory of Machine Tools, Cassell and Co. Ltd., London, p. 374, 1965.

Homer, D. E.: Kinematic Design of machines and mechanisms, McGraw-Hill, New York, p. 661, 1998.

Jomartov, A.: Dynamics of machine-automaton jointly with cyclegram, World Congress on Engineering, Lîndon, UK, 1224–1229, 2010.

Jomartov, A.: Multi-Objective Optimization Of Cyclogram Mechanisms Machine-Automaton, World Congress on Engineering, Lîndon, UK, 2562–2565, 2011.

Levner, E. and Kats, V. D.: Cyclic Scheduling in Robotic Cells: An Extension of Basic Models in Machine Scheduling Theory, Multiprocessor Scheduling: Theory and Applications, Vienna, Austria, p. 436, 2007.

Natale, C.: Interaction control of robot manipulators: six degrees-of-freedom tasks, Springer-Verlag, Berlin, p. 108, 2003.

Niir Board: Complete technology book on textile, spinning, weaving, finishing and printing, National Institute of Industrial Re, Delhi, p. 814, 2009.

Norton, R. L.: Cam Design and Manufacturing Handbook, Industrial Press Inc., p. 591, 2009.

Novgorodtsev, V. A.: Presentation of the machine timing diagram as a graph, Theory of mechanisms and machines, Kharkov, 33, 57–60, 1982.

Petrokas, L. V.: Reviews of timing diagram manufacturing machines and automatic production lines, J. Theory of automatic machines and pneumatic, Moscow, 22–36, 1970.

Sandler, B. Z.: Robotics: designing the mechanisms for automated machinery, Academic Press, San Diego, p. 433, 1999.

Singh, B. and Bhattacharya, S. K.: Control of machines, New Age International, New Delhi, p. 338, 2006.

Topalbekiroglu, M. and Celik, H. I.: Kinematic analysis of beat-up mechanism used for handmade carpet looms, Indian J. Fibre Textile Res., 34, 129–136, 2009.

Wolfson, I. I.: Dynamics calculations of cycle mechanisms, Mashinostroenie, Leningrad, p. 328, 1976.

Youssef, H. A. and El-Hofy, H.: Machining technology: machine tools and operations, Taylor & Francis Group, London, p. 672, 2008.

13

ADLIF: a new large-displacement beam-based flexure joint

X. Pei and J. Yu

School of Mechanical Engineering and Automation, Beihang University, Beijing 100083, China

Abstract. A flexure joint is an important component in flexure mechanisms. Most of well known flexure joints have always a trade-off among such performances as precision, stiffness, and stroke, which heavily affect the overall performances of flexure mechanisms. In this paper, a new flexure joint, named an anti-symmetric double leaf-type isosceles-trapezoidal flexure joint (ADLIF), is introduced. The joint is constructed by two leaf-type isosceles-trapezoidal flexure (LITF) building blocks in an anti-symmetrical form. In order to investigate such characteristics as precision, stiffness and stroke, two ADLIFs with different structural parameters are compared with a cartwheel hinge. In addition, a simple and accurate pseudo-rigid body (PRB) joint model of the ADLIF is formulated to simplify the parametric model and achieve the structural optimization. The results show that the ADLIF can gain a great improvement in precision as well as maintain other characteristics such as stiffness and ranges of motion similar. Even the ADLIF gets more than 16 times improvement in precision in the case that the rotational angle is less than five degrees (5°). The ADLIF can thus be used for the replacement of the cartwheel joint in some precision application fields.

1 Introduction

Flexure mechanisms are devices that attain motion by means of elastic deformation of flexures. They have been utilized in many applications, in particular as precise instruments (Her and Chang, 1994; Kota et al., 1999; Onillon et al., 2003; Pernette et al., 1997; Slocum, 1992) due to a number of advantages including low cost, reduced weight and smooth motion. Besides, simplified, especially monolithic manufacturing for the flexure mechanisms can cut down the number of work pieces; sequentially diminish the errors brought by assembly.

As one of the most important elements in flexure mechanisms, flexure joints transfer the motion and energy, and ensure the movement and precision that the specified applications required. One of the commonly-used flexure joints is notch hinges (Lobontiu, 2003), which provide a high precision and a large stiffness but a very limited stroke. One the contrary, another commonly-used flexure joint is a leaf spring or a compliant beam, the stroke of which is much larger due to its distributed-compliance characteristic. However, it lacks in precision and stiffness. In order to overcome such disadvantages, some complex flexure joints were inves-

tigated recently (Tseytlin, 2006; Trease et al., 2005). The typical examples are the cross-axis flexural pivots (Jensen and Howell, 2002), the cartwheel hinges (Smith, 2000), the split-tube pivots (Goldfarb and Speich, 1999), the butterfly flexural pivot (Henein et al., 2003) and so on. Most of them consist of two and more flexural beams: a cross-axis flexural pivot is assembled by two beams; a cartwheel hinge is constructed by four beams; and butterfly pivot is up to eight beams.

Some complex flexure joints, such as the cross-axis flexural pivot, are indeed the spatial structures; therefore they are hard to be machined out from a monolithic block of material. Therefore, they are generally unsuitable for being used in the high precision systems because assembly error could be easily introduced.

A cartwheel hinge, on the contrary, has a planar nature. It provides not only a large-deflection stroke, but also overcomes such disadvantages of a cross-axis pivot as an unavoidable assembly and a relatively low rotational precision. As a result, by comparison with the cross-axis pivot, the cartwheel hinge gets a five times improvement in the stability of the center of rotation, but as a cost of nearly four times lose in the rotational stroke (Smith, 2000).

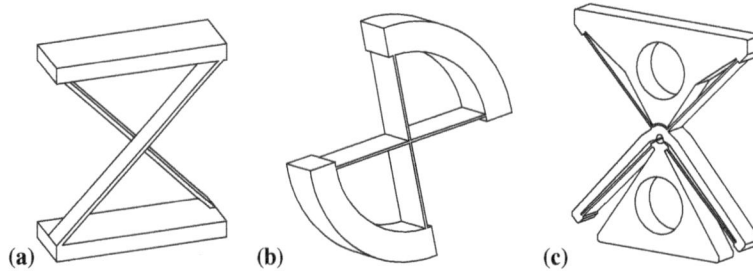

Figure 1. Leaf-Type Flexure joints: (**a**) a cross-axis pivot; (**b**) a cartwheel hinge; (**c**) an ADLIF.

Another type of planar flexure joint is the butterfly flexural pivot. Although there exists 8 beams in the architecture, it is a nearly perfect combination of a large-deflection stroke and a high precision (much better than a cartwheel hinge).

Whether for a cartwheel hinge or a butterfly flexural pivot, we notice that both of them can be regarded as the combination of two and more LITF pivots (Pei et al., 2008a), in which the remote-center-motion (RCM) enable them to be easily superimposed to construct some new complex flexure joints. Just inspired by this design phylosiphy, in this paper, we present a new type of a planar leaf-type flexure joint, i.e. an anti-symmetric double leaf-type isosceles-trapezoidal flexure joint (ADLIF), which is constructed by four beams. The ADLIF possesses almost all the merits of the cartwheel hinge and meantime can gain much higher precision. In the later sections, Finite Element Analysis (FEA) is used to validate such an improvement. In addition, in order to provide a simple tool to design or optimize the ADLIF, an improved pseudo-rigid-body (PRB) method is also formulated. The proposed PRB model is not only simple and accurate, but also intuitive to designers.

2   Conceptual design

As shown in Fig. 1c, an ADLIF consists of four leaf-type flexures, and the extended lines of the four leaves intersected at a point called the virtual pivot point. Both of the ADLIF and the cartwheel hinge can be regarded as two LITF building blocks connected in series (Pei et al., 2008a). The major difference between them is that whether the two LITF modules are arranged symmetrically or anti-symmetrically. In the ADLIF as shown in Fig. 2b, the two LITF modules are connected by an intermediate body. In a cartwheel hinge, the intermediate body is reduced to zero. When either of two rims (Rim 1 and Rim 2) is stationary, the other one becomes movable.

Three parameters are needed to determine the configuration of the ADLIF: (1) h_f denotes the distance between the bottom end of the leaves and the pivot point O; (2) H denotes the distance between the upper end of the leaves and

the pivot point O; (3) ϕ denotes half of the angle between two leaves. The length of each leaf is thus written as

$$l = \overline{DA} = (H - h_f)/\cos\phi \tag{1}$$

Assuming one rim (Rim 1 or Rim 2) is stationary, when the joint deflects, θ is used to denote the rotation angle of the other rim.

3   Pseudo-rigid-body model

Although the FEA is an effective and credible way to analyze a flexible body, the commercial software is usually expensive; what is more important, the modelling and simulation processes are rather time-consuming. In the early design phases of flexures, a PRB is instead a useful tool. It is not only intuitive to designers, but also a resultant parameterized model that facilitates structural optimization. In this section, based on the results for the LITF model (Pei et al., 2008b, c), a simple PRB model of the ADLIF is formulated. The accuracy of the model is relatively high, and it can be confirmed in Sect. 5 by comparison with the results of FEA.

An improved PRB bar model of the ADLIF is shown in Fig. 2b, where the bold lines denote the rigid segments; eight ideal rigid pivots are added to predict the deflected path; four torsional springs are attached at the inner four pivots to reveal the force-deflection relationships in the ADLIF.

The vertical distances between all four pivots A, B, A', B' and the center O are equal to h, which is given by

$$h = \gamma h_f + (1 - \gamma)H \tag{2}$$

where the characteristic radius factor γ is defined using the equation (Pei et al., 2008b)

$$\gamma = \frac{15}{2n_f^2 - n_f + 17}. \tag{3}$$

$$n_f = h_f/H \tag{4}$$

Using a rough estimation, γ is approximately equal to 8/9.

The stiffness of the spring K can be calculated as (Pei et al., 2008b)

$$K = \frac{4EI\gamma^2(1 + n_f + n_f^2)}{l_f} \tag{5}$$

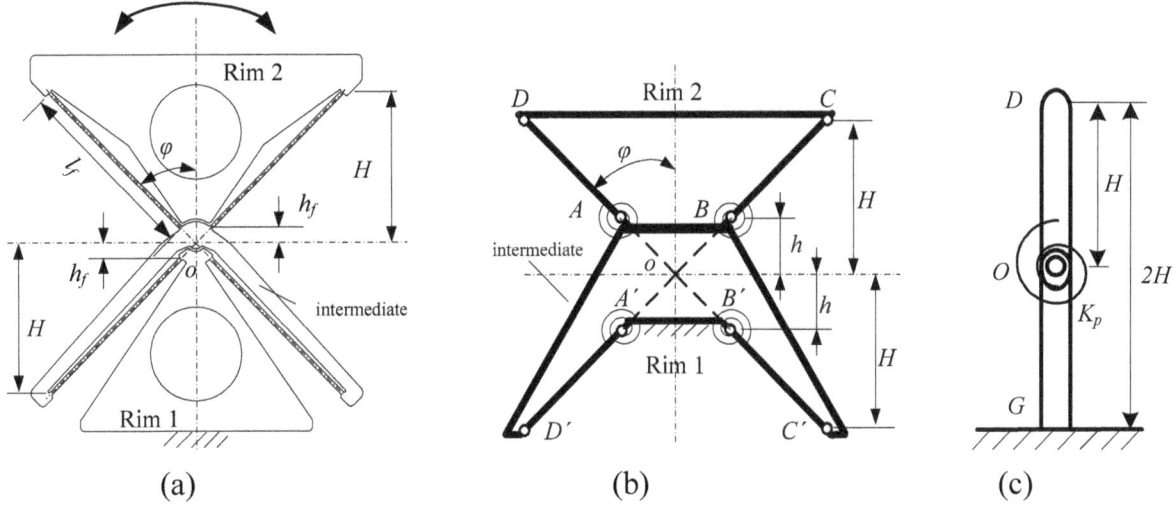

Figure 2. The PRB model of an ADLIF: (**a**) the ADLIF; (**b**) the bar model; (**c**) the pin-joint model.

3.1 Stiffness

With aid of the PRB bar model shown in Fig. 2b, the force-displacement relationship of the ADLIF can be found correspondingly. In addition, a pin-joint model of ADLIF the shown in Fig. 2c is proposed by simplifying the PRB bar model. The more detailed modeling process has been demonstrated extensively in the literature (Pei et al., 2008b). The stiffness value of the pin-joint model can be obtained by

$$K_\mathrm{p} = \frac{4EI(H^2 + Hh_\mathrm{f} + h_\mathrm{f}^2)\cos\phi}{(H - h_\mathrm{f})^3} \tag{6}$$

$$M = K_\mathrm{p} \cdot \theta \tag{7}$$

3.2 Center-shift

The developed pin-joint model only can be used to determine the moment-displacement nature. When the detail feature of the joint needs to be investigated, such as the accurate motion of the Rim 2, the bar model prefers to be used in this case. The center-shift δ is an important criterion for evaluating the rotational precision of a general flexure joint. It can be obtained in a way proposed in literature (Pei et al., 2008a). The vector-form result is given by

$$\delta = -\delta_1 \cos\frac{\theta}{2} + \delta_1 \sin\frac{\theta}{2} \cdot i \tag{8}$$

$$\|\delta\| = \delta_1 \sqrt{2\left(1 - \cos\frac{\theta}{2}\right)} = \delta_1 \left|\frac{\theta}{2}\right| \tag{9}$$

where δ_1 is the center-shift of a single LITF joint, and

$$\delta_1 = \delta_{1x} + i\delta_{1y} \tag{10}$$

$$\frac{\|\delta_1\|}{H} = B_3 \sqrt{B_1} \tag{11}$$

where δ_{1x} and δ_{1y} are the components of the center-shift. They can be written as

$$\frac{\delta_{1x}}{H} = B_3 \cdot \sin\frac{\theta}{2} \tag{12}$$

$$\frac{\delta_{1y}}{H} = \left(n - \cos\frac{\theta}{2}\right) \cdot B_3 \tag{13}$$

where

$$B_1 = 1 + n^2 - 2n\cos\frac{\theta}{2} \tag{14}$$

$$B_2 = (1 - n)^2 / \sin^2\varphi \tag{15}$$

$$B_3 = 1 - \tan\varphi \sqrt{\frac{B_2}{B_1} - 1} \tag{16}$$

$$n = h/H \tag{17}$$

3.3 Stress analysis

Stress always occurs as the hinge deflects. It is therefore another important criterion to be considered, which can also be used to decide the maximum deflection for the hinge.

For a single flexible segment, because the length l is much larger than the thickness t, thus the shear deflection can be neglected, and only the bending deflection is considered. In this case, the stress is given by

$$\sigma_\mathrm{max} = \frac{Et(2H + h_\mathrm{f})\cos\phi}{2(H - h_\mathrm{f})^2}\theta \tag{18}$$

The maximum deflection of the ADLIF may be found by replacing σ with the yield strength (S_y).

$$\theta_\mathrm{max} = \frac{2(H - h_\mathrm{f})^2}{Et(2H + h_\mathrm{f})\cos\phi}S_\mathrm{y} \tag{19}$$

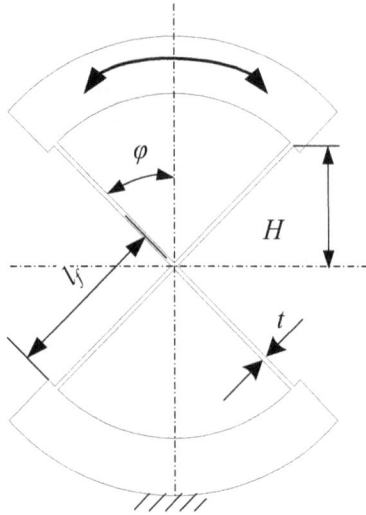

Figure 3. Parametric model of the cartwheel hinge.

3.4 Model validation

To validate the proposed PRB model, two ADLIFs with different structural parameters are adopted in thus study, and the results are plotted in Fig. 4 to compare with FEA simulation results. Figure 4 shows that the curves of the PRB and FEA results are very close to each other.

4 Case study

In order to validate the improvement in performances, two ADLIFs with different structural parameters are selected to make a comparison with a cartwheel hinge. Figure 3 illustrates the parametric model for the cartwheel hinge, and all structural parameters in the ADLIF are described in Fig. 2a. The concrete parameter values used for comparison are listed in Table 1. The length of leaves of the first ADLIF (ADLIF I) is same as that of the cartwheel joint, while the stiffness specification of the second ADLIF (ADLIF II) is approximate to that of the cartwheel joint. In Table 1, t is the thickness of the flexible segment; b is the width of the flexible segment; E is the Young's modulus.

The commercial FEA software (ANSYS 9.0) capable of making a large-deflection nonlinear analysis is used. The selected material in these cases is aluminum alloy. The Young's modulus, E, is thus 71 GPa and the Poisson's ratio, μ is 0.33. The BEAM3 elements have been chosen with the large displacement option turned on. The options of non-linear computation and stress stiffening are also turned on. A moment is loaded in the middle of moveable rim for each joint; while the other rim is fixed on the ground.

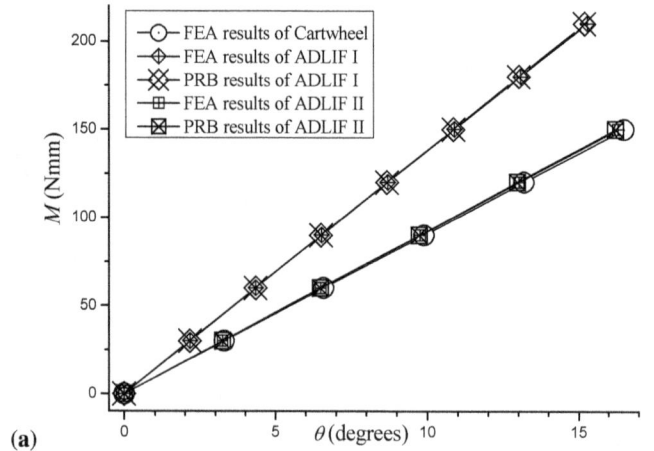

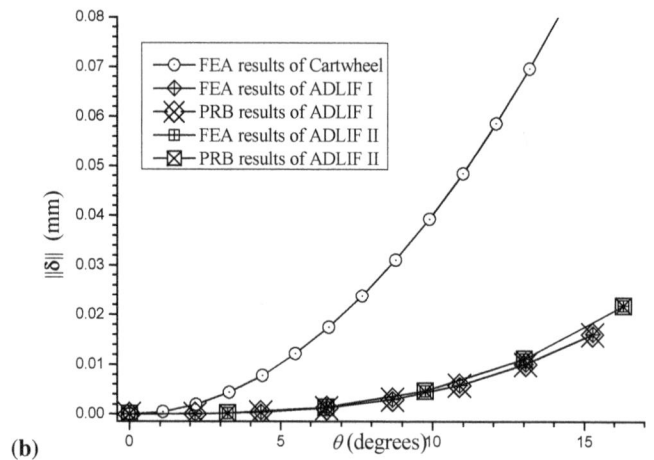

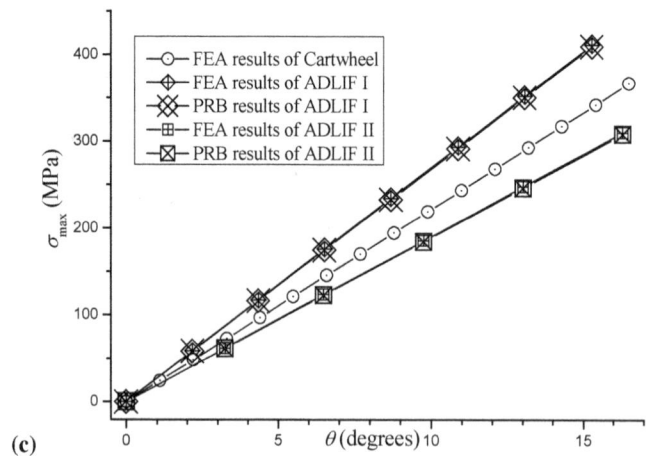

Figure 4. Performance comparisons of a cartwheel hinge and two ADLIFs: **(a)** Stiffness; **(b)** Center-shift; **(c)** Stress.

Table 1. Structural parameters of three flexure joints.

	H (mm)	h_f (mm)	ϕ (degree)	l_f (mm)	t (mm)	b (mm)	E (Mpa)
Cartwheel	20	–	45	28.28	0.5	5	71 000
ADLIF I	23	3	45	28.28	0.5	5	71 000
ADLIF II	30	3	45	38.18	0.5	5	71 000

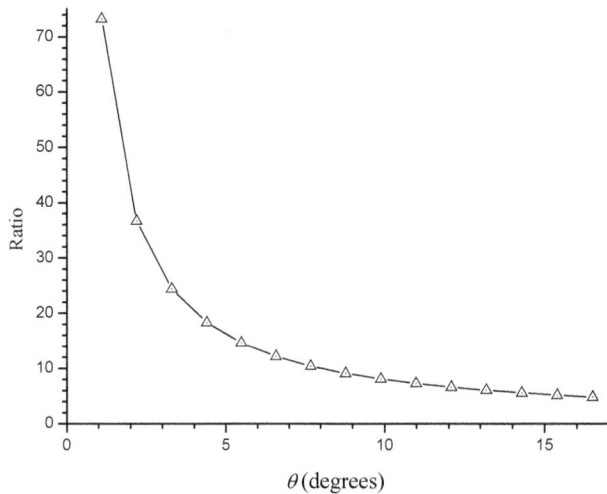

Figure 5. Center-shifts ratio of the cartwheel hinge to the ADLIF II.

5 Results and discussion

The moment-displacement characteristics, center-shifts, and stresses of the three flexure joints are measured and evaluated through the FEA program, and the results are ploted in Fig. 4.

h_f of ADLIF cannot be too small due to the limit of the manufacture capability. This makes the stiffness value of the ADLIF I higher than the cartwheel hinge (shown in Fig. 4a). When the length of all leaves is identical to each other, increase in the leaves' length of the ADLIF (see ADLIF II) may lead to the reduced stiffness. Meanwhile, the center-shift of the ADLIF II increases by a little, as shown in Fig. 4b, and it is still much smaller than that of the cartwheel hinge.

As illustrated in Fig. 5, the center-shift ratio of the cartwheel hinge to the ADLIF II is decreasing when the rotational angle becomes larger. When the rotational angle is up to 5°, the center-shift of the cartwheel joint (about 0.01 mm) is about 16 times larger than that of ADLIF II (about 0.0006 mm). In other words, by comparison with the cartwheel hinge, the ADLIF II can gain more than 16 times improvement in precision when rotational angle is less than 5°.

The maximum deflections of all three joints can be found in Fig. 4c. If the yield strength S_y is 250 MPa, the ranges of motion corresponding to the cartwheel hinge, the ADLIF I and the ADLIF II are about 11.4°, 9.3° and 13.2°, respectively.

6 Conclusions

A novel large-displacement flexure joint, named an ADLIF, is proposed in this paper. It consists of two LITF building blocks, but these two blocks are arranged asymmetrically. Compared with the cartwheel hinge commonly used in precision engineering and characterized as two LITF building blocks arranged symmetrically, the ADLIF can gain a great improvement in precision as well as keeping other characteristics such as stiffness and ranges of motion at the same level. In order to quantitatively evaluate these performances, A PRB model of the ADLIF is developed, and the moment-displacement characteristics, center-shifts, and stresses of two cases are calculated correspondingly. By aid of validation by the FEA result, the PRB model is proved accurate. According to both the theoretical and simulation results, the ADLIF can get more than 16 times improvement in precision as the rotational angle is less than 5°. Therefore, the ADLIF is suitable for a replacement of the cartwheel hinge in precision some applications.

Acknowledgements. The authors would like to acknowledge the support of National Natural Science Foundation of China, through Grant No. 50905005, 50875008.

Edited by: N. Tolou

References

Goldfarb, M. and Speich, J.: A Well-Behaved Revolute Flexure Joint for Compliant Mechanism Design, ASME J. Mech. Des., 121, 424–429, 1999.

Henein, S., Droz, S., Myklebust, L., and Onillon, E.: Flexure pivot for aerospace mechanisms, Proc. 10th European Space Mechanisms and Tribology Symposium, 24–26 September 2003, San Sebastian, Spain, 1–4, 2003.

Her, I. and Chang, J. C.: A linear scheme for the displacement analysis of micropositioning stages with flexure hinges, J. Mech. Design, 116, 770–776, 1994.

Jensen, B. D. and Howell, L. L.: The modeling of cross-axis flexural pivots, Mech. Mach. Theory, 37, 461–476, 2002.

Kota, S., Hetrick, J., Li, Z., and Saggere, L.: Tailoring unconventional actuators using compliant transmissions: design methods and applications, IEEE/ASME Transactions on Mechatronics, 4, 396–408, 1999.

Lobontiu, N.: Compliant Mechanisms: Design of Flexure Hinges, CRC Press, Boca Raton, FL, 2003.

Onillon, E., Henein, S., and Theurillat, P.: Small scanning mirror mechanism, 2003 IEEE/ASME International Conference on Advanced Intelligent Mechatronics, 1129–1133, 2003.

Pei, X., Yu, J. J., Zong, G. H., and Bi, S. S.: A Novel Family of Leaf-Type Compliant Joints: Combination of Two Isosceles-Trapezoidal Flexural Pivots in Series, Journal of Mechanisms and Robotics, 1, 021005, 1–6, 2008a.

Pei, X., Yu, J. J., Zong, G. H., and Bi, S. S.: The stiffness model of leaf-type isosceles-trapezoidal flexural pivots, J. Mech. Design, 130, 082303, doi:10.1115/1.2936902, 2008b.

Pei, X., Yu, J. J., Zong, G. H., Bi, S. S., and Yu, Z. W.: Analysis of rotational precision for an isosceles-trapezoidal flexural pivot, J. Mech. Design, 130, 052302 doi:10.1115/1.2885507, 2008c.

Pernette, E., Henein, S., Magnani, I., and Clavel, R.: Design of parallel robots in microrobotics, Robotica, 15, 417–420, 1997.

Slocum, A. H.: Precision Machine Design, Society of Manufacturing Engineers, 1992.

Smith, S. T.: Flexures: elements of elastic mechanisms, Gordon and Breach Science, New York, 153–230, 2000.

Trease, B. P., Moon, Y. M., and Kota, S.: Design of Large-Displacement Compliant Joints, J. Mech. Design, 127, 788–798, 2005.

Tseytlin, Y. M.: Structural Synthesis in Precision Elasticity, Springer, New York, 2006.

14

Review Article: Inventory of platforms towards the design of a statically balanced six degrees of freedom compliant precision stage

A. G. Dunning, N. Tolou, and J. L. Herder

Faculty of Mechanical, Maritime and Materials Engineering, Department of Biomechanical Engineering,
Delft University of Technology, Delft, The Netherlands

Abstract. For many applications in precision engineering, a six degrees of freedom (DoF) compliant stage (CS) with zero stiffness is desirable, to deal with problems like backlash, friction, lubrication, and at the same time, reduce the actuation force. To this end, the compliant stage (also known as compliant mechanism) can be statically balanced with a stiffness compensation mechanism, to compensate the energy stored in the compliant parts, resulting in a statically balanced compliant stage (SBCS). Statically balanced compliant stages can be a breakthrough in precision engineering. This paper presents an inventory of platforms suitable for the design of a 6 DoF compliant stage for precision engineering. A literature review on 3–6 DoF compliant stages, static balancing strategies and statically balanced compliant mechanisms (SBCMs) has been performed. A classification from the inventory has been made and followed up by discussion. An obviously superior architecture for a 6 DoF compliant stage was not found. All the 6 DoF stages are either non-statically balanced compliant structures or statically balanced non-compliant structures. The statically balanced non-compliant structures can be transformed into compliant structures using lumped compliance, while all SBCMs had distributed compliance. A 6 DoF SBCS is a great scope for improvements in precision engineering stages.

1 Introduction

Many applications in precision engineering, including lithography, electron beam microscopy, micro assembly, aerospace, medical applications, require ultra precision positioning to manipulate an object in a vacuum or wet environment. For instance, in lithography the electrical circuits written on a wafer will have a resolution smaller than 20 nm (Willson and Roman, 2008). In the medical field, precise surgical tools with good force feedback are required to avoid tissue damage during operation (Sjoerdsma et al., 1997). All the named applications are situated inside a vacuum or wet environment. Therefore it is difficult to use conventional bearings, due to the need of lubrication. The backlash in conventional joints also has been an issue in high precision engineering. To overcome these problems, compliant mechanisms can be used.

A compliant mechanism is a mechanism that transfers force, motion or energy by using the elastic deformation of its flexible components rather than using rigid-body joints only. An advantage of compliant mechanisms is that it can easily be manufactured as a monolithic structure due to its hingeless nature of the design. This absence of movable joints reduces wear, friction and backlash in the mechanism and correspondingly increases precision, which is an important factor in the design of high-precision instrumentation. There is also no need for lubrication and the mechanism is insensitive to dust, which is an important advantage in instruments under vacuum (Howell, 2001).

However, the compliant mechanisms rely on the deflection of flexible members, which introduces positive stiffness and requires energy to deform. Therefore, the energy storage in the flexible members is distorting the input-output relationship and challenges the mechanical efficiency. When the deformation of the flexible members is large, non-linearities are introduced, which increases the complexity of the design (Herder and van den Berg, 2000; Morsch and Herder, 2010).

In many of the mentioned fields, it is required to manipulate an object in six degrees of freedom (DoF). In particular, in lithography and electron beam microscopy, the actuation of the 6 DoF positioning stage produces too much heat, mainly caused by the stiffness of the stage, which can affect the precision of the application (Nieuwenhuis, 2010). In medical instruments, the force feedback is not optimal, due to the stiffness and friction introduced in compliant and contact members (Sjoerdsma et al., 1997).

To overcome these problems a stiffness compensation mechanism can be added to the compliant mechanism, resulting in a statically balanced compliant mechanism (SBCM) with nearly zero stiffness. A statically balanced mechanism (SBM) is a mechanism on which the forces of one or more potential energy storage elements are acting, such that the mechanism is in static equilibrium and therefore has zero stiffness. The total potential energy should be constant in every position of the mechanism (Herder, 2001). To create static balancing a positive stiffness of the mechanism should be balanced with a negative stiffness compensation device. Therefore, it can be very advantageous to integrate a 6 DoF SBCM into an available application and replace the conventional positioning system.

The purpose of this literature survey consists of (1) to provide an overview of the state of the art of 6 DoF compliant stages. Interesting stages with less degrees of freedom, where translations are combined with rotations have also been investigated. A classification is made to compare the available stages to investigate whether there is a superior design for 6 DoF compliant stages. Thereafter, (2) an inventory on balancing strategies for compliant mechanisms is made. Finally, (3) possibilities to combine a 6 DoF compliant stage with static balancing will be investigated.

In Sect. 2, the method, including search method, search criteria, and the method to classify the results, is explained. The results of the literature survey are briefly described in Sect. 3. In Sect. 3.1 the results of the 6 DoF compliant stages are presented. It presents the type and classification of flexures, serial and planar positioning structures. Section 3.2 describes the balancing strategies with existing SBCMs and structures combining 6 DoF with static balancing. Section 4 interprets and discusses the results of each goal. Conclusions are presented in Sect. 5.

2 Method

2.1 Search method

The literature survey is separated into two parts. In the first part a literature search is conducted for 6 DoF compliant precision stages. This part also considers stages with fewer DoFs that may be converted into 6 DoF. These are stages with 3, 4 or 5 degrees of freedom, where translational degrees of freedom were combined with rotational degrees of freedom.

The second part is to examine the static balancing strategies for compliant mechanisms and make a classification.

By analyzing the topics a search plan was made. The key subjects and constraints were determined, particularly in the field of precision engineering. Only stages with a motion smaller than 1 mm were searched for. Subsequently, key subjects were transformed into search terms, comprising synonyms and related terms. These search terms were used in the set of keywords in the search engines.

In total five different sets of keywords have been used, concerning keywords defining (1) compliant mechanisms, (2) the field of precision engineering, (3) 6 DoF stages, (4) static balancing and (5) zero stiffness.

In order to optimize the search, all sets of keywords were combined and narrowed. Also the references of the articles were checked for useful articles in the same subject. The results were first filtered by inspecting the article titles. Subsequently, the reduced results were filtered by reading the abstracts and looking to the images in the article. From the abstract or the images the working principle needed to be clear. Otherwise the papers were discarded.

The literature search was conducted using two search engines (Scopus; Espacenet). SCOPUS was used for journal articles and conference proceedings, while Espacenet was used to search for patents. All five sets of keywords were used in SCOPUS. Espacenet is the search engine of the European Patent Office and searches patents from all over the world. This engine is able to search patents with a set of keywords, instead of a classification system. Only patents of 6 DoF compliant stages and SBCMs were of interest for this literature survey, only specific combinations of sets of keywords were used. An overview of the sets of keywords can be found in Table 1.

2.2 Classification

A classification was made to compare the results of the compliant mechanisms within the field of 6 DoF stages and precision engineering. The following strategy and criteria have been used for classification.

The first and second level of classification, indicated the architecture of the mechanism. In the first level, a distinction was made between planar and spatial geometry of the structure. In a planar structure, in contrast to spatial structures, flexible elements to perform a 6 DoF motion are in the same plane, so for some motion out-of-plane motion is required. The second level described the configuration of the kinematic chain mechanism. This can be a parallel or a serial configuration (Lobontiu, 2003). In a parallel configuration, also called a closed-loop configuration, the fixed base is connected to the movable end-effector through multiple kinematic chains. A good example of a parallel mechanism is the Stewart platform (Stewart, 1965). Serial mechanisms use an open loop serial chain of links to connect the base with the end-effector. A robot arm is a good example of a serial mechanism.

Table 1. Overview of the sets of keywords used in SCOPUS (1–5) and Espacenet (1, 3, 4, 5).

Sets	Keywords
(1) Compliant mechanisms	– Compliant, flexible, flexure, monolithic – Mechanism, structure, design
(2) Precision engineering	– Precision, micro, nano, sensible – Stage
(3) 6 DoF stage	– Six degrees of freedom, six axis – Stage
(4) Static balancing	– Static balancing, neutral equilibrium
(5) Zero stiffness	– Zero/neutral/eliminate/remove/cancel stiffness – Constant potential energy, pre-stressed – Neutral stability – Gravity compensation

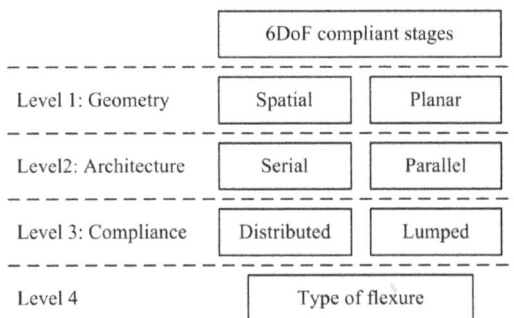

Figure 1. Schematic representation of the classification levels to compare the 6 DoF compliant stages.

The third level of classification described the types of stress distribution in the mechanism, which are lumped compliance and distributed compliance (Ananthasuresh and Kota, 1995).

In the fourth level the type of flexures used in the mechanism was distinguished.

In Fig. 1, a schematic representation of the classification is provided. Quantitative data found, involving size (S), working range (WR), will be noted.

To compare the stages, the ratios between translations, rotations and the size of the stages were investigated.

The SBCMs were classified according to the balancing principle, using (1) counterweights or (2) elastic elements, to compensate gravity forces or strain energy inside the mechanism (Herder, 2001). The mechanisms in these categories can be classified further according to the type of compensation mechanism. If reported in the article, the remaining stiffness after balancing, the statically balanced stroke and the size of the balancing mechanism is mentioned.

3 Results

3.1 State of the art in 6 DoF compliant stages

In the field of precision engineering the demand for 6 DoF stages is high. These stages have to be very accurate, with a resolution in the order of nanometers (Willson and Roman, 2008). In literature, precision compliant stages, which combine translations and rotations, with 3, 5 and 6 DoF were found. All the 6 DoF stages had three translational (x, y, z) and three rotational (θx, θy, θz) degrees of freedom. One 5 DoF stage (Wang et al., 2005) was found, which had no degree of freedom in rotation around the z-axis, and the 3 DoF stages had all two translational (x, y) and one rotational (θz) degrees of freedom. All the designs found in literature were fully compliant. In other words, no conventional joints were used for transferring motion. Besides, all the designs were highly symmetric, otherwise it is mentioned.

An overview of all the available results, including flexure type, size (S) and working range (WR) is shown in Table 2.

3.1.1 Type of flexures

Different flexures were found in the compliant mechanisms. Depending on the characteristics of the flexure it can have single or multiple deflection axes, which can be translational or rotational. Two rotational deflection axes in a joint create a universal joint and a combination of three rotational joints creates a spherical joint.

The flexible components could be classified in two groups, with flexures having (1) lumped compliance and (2) distributed compliance. With lumped compliance the flexion concentrates around a distinct number of flexures, causing high stress concentrations in the mechanism. These flexible elements have also low static and fatigue strength, usually undergoes small displacements, and manufacturing

Table 2. Overview of the results of the compliant stages, mentioned flexure type (mentioned with •), size and working range. Data not available identified with –.

		Flexure type									Size (mm)			Working range					
														Translation (μm)			Rotation (mrad)		
	Reference	Leaf spring	Pin flexure	Small-length pin flexure	Small-length plate flexure	Corner filleted notch	Circular notch	Parabolic notch	Spherical notch	Monolithic	ΔX	ΔY	ΔZ	ΔX	ΔY	ΔZ	Δθx	Δθy	Δθz
Spatial parallel structures	Brouwer et al. (2010)	•									6.2	6.2	0.5	20	20	20	52.36	52.36	52.36
	Seugling et al. (2002)	•									100	100	100	0.93	0.93	0.93	38e-3	38e-3	38e-3
	Moon and Kota (2002)	•									–	–	–	–	–	–	–	–	–
	Helmer et al. (2004)					•					164	147	255	4000	4000	4000	69.8	69.8	69.8
	Hu et al. (2008b)						•				±Ø95.2		21.6	50	50	50	8.73	8.73	8.73
	Liu et al. (2001)							•			–	–	–	–	–	–	–	–	–
	Sun et al. (2003)							•			–	–	–	–	–	–	–	–	–
	Wang et al. (2003)							•			Ø130		98.3	–	–	–	–	–	–
	Wang et al. (2007)		•						•	•	–	–	–	5.8	5.7	1	–	–	–
	Sun (2007)			•					•		–	–	–	1023	1023	1023	–	–	–
	Yun and Li (2010)				•						250	250	250	9700	9700	9700	240	240	240
Spatial serial structures (2 parallel mechanisms in serie)	Choi and Lee (2005)	•								•	Ø258		10	–	–	–	–	–	–
	Hu et al. (2008a)					•					Ø240		31.26	77.42	67.45	24.56	0.93	0.95	3.1
	Chao et al. (2005)						•				–	–	–	130	140	18	–	–	–
	Xiaohui et al. (2010)			•		•	•				–	–	–	–	–	–	–	–	–
	Xuchu and Qianfeng (2009)			•		•					–	–	–	–	–	–	–	–	–
	Liang et al. (2011)	•		•		•					–	–	–	0.034	0.034	0.034	–	–	–
	Gao and Swei (1999)			•			•	•	•		–	–	–	–	–	–	–	–	–
	Wang et al. (2005)					•			•		–	–	–	–	–	–	–	–	–
	Chang et al. (1999a, b)	•								•	200	200	50	17.9	17.9	–	–	–	0.585
Planar parallel structures	Anderson (2003), Culpepper (2006), Culpepper and Anderson (2004)		•							•	±Ø180		3	100	100	100	4	4	4
	Chen and Culpepper (2006)		•							•	Ø3		5.18	8.4	12.8	8.8	19.2	17.5	33.2
	Zhang et al. (2005)				•					•	14	14	0.8	2	2	2	0.25	0.25	0.25
	Park and Yang (2005)					•				•	–	–	–	7	7.1	10	0.25	0.23	0.26
	Lu et al. (2004)					•				•	–	–	–	14	13	–	–	–	0.756
	Ryu et al. (1997)					•				•	±Ø115		–	41.5	47.8	–	–	–	1.565
	Tian et al. (2010)					•				•	–	–	–	–	–	–	–	–	–
	Wang and Zhang (2008)					•				•	±Ø150		18.5	–	–	–	–	–	–
	Yi et al. (2003)					•				•	±Ø120		–	100	100	–	–	–	17.5
	Jong de, et al. (2010)	•								•	5.5	5.5	–	10	10	–	–	–	34.9
	Lee and Kim (1997)					•				•	–	–	–	–	–	–	–	–	–

these elements can give difficulties, due to very thin sections (Ananthasuresh and Kota, 1995; Gallego and Herder, 2009). In this group, notch-type flexures and small-length plate and pin flexures could be found. The notch profile could be a (1) rectangular corner-filleted, (2) circular, (3) parabolic, or (4) spherical section (Fig. 2). The small-length plate flexure could bend in one degree of freedom and the pin flexure could bend in all three rotational degrees of freedom (Gallego and Herder, 2009).

For distributed compliant flexures, the flexibility is distributed equally over the entire flexible element. The flexible element has a constant cross-section, which prevent stress concentration around a point. Distributed compliance offers better performance and reliability compared to lumped compliance (Ananthasuresh and Kota, 1995). The pin flexure could bend in all three rotational degrees of freedom and

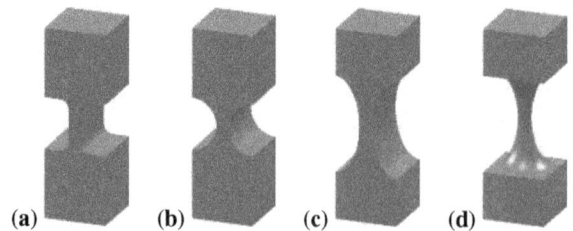

(a)   **(b)**   **(c)**   **(d)**

Figure 2. Notch-type flexures with lumped compliance. The notch profile is (**a**) rectangular corner-filleted, (**b**) circular, (**c**) parabolic, or (**d**) spherical.

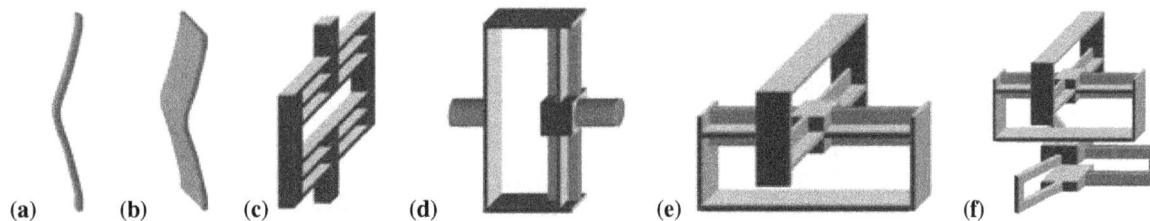

Figure 3. Flexures with distributed compliance. The flexures could be a (**a**) pin, (**b**) chevron, (**c**) translational, (**d**) rotational, (**e**) universal, or (**f**) spherical flexure. Reproduced from Gallego and Herder (2009).

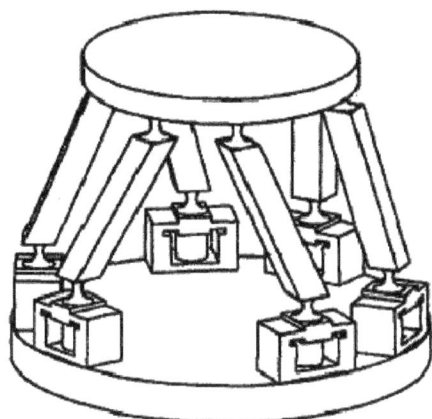

Figure 4. Typically example of a spatial parallel compliant stage (Liu et al., 2001). The platform is supported by legs, with compliant joints at both ends.

a chevron flexure, also called a leaf spring, could bend in one direction and take up torsion. Almost all of these flexures were built up from combining several chevron flexures in such a way that joints with different degrees of freedom are possible (Fig. 3) (Gallego and Herder, 2009).

3.1.2 Spatial compliant stages

The results for spatial compliant stages were separated into a group with a parallel and a serial kinematic chain. First the parallel designs will be described (Fig. 4).

In Brouwer et al. (2010) in-plane leaf springs form prismatic joints and three slanted leaf springs for out-of-plane motion form three universal joints. The flexures, arranged by 120°, create a monolithic spatial parallel platform stage. The same kind of flexures are used in Seugling et al. (2002) and Moon and Kota (2002). In the latter article, the leaf springs were combined such that they form a prismatic, rotational and spherical joint, respectively.

A large non-symmetric stage with corner-filleted notches was developed in Helmer et al. (2004).

Circular notch-type flexures are used in Hu et al. (2008b). Here six slanted trapeziform displacement amplifiers form a

spatial stage. Each trapeziform amplifier can be modeled as two prismatic joints.

Spherical notches were found in mechanisms based on the Stewart platform. In Liu et al. (2001), Sun et al. (2003), and Wang et al. (2003) the platform is supported by 6 legs, that is the compliant equivalent of a 6-spherical-prismatic-spherical manipulator. In Wang et al. (2007) the platform is supported by 3 legs. Each leg is the compliant equivalent of a rotational-spherical manipulator. The legs are placed on small compliant mechanisms, which enables translational motion in 2 DoF with leaf springs and are placed 120° of each other.

Sun (2007) used a non-symmetric stage with spherical notch-type flexures in series with small-length plate flexures (prismatic joints) to create the desired degrees of freedom.

In Yun and Li (2010) small-length pin flexures on both sides of an actuator are used to move a platform. In total eight non-symmetrically placed actuators are used, which makes the stage the compliant equivalent of a 8-prismatic-spherical-spherical/spherical-prismatic-spherical manipulator.

All stages with a serial kinematic chain were constructed as two parallel mechanisms in series, a so-called serial-parallel mechanism (Fig. 5). All stages consist of a parallel monolithic mechanism, which could perform the motion in x, y and θz direction (further mentioned as in-plane motion), and a parallel mechanism performing motion in z, θx, θy direction (further mentioned as out-of-plane motion). The flexures are all arranged 120° of each other.

Choi and Lee (2005) designed a stage where the motion is enabled by leaf springs. The x, y and θz motions are transferred by six L-shaped leaf springs and the z, θx, θy motions are transferred by wide leaf springs.

In Hu et al. (2008a) the flexures are cornered-filleted notches. The in-plane mechanism is the compliant equivalent of a traditional 3-revolute-revolute-revolute manipulator. The out-of-plane mechanism is an equivalent of a traditional 3-universal-prismatic-universal manipulator.

Chao et al. (2005) used a 3-revolute-revolute-revolute compliant mechanism with circular notches for the in-plane motion. For the out-of-plane motion a 3-revolute-prismatic-spherical compliant mechanism with circular notches is used to form 3 legs, supporting the moving platform. The stage

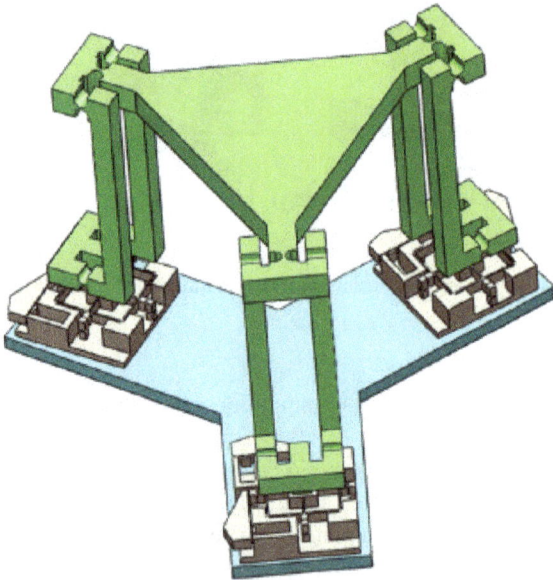

Figure 5. Typically example of a spatial serial compliant stage (Liang et al., 2011). Three legs forms a parallel compliant mechanism performing motion in z, θx, θy, θz. The legs are supported by parallel 2 DoF compliant mechanisms. Both parallel mechanisms in serie forms the spatial serial compliant stage.

from Xiaohui et al. (2010) has the same compliant equivalent structure as Chao et al. (2005) for in-plane motion. The out-of-plane motion is performed by 3 parabolic notch-type flexures. In Xuchu and Qianfeng (2009) a 3-revolute-revolute compliant mechanism with circular notches is used for in-plane motion. Small-length plate flexures are used for the out-of-plane motion.

Liang et al. (2011) used 3 legs, each consisting of two universal joints, supporting a platform for out-of-plane motion with 4 DoF (z, θx, θy, θz). These universal joints were manufactured with circular notch-type flexures. The in-plane motion (x, y) is provided by a spatial mechanism consisting of small-length plate flexures and leaf springs.

In Gao and Swei (1999) the compliant equivalent of a 3-revolute-prismatic-revolute manipulator is used for in-plane motion and a 3-revolute-prismatic-spherical manipulator for the out-of-plane motion. Three legs, with a parabolic and a spherical notch-type flexure, support the platform. The in-plane motion is provided by small-length plate flexures.

Wang et al. (2005) developed a 5 DoF compliant stage made with circular notch-type flexures, having a monolithic mechanism to provide translation along the x-axis and y-axis and a 4-revolute-revolute compliant mechanism to provide translation along the z-axis and rotations in all directions. The flexures in this stage are not arranged 120° of each other.

Chang et al. (1999a, b) designed a 3 DoF stage with leaf springs and small-length plate flexures, consisting of a 2 DoF (x, y) stage and a 1 DoF (θz) stage on top of it, which makes it also a serial-parallel structure.

Figure 6. Typically example of a planar compliant stage (Anderson, 2003; Culpepper, 2006; Culpepper and Anderson, 2004). The flexures to perfom motion are in the same plane.

3.1.3 Planar compliant stages

Only a few stages have a planar structure (Fig. 6). The main advantage of planar structures is that the whole mechanism can be manufactured monolithic and have a high stiffness, but usually a small workspace, compared to serial mechanisms. All the planar designs found in the articles were monolithic, and had a parallel kinematic chain. The differences in each design were the used flexure type.

In Anderson (2003), Culpepper (2006), and Culpepper and Anderson (2004) a nano-manipulator, called the HexFlex, which use 3 long pin flexures, placed 120° to each other, to enable 6 DoF is presented. Each flexure enables in-plane and out-of-plane motion. In Chen and Culpepper (2006) and Culpepper and Golda (2007) two different types of mirco-scaled versions of the HexFlex are made. In Zhang et al. (2005) the 6 DoF motion is enabled by four parallelograms. With small-length pin flexures the parallelograms can move in-plane and out-of-plane. In Park and Yang (2005) a set of circular notches arranged by 120° creates in-plane motion, and inclined circular notches placed 45° with respect to the plane enables out-of-plane motion.

Planar monolithic 3 DoF stages were found in Lu et al. (2004), Ryu et al. (1997), Tian et al. (2010), Wang and Zhang (2008), and Yi et al. (2003). The circular notch flexure groups are arranged 120° of each other. All the designs are modeled with a 3-revolute-revolute-revolute manipulator. Almost the same structure was found in a MEMS-based manipulator, produced by Jong de, et al. (2010), but the flexures are leaf springs and the compliant equivalent of a 3-prismatic-revolute-revolute manipulator is used. Lee and Kim (1997) designed an ultra-precision micro stage, with circular notch flexures, to correct the errors of a global stage.

Table 3. Overview of the results of the statically balanced compliant mechanisms (SBCM) and 6 DoF statically balanced mechanisms (SBM). The balancing mechanism is either with counterweights (C) or elastic elements, using springs (S), zero-free-length springs (ZFLS) or compliant flexures (CF), which are categorized into the use of buckling plates (BP), preloaded plates (PP), to balance strain energy (E) or gravity forces (G). Data not available identified with –.

	Reference	Leaf spring	Circular notch	Parabolic notch	Spherical notch	Monolithic	Balancing mechanism	Preloading	Category of SBCM	Stiffness/force compensation (%)	Compensated stiffness/force upper bound	Statically balanced stroke (mm)	Size of the balancing mechanism (mm³)
SBCM	Eijk van, and Dijksman (1979)	•					BP	E	–	100 %	–	–	–
	Herder and van den Berg (2000)	•					S	E	1	99.9 %	12.9 N	1	±49×10³
	Stapel and Herder (2004)	•				•	PP	E	3	100 %	±50 N mm⁻¹	0.3	±4280
	Tolou and Herder (2009)	•					PP	E	3	100 %	19 N	4.17	±720
	Lange de, et al. (2008)	•				•	BP	E	3	90 %	300 N	0.65	±980
	Powell and Frecker (2005)	•					S	E	1	100 %	–	–	–
	Hoetmer et al. (2009)	•				•	BP	E	3	120 %*	1 N mm⁻¹	1.7	±1850
	Morsch and Herder (2010)	•					PP	E	3	70 %	6.5 N**	23.6***	±4×10⁵
	Trease and Dede (2004)	•					CF	G	3	100 %	±5 N	–	–
	Tolou and Herder (2010) (case I)	•				•	BP	E	3	99 %	60 mN	0.05	9.6
	Tolou and Herder (2010) (case II)	•				•	BP	E	3	86 %	40 mN	0.06	1.6
6 DoF SBM	Streit (1991)					•	ZFLS	G					
	Ebert-Uphoff and Johnson (2002), Ebert-Uphoff et al. (2000)					•	S	G					
	Gosselin and Wang (2000)		•	•			C, S	G					
	Leblond and Gosselin (1998)					•	S	G					
	Shekarforoush et al. (2010)					•	ZFLS	G					

* This mechanism is overcompensated.
** Compensated force is calculated from given compensated moment.
*** Stroke is calculated from stroke given in radian.

3.2 Static balancing strategies for compliant mechanisms

Static balancing can be classified according to the balancing principle (Herder, 2001). These balancing principles are: (1) the addition of counterweights and (2) the use of elastic elements, to compensate gravity forces or strain energy inside the mechanism.

With the use of counterweights, the system is in equilibrium in any position. This method adds extra mass and inertia to the system, relative to springs or other elastic elements. The total potential energy of all gravity and elastic elements must be constant for perfect static balance.

There are several categories of SBCMs. These include (1) a compliant part balanced with a non-compliant compensation mechanism, (2) a compliant part with a compliant balancing mechanism, where the energy is stored in a separate spring, (3) the compensation energy is stored in a compliant part of the mechanism, rather than in a separate spring, and (4) adaptive balancing, taking into account that compliant mechanisms behave different under loaded and unloaded situations (Herder and van den Berg, 2000).

In Table 3 an overview of the results can be found.

3.2.1 Statically balanced compliant mechanisms

In literature, examples of SBCMs using elastic elements are very rare. In Eijk van, and Dijksman (1979) a mechanism with a constant negative stiffness, using a buckled plate spring, has been studied. Herder and van den Berg (2000) compensate the undesired stiffness in a laparoscopic grasper with a rolling-link mechanism and conventional helical springs (category 1). The reduced stiffness is in the order of 0.1 % of the stiffness of the gripper. In Stapel and Herder (2004) a fully compliant compensation device, based on a slider-rocker mechanism, for the laparoscopic grasper is developed (category 3). The total potential energy in the system is almost constant. In Tolou and Herder (2009), the gripper of Herder and van den Berg (2000) is balanced with a partially compliant mechanism, consisting of pairs of pre-stressed pinned-pinned initially curved beams, arranged perpendicular to the link driving the grasper and placed inside the tip of the grasper (category 3). This resulted in force of almost 0N to operate the grasper. Lange de, et al. (2008) used topology optimization to design a fully compliant grasper with a bistable balancing mechanism, with an actuation force reduction of 90 %, but due to calculated high stresses, a prototype is never fabricated (category 3). Powell and Frecker (2005) balanced a compliant forceps with a rigid link slider-crank mechanism with a non-linear spring, optimized with potential energy analysis with finite element analysis (category 1).

Hoetmer et al. (2009) used the Building Block Approach to balance a gripper. With the use of a new balanced building block, consisting of buckling plates, the stiffness was reduced from 1 N mm^{-1} to −0.2 N mm^{-1} (category 3).

In Morsch and Herder (2010), the joint of a conventional balanced mechanism (Herder, 2001) is replaced by a cross-axis flexural pivot, and the zero-free-length springs by compliant leaf springs (category 3). This resulted in a fully compliant joint with a moment reduction of 70 %, measured from experiments.

Trease and Dede (2004) designed a partially compliant four bar mechanism with novel "open-cross" compliant joints to form a torsion-spring-based statically balanced gravity compensator (category 3). The potential energy of the system was balanced over ±45° from horizontal plane within a 3 % error.

In Tolou and Herder (2010), two different statically balanced compliant micro mechanisms were designed (category 3) where the preloading force and stroke are either perpendicular or collinear. The first type compensated the force for 99 % in the beginning of the travel path, due to external preloading force. But the collinear-type has been internally balanced without separated external preloading force, by using a bi-stable mechanism, compensating the force for 86 % at the end of the stroke.

All the above-mentioned SBCMs had one degree of freedom and had distributed compliance. The design methods may well be used to implement in a 6 DoF structure.

3.2.2   6 DoF statically balanced mechanisms

In literature 6 DoF SBCMs is not readily available. An investigation of the possibilities to combine compliant mechanisms with static balancing some 6 DoF SBMs found in literature are discussed here. All the structures discussed here are spatial parallel platform mechanisms.

Streit (1991) presented the first 6 DoF SBM. He presented a parallel platform mechanism consisting of three legs, where each leg has three degrees of freedom. The legs are parallelograms connected to the platform with spherical joints, and balanced with zero-free-length springs. Static balancing is only achieved when the centre of mass of the platform is close to the plane of the spherical joints. In Ebert-Uphoff and Johnson (2002) and Ebert-Uphoff et al. (2000) this condition is removed by introducing pulling and pushing legs connected to the platform with spherical joints. The mechanism has three active pushing legs, which tilt the platform upwards, and one passive pulling leg, attached in slightly off-centre of the platform and pulling the platform down to a static balanced condition.

Gosselin and Wang (2000) used six legs with revolute actuators to balance a platform, using both the counterweights method and the spring method.

Leblond and Gosselin (1998) showed different ways to balance existing spatial parallel mechanisms, such as the Gough-Stewart platform, with additional elements.

Shekarforoush et al. (2010) balanced two types of 6 DoF tensegrity systems, with passive zero-free-length springs and with an adjustable cable-spring combination. The connection between legs and the platform are all ball-socket joints, which could be represented as spherical joints.

In Table 3 the results are shown for balancing principle and which compliant flexure type could represent the joints in the mechanisms.

4   Discussion

In this part, the results are compared and discussed with each other based on criteria. Many articles did not mention size or working range, which makes it a challenge to compare all stages with each other. Besides, not every stage had the same structure to make a good comparison. Therefore, a comparison between all planar structures is made and finally the spatial stages are compared.

To make a good comparison, the ratios between translations, rotations and the size of the stages are compared. The ratios are normalized to the largest in the group, as shown in Fig. 7.

First, the ratios of translations (in μm) in the XY-plane relative to the size (in mm) of the XY-plane of planar structures (WRx·y/Sx·y) are compared. It is noteworthy, that in Chen and Culpepper (2006) the largest ratio is reached. Considering the ratios between rotations (in mrad) around the z-axis and the size (in mm) in the XY-plane (WRθz/Sx·y), again the largest ratio has been reached in Chen and Culpepper (2006). Also in Jong de, et al. (2010) and Ryu et al. (1997) a relative large ratio is found, compared to the other stages. The results showed that there is no clear relation between flexure type and translation/size or rotation/size ratio in XY-plane. Both Chen and Culpepper (2006) and Chang et al. (1999a) used leaf springs, but had the largest and the smallest ratios, respectively. Also the notch-type flexures did not showed ratios in the same order.

For the spatial stages the ratios of working range of the translations (in μm) relative to the size (in mm) of the stage (WRx·y·z/Sx·y·z) shows that the stage from Seugling et al. (2002) has a very small working range with respect to the size. In Brouwer et al. (2010), Culpepper and Anderson (2004), and Chen and Culpepper (2006) the ratios are high, due to the almost planar structure of the stages, which are able to perform 6 DoF motion. But the largest ratio is reached by a spatial structure (Yun and Li, 2010). Comparing the ratios between rotations (in mrad) and size (in mm) (WRθx·θy·θz/Sx·y·z) shows high ratios in Brouwer et al. (2010) and Chen and Culpepper (2006). This is also due to their planar structure. Remarkably, the ratio of the planar stage in Culpepper and Anderson (2004) is not as high as

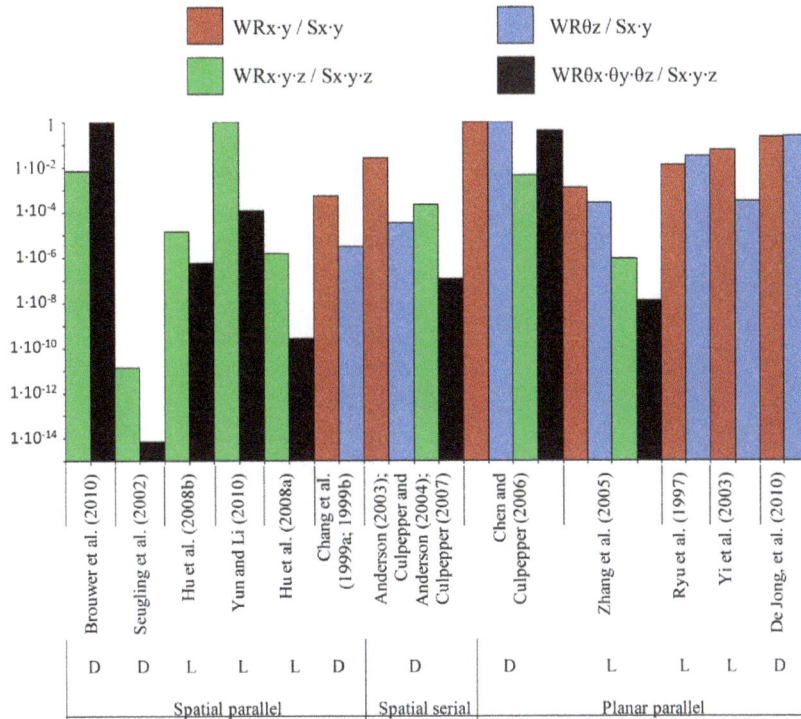

Figure 7. The ratios between translation, or rotation, and size for each compliant stage, if data was available. The ratios were normalized to the largest in the group, shown in logarithmic scale. The mechanisms use distributed compliance (D) or lumped compliance (L).

expected. Also in spatial structures there is no clear relation between working range and flexure type.

In theory, flexures with distributed compliance have a larger range of motion than flexures with lumped compliance. But also lumped compliant flexures were designed such that the complete stage had a large range of motion, using amplifiers in the stage (e.g. the legs in the spatial stages or the 3-revolute-revolute-revolute structure in planar stages act as amplifiers). Most of the stages with lumped compliance are based on these kinds of structures.

In many designs the groups of flexures are arranged 120° of each other. With a minimum of three equally distributed compliant structures, it is possible to create both translation and rotation of the whole stage, using only translation actuation. In other words, with minimal three 1 DoF compliant structures it is possible to create a 3 DoF stage. Due to this arrangement many stages were highly symmetric. This is to decrease the effect of the temperature gradient on accuracy of the design (Ryu et al., 1997).

From the results it appears that most of the 6 DoF spatial compliant structures are non-monolithic. Some 3 DoF planar structures are promising when implemented in a 6 DoF stage.

All the SBCMs, except one, have distributed compliance and use elastic elements to balance strain energy in the mechanism. The elastic elements (springs and compliant flexures) have been preloaded to store the strain energy, creating zero stiffness. However, pre-stressing of the elastic elements is

a challenge and gives difficulties in the design of statically balanced monolithic structures.

For further illustration, the ratios of the statically balanced stroke and compensated force relative to the size of the balancing mechanism is shown in Fig. 8. The compliant micro mechanisms (category 3 of SBCMs) have the largest ratios for statically balanced stroke relative to the size, while this ratio for compensated force relative to the size is still above the average of the other works. The largest ratio for compensated force relative to the size of the balancing mechanism is again for the category 3 of SBCMs. It may be concluded that a balancing mechanism based on buckling plates have great advantages to compensate relative large forces in a relative large stroke, compared to the size. The design with the non-compliant balancing mechanism (category 1 of SBCMs) has the smallest ratio for balanced stroke relative to the size. The preloaded plates shows less efficiency in terms of compensated force and balanced stroke relative to the average, however in all above case, further research is needed as only a few designs were available.

There are few examples of 6 DoF SBMs, but these are all spatial structures, which could be modeled with lumped compliance, balancing gravity forces. No example is available for SBCMs with lumped compliance. Combining SBCMs with lumped compliance, or redesigning an existing 6 DoF SBM, using distributed compliance and balancing strain energy, needs further research and will probably results in a complete new stage design.

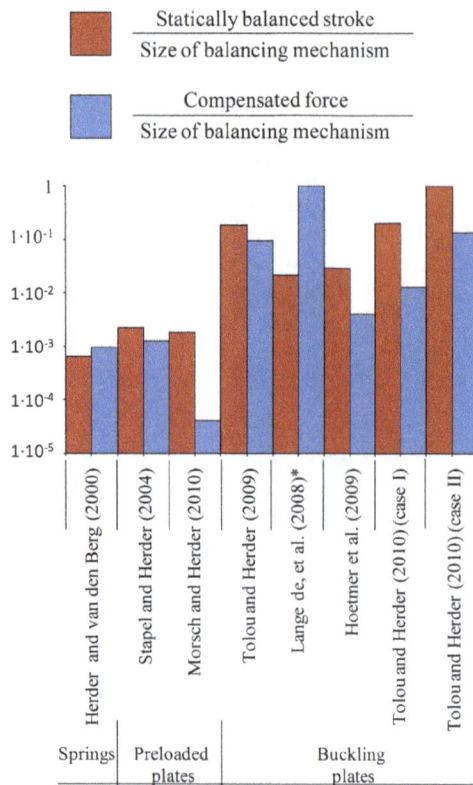

Figure 8. The ratios of statically balanced stroke and compensated force relative to the size of the balancing mechanism. Note that the ratios were normalized to the largest in the group and shown in logarithmic scale. The balancing mechanism used springs, preloaded plates or buckling plates to balance the mechanism.
* This design has an exceptionally high compensated force, but was never fabricated due to calculated high stresses.

5 Conclusions

An overview of existing compliant stages, combining translations and rotations (3–6 DoF), classification and discussion, comparing the ratios between translations, rotations and the size, has been made towards the design of 6 DOF statically balanced compliant stage.

It was found that different types of flexures are used in the planar and spatial stages. From the results there is no clear relation between the range of motion and the type of flexure. Where distributed compliance should have a larger range of motion, the lumped compliance stages use different kind of amplifiers to create a large range of motion. Consequently, it can be concluded that effectively each architecture for 6 DoF compliant stages performed equally well.

Different balancing strategies have been studied, as well as the possibilities to combine 6 DoF compliant stages with static balancing.

The compliant balancing mechanisms using buckling plates (either in micro- or mesoscale) shows the better per-

formance in terms of force compensation and stroke of static balancing relative to the size of the balancing mechanism.

It is shown that no 6 DoF statically balanced compliant stage is readily available. The existing statically balanced compliant mechanisms have 1 DoF, use pre-stressed elastic elements as balancing mechanism, and have distributed compliance, while all existing non-compliant 6 DoF statically balanced stages can be modeled with lumped compliance. Combining static balancing with a 6 DoF compliant stage needs either a new 6 DoF distributed compliant stage, balanced according to the method for balancing distributed compliance, or a new method to balance a lumped compliant 6 DoF stage.

A promising direction for future research would be to find a strategy to combine a 6 DoF monolithic compliant stage with static balancing.

Acknowledgements. This research is part of VIDI Innovational Research Incentives Scheme grant for the project "Statically balanced compliant mechanisms", NOW-STW 7583.

The authors thank Jerry Peijster for his good revisions on this work.

Edited by: C. Kim

References

Ananthasuresh, G. K. and Kota, S.: Designing compliant mechanisms, Mech. Eng., 117, 93–96, 1995.

Anderson, G. A. B.: A six degree of freedom flexural positioning stage, M.S. thesis, Massachusetts Institute of Technology, Cambridge, USA, 136 pp., 2003.

Brouwer, D. M., de Jong, B. R., and Soemers, H. M. J. R.: Design and modeling of a six DOFs MEMS-based precision manipulator, Precis. Eng., 34, 307–319, 2010.

Chang, S. H., Tseng, C. K., and Chien, H. C.: An ultra-precision XYθz piezo-micropositioner part I: Design and analysis, IEEE T. Ultrason. Ferr., 46, 897–905, 1999a.

Chang, S. H., Tseng, C. K., and Chien, H. C.: An ultra-precision XYθz piezo-micropositioner part II: Experiment and performance, IEEE T. Ultrason. Ferr., 46, 906–912, 1999b.

Chao, D., Zong, G., and Liu, R.: Design of a 6-DOF compliant manipulator based on serial-parallel architecture, in: Proceeding of the 2005 IEEE/ASME International Conference on Advanced Intelligent Mechatronics, Monterey, California, USA, 24–28 July 2005, 765–770, 2005.

Chen, S. C. and Culpepper, M. L.: Design of a six-axis micro-scale nanopositioner-μHexFlex, Precis. Eng., 30, 314–324, 2006.

Choi, K. B. and Lee, J. J.: Passive compliant wafer stage for single-step nano-imprint lithography, Rev. Sci. Instrum., 76, 075106, doi:10.1063/1.1948401, 2005.

Culpepper, M. L.: Multiple degree of freedom compliant mechanism, patent US2006252297, 2006.

Culpepper, M. L. and Anderson, G.: Design of a low-cost nanomanipulator which utilizes a monolithic, spatial compliant mechanism, Precis. Eng., 28, 469–482, 2004.

Review Article: Inventory of platforms towards the design of a statically balanced six degrees of freedom...

121

Culpepper, M. L. and Golda, D.: 6-Axis electromagnetically-actuated meso-scale nanopositioner, patent US2007220882, 2007.

Ebert-Uphoff, I. and Johnson, K.: Practical considerations for the static balancing of mechanisms of parallel architecture, P. I. Mech. Eng. K.-J. Mul., 216, 73–85, 2002.

Ebert-Uphoff, I., Gosselin, C. M., and Laliberté, T.: Static balancing of spatial parallel platform mechanisms – revisited, Mech. Des.-T. ASME, 122, 43–51, 2000.

Eijk van, J. A. and Dijksman, J. F.: Plate spring mechanism with constant negative stiffness, Mech. Mach. Theory, 14, 1–9, 1979.

Espacenet: http://www.espacenet.com/, last acces: 25 January 2011.

Gallego, J. A. and Herder, J. L.: Synthesis methods in compliant mechanisms: an overview, in: Proceedings of the ASME 2009 International Design Engineering Technical Conferences & Computers and Information in Engineering Conference, San Diego, California, USA, 30 August–2 September 2009, DETC2009-86845, 2009.

Gao, P. and Swei, S. M.: Six-degree-of-freedom micro-manipulator based on piezoelectric translators, Nanotechnology, 10, 447–452, 1999.

Gosselin, C. M. and Wang, J.: Static balancing of spatial six-degree-of-freedom parallel mechanisms with revolute actuators, J. Robotic Syst., 17, 159–170, 2000.

Helmer, P., Mabillard, Y., Clavel, R., and Bottinelli, S.: High precision apparatus for imposing or measuring a position or a force, patent US20040255696, 2004.

Herder, J. L.: Energy-free systems: theory, conception, and design of statically balanced spring mechanisms, Ph.D. thesis, Delft University of Technology, Delft, The Netherlands, 248 pp., 2001.

Herder, J. L. and van den Berg, F. P. A.: Statically balanced compliant mechanisms (SBCM's), an example and prospects, in: Proceedings of ASME 2000 Design Engineering Technical Conferences and Computers and Information in Engineering Conference, Baltimore, Maryland, USA, 10–13 September 2000, DETC2000/MECH-14144, 2000.

Hoetmer, K., Herder, J. L., and Kim, C. J.: A building block approach for the design of statically balanced compliant mechanisms, in: Proceedings of ASME 2009 International Design Engineering Technical Conferences & Computers and Information in Engineering Conference, San Diego, California, USA, 30 August–2 September 2009, DETC2009-87451, 2009.

Howell, L. L.: Compliant mechanisms, John Wiley & Sons, New York USA, 459 pp., 2001.

Hu, K., Kim, J. H., Schmiedeler, J., and Menq, C. H.: Design, implementation, and control of a six-axis compliant stage, Rev. Sci. Instrum., 79, 025105, doi:10.1063/1.2841804, 2008a.

Hu, Y. H., Lin, K. H., Chang, S. C., and Chang, M.: Design of a compliant micromechanism for optical-fiber alignment, Key Eng. Mat., 381–382, 141–144, 2008b.

Jong de, B. R., Brouwer, D. M., de Boer, M. J., Jansen, H. V., Soemers, H. M. J. R., and Krijnen, G. J. M.: Design and fabrication of a planar three-DOFs MEMS-based manipulator, J. Microelectromech. S., 19, 1116–1130, 2010.

Lange de, D. J. B. A., Langelaar, M., and Herder, J. L.: Towards the design of a statically balanced compliant laparoscopic grasper using topology optimization, in: Proceedings of the ASME 2008 International Design Engineering Technical Con-

ferences & Computers and Information in Engineering Conference, Brooklyn, New York, USA, 3–6 August 2008, DETC2008-49794, 2008.

Leblond, M. and Gosselin, C. M.: Static balancing of spatial and planar parallel manipulators with prismatic actuators, in: Proceedings of the 1998 ASME Design Engineering Technical Conferences, Atlanta, Georgia, USA, 13–16 September 1998, DETC98/MECH-5963, 1998.

Lee, C. W. and Kim, S. W.: An ultraprecision stage for alignment of wafers in advanced microlithography, Precis. Eng., 21, 113–122, 1997.

Liang, Q., Zhang, D., Chi, Z., Song, Q., Ge, Y., and Ge, Y.: Six-DOF micro-manipulator based on compliant parallel mechanism with integrated force sensor, Robot. CIM-Int. Manuf., 27, 124–134, 2011.

Liu, X. J., Wang, J., Gao, F., and Wang, L. P.: On the design of 6-DOF parallel micro-motion manipulators, in: Proceedings of the 2001 IEEE/RSJ International Conference on Intelligent Robots and Systems, Maul, Hawaii, USA, 29 October–3 November 2001, 343–348, 2001.

Lobontiu, N.: Compliant mechanisms: design of flexure hinges, CRC Press LLC, Boca Raton, Florida, USA, 447 pp., 2003.

Lu, T. F., Handley, D. C., Yong, Y. K., and Eales, C.: A three-DOF compliant micromotion stage with flexure hinges, Ind. Robot, 31, 355–361, 2004.

Moon, Y. M. and Kota, S.: Design of compliant parallel kinematic machines, in: Proceedings of the ASME 2002 Design Engineering Technical Conferences and Computer and Information in Engineering Conference, Montreal, Canada, 29 September–2 October 2002, DETC2002/MECH-34204, 2002.

Morsch, F. M. and Herder, J. L.: Design of a generic zero stiffness compliant joint, in: Proceedings of the ASME 2010 International Design Engineering Technical Conferences & Computers and Information in Engineering Conference, Montreal, Quebec, Canada, 15–18 August 2010, DETC2010-28351, 2010.

Nieuwenhuis, C.: Thermal behavior Mapper Short Stroke, restricted internal report, Demcon Advanced Mechatronics B.V., Oldenzaal, The Netherlands, 16 pp., 2010.

Park, S. R. and Yang, S. H.: A mathematical approach for analyzing ultra precision positioning system with compliant mechanism, J. Mater. Process. Tech., 164–165, 1584–1589, 2005.

Powell, K. M. and Frecker, M. I.: Method for optimization of a nonlinear static balance mechanism, with application to ophthalmic surgical forceps, in: Proceedings of the ASME 2005 International Design Engineering Technical Conferences & Computers and Information in Engineering Conference, Long Beach, California, USA, 24–28 September 2005, DETC2005-84759, 2005.

Ryu, J. W., Gweon, D. G., and Moon, K. S.: Optimal design of a flexure hinge based $Xy\theta$ wafer stage, Precis. Eng., 21, 18–28, 1997.

Scopus: http://www.scopus.com/, last acces: 25 January 2011.

Seugling, R. M., Lebrun, T., Smith, S. T., and Howard, L. P.: A six-degree-of-freedom precision motion stage, Rev. Sci. Instrum., 73, 2462–2468, 2002.

Shekarforoush, S. M. M., Eghtesad, M., and Farid, M.: Design of statically balanced six-degree-of-freedom parallel mechanisms based on tensegrity system, in: Proceedings of the ASME 2009 International Mechanical Engineering Congress & Exposition, Lake Buena Vista, Florida, USA, 13–19 November 2009,

IMECE2009-12625, 2010.

Sjoerdsma, W., Herder, J. L., Horward, M. J., Jansen, A., Bannenberg, J. J. G., and Grimbergen, C. A.: Force transmission of laparoscopic grasping instruments, Minim. Invasiv. Ther., 6, 274–278, 1997.

Stapel, A. and Herder, J. L.: Feasibility study of a fully compliant statically balanced laparoscopic grasper, in: Proceedings of the ASME 2004 Design Engineering Technical Conferences and Computers and Information in Engineering Conference, Salt Lake City, Utah, USA, 28 September–2 October, DETC2004-57242, 2004.

Stewart, D.: A platform with six degrees of freedom, Proceedings of the Institution of Mechanical Engineers, 180, 371–386, 1965.

Streit, D. A.: Spatial manipulator and six-degree-of-freedom platform spring equilibrator theory, in: Proceedings of the Second National Conference on Applied Mechanisms and Robotics, 1991.

Sun, L., Wang, J., and Wang, Z.: Six-freedom precision paralleled robot, patent CN2576434, 2003.

Sun, S.: Research of nanometer positioning stage with six degree of freedom based on binary actuation principle, in: Proceeding of MircoNanoChina07, Sanya, Hainan, China, 10–13 January 2007, 1639–1647, 2007.

Tian, Y., Shirinzadeh, B., and Zhang, D.: Design and dynamics of a 3-DOF flexure-based parallel mechanism for micro/nano manipulation, Microelectron. Eng., 87, 230–241, 2010.

Tolou, N. and Herder, J. L.: Concept and modeling of a statically balanced compliant laparoscopic grasper, in: Proceedings of ASME 2009 International Design Engineering Technical Conferences & Computers and Information in Engineering Conference, San Diego, California, USA, 30 August–2 September 2009, DETC2009-86694, 2009.

Tolou, N. and Herder, J. L.: Statically balanced compliant micro mechanisms (SM-MEMS): concepts and simulation, in: Proceedings of the ASME 2010 International Design Engineering Technical Conferences & Computers and Information in Engineering Conference, Montreal, Quebec, Canada, 15–18 August 2010, DETC2010-28406, 2010.

Trease, B. Dede, E.: Statically-balanced compliant four-bar mechanism for gravity compensation, in: 2004 ASME Student Mechanism Design Competition, 2004.

Wang, H. and Zhang, X.: Input coupling analysis and optimal design of a 3-DOF compliant micro-positioning stage, Mech. Mach. Theory, 43, 400–410, 2008.

Wang, L., Rong, W., Sun, L., and Jiao, J.: Analysis and design of a three-limb six degree-of-freedom parallel micromanipulator with flexure hinges, in: Proceeding of MircoNanoChina07, Sanya, Hainan, China, 10–13 January 2007, 1561–1566, 2007.

Wang, S. C., Hikita, H., Kubo, H., Zhao, Y. S., Huang, Z., and Ifukube, T.: Kinematics and dynamics of a 6 degree-of-freedom fully parallel manipulator with elastic joints, Mech. Mach. Theory, 38, 439–461, 2003.

Wang, Y., Liu, Z., Bo, F., and Zhu, J.: Design and research of 5-DOF Integrated nanopositioning stage, Zhongguo Jixie Gongcheng/China Mechanical Engineering, 16, 1317–1321, 2005.

Willson, C. G. and Roman, B. J.: The future of lithography: Sematech litho forum 2008, ACS Nano, 2, 1323–1328, 2008.

Xiaohui, J., Yanling, T., and Dawei, Z.: Six-freedom-degree precision positioning table for nano-imprint lithography system, patent CN101726997, 2010.

Xuchu, J. and Qianfeng, Q.: Precise positioning platform with six freedom of motion, patent CN101488371, 2009.

Yi, B. J., Chung, G. B., Na, H. Y., Kim, W. K., and Suh, I. H.: Design and experiment of a 3-DOF parallel micromechanism utilizing flexure hinges, IEEE T. Robotic. Autom., 19, 604–612, 2003.

Yun, Y. and Li, Y.: Design and analysis of a novel 6-DOF redundant actuated parallel robot with compliant hinges for high precision positioning, Nonlinear Dynam., 61, 829–845, 2010.

Zhang, D. Y., Ono, T., and Esashi, M.: Piezoactuator-integrated monolithic microstage with six degrees of freedom, Sensor. Actuat. A-Phys., 122, 301–306, 2005.

Fabrication of compliant mechanisms on the mesoscale

G. R. Hayes, M. I. Frecker, and J. H. Adair

The Pennsylvania State University, University Park, PA, 16802, USA

Abstract. The fabrication of compliant mechanisms on the mesoscale requires collaboration of mechanical engineering design, with materials science and engineering fabrication approaches. In this paper, a review of current fabrication approaches to produce mesoscale devices is given, highlighting the benefits and limitations of each technique. Additionally, a hierarchy is provided, eliminating fabrication techniques that do not completely satisfy the mechanical design requirements of the compliant mechanisms. Furthermore, the lost mold-rapid infiltration forming process (LM-RIF) is described, and compared to existing fabrication approaches. Finally, prototype mesoscale compliant mechanisms are fabricated, demonstrating the versatility of the LM-RIF process to produce both metal and ceramic devices, as well as ability of a fabrication process to work in collaboration with mechanical design.

1 Introduction

As engineering applications become increasingly complex, the need for collaboration between mechanical engineering design and materials science engineering becomes increasingly apparent. Just as advances in mechanical design have motivated materials scientists to develop new materials with tailored properties, breakthroughs in materials science have, in turn, motivated mechanical engineers to design new and improve existing devices. This exchange of engineering knowledge can be found in the development of almost every present day device and component, ranging from large scale applications such as composite materials used in the automotive and aerospace industries, to small scale applications such as microelectromechanical systems.

In this paper, a collaborative effort between mechanical engineering and materials science engineering is utilized to fabricate and further the development of two mesoscale compliant mechanisms: (1) a compliant forceps for minimally invasive surgery, and (2) a contact-aided compliant cellular mechanism (C3M). In addition, this paper presents the lost mold-rapid infiltration forming (LM-RIF) process, in comparison to other mesoscale fabrication techniques, as a viable mesoscale compliant mechanism fabrication route. This process is based on the initial work of Antolino et al. (2009a, b), to manufacture large arrays of meso-scale devices colloidal science with ultra thick photoresist molding methods.

2 Micro scale fabrication techniques

Many compliant mechanism designs, such as those designed by Aguirre (2011) and Mehta et al. (2010), result in part sizes on the millimeter scale or larger, with feature sizes on the micrometer scale. While these devices show enhanced performance over traditionally designed devices, the fabrication method used to manufacture the compliant mechanisms needs to be considered. To this end, a summary of fabrication approaches applicable to compliant mechanism device manufacturing have been summarized and explained in this section. The mesoscale is defined in Fig. 1 as a component with dimensions on the millimeter scale with feature sizes on the micrometer scale. The part size and feature size of these mesoscale devices falls between traditional large and small scale fabrication approaches. Therefore, mesoscale compliant mechanisms require new or modified fabrication techniques for successful prototype manufacturing. To date, many microfabrication techniques have been explored to create free standing parts on the mesoscale, consisting of top down, and bottom up approaches (Heule et al., 2003). Top-down approaches consist of processes typical to semi-conductor processing in which, for example, a film is deposited via vapor deposition techniques followed by chemical or reactive ion etching. Additionally, small scale machining technologies, such as direct ceramic machining, wire electrical discharge machining (EDM) (Yan et al., 2004), and low-temperature-co-fired-ceramics machining can cut and grind devices from bulk materials, but can only produce a few parts at a time, generally have significant

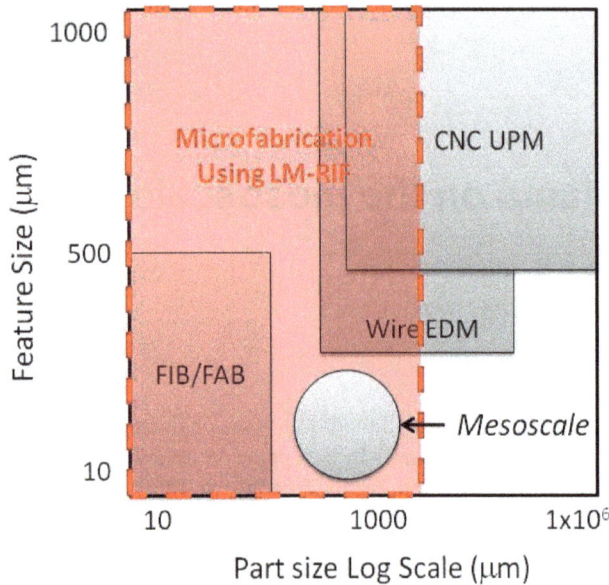

Figure 1. The feature size versus part size comparison are presented for various types of fabrication approaches; focused ion beam and fast atom beam (FIB and FAB), wire electrical discharge machining (Wire EDM), and computer numerical control ultra precision machining (CNC UPM). The lost mold-rapid infiltration forming (LM-RIF) process encompasses the mesoscale while adding the ability to simultaneously manufacture large arrays of parts. The mesoscale lies in the gap in manufacturing size regime above semiconductor fabrication sizes in the sub-10 micron range, and below traditional bulk machining near 300 micron features. Feature size is given on a linear scale, while a logarithmic scale is used to show the wide range in part sizes possible with the fabrication approaches listed.

surface flaws due to the mechanical approach, and yield large quantities of particulate debris (Heule et al., 2003). Machining technologies are also limited by the cutting tool size used to shape the part (Frazier et al., 1995).

Bottom-up techniques consist of the assembly of particulate elements via directed assembly or self-assembly. The additive processes provided by bottom-up approaches are attractive because of a more efficient use of materials and resources, while minimizing manufacturing debris, and avoiding size restrictions due to tool size. Self assembly (Clark et al., 2001) can be used to create arrays of small building blocks, which assemble due to specifically functionalized surfaces, and without the need for external intervention. Alternatively, directed assembly of particulates, via a molding (Muller et al., 2009; Fu et al., 2005), printing (Heule et al., 2003), stereolithography (Heule et al., 2003), or selective laser sintering (Nelson et al., 1995) process can be easily implemented to manufacture devices from a variety of materials available in particulate form. As shown by Bowden and Whitesides (Bowden et al., 1997; Terfort et al., 1997; Xia and Whitesides, 1998), elegant and interesting shapes can be

produced with self assembly strategies. The drawbacks of self assembly include arbitrary shape fabrication and the formation of shapes with a substantial thickness (Klajn et al., 2007). Therefore, in the LM-RIF process, directed assembly methods are utilized to form complex particulate bodies with thicknesses of 20 to 300 microns (Antolino, 2010; Antolino et al., 2009a, b).

For directed assembly approaches, there are two main areas of research and development: direct writing methods, and lithography mold-based methods. Direct writing methods, such as direct ink writing (Lewis et al., 2006) of ceramic slurries, do not have the edge control and subsequent edge resolution required for surgical instrumentation, and normally can only fabricate one structure at a time, making these processes relatively inefficient and time consuming. Lost mold processes, such as injection molding of polymer molds (Knitter et al., 2001) and filling of photoresist molds (Schonholzer et al., 2000) offer the ability to create free standing parts large enough, with the desired edge resolution, to be viable options for microfabrication. Furthermore, mold fabrication via lithography is one of the least expensive microfabrication techniques (Lawes, 2007). New advances in ultra thick photoresist techniques permit the fabrication of single layer lithographic molds up to 1mm thick, while maintaining good edge resolution (Lin et al., 2002). Additionally, aqueous gel-casting (Janney et al., 1998; Christian and Kenis, 2007), aqueous tape-casting (Hotza and Greil, 1995), and non-aqueous colloidal suspension formulation (Imbaby and Jiang, 2009), have shown the capability to produce uniform green bodies via colloidal slurries that can be cast into the lithographic molds. Finally, filling photoresist molds with a particulate-based suspension opens the possibilities of processing a wide range of materials including metals, ceramics, composite materials, and multilayer laminated materials. Manufacturing methods for micro components are summarized by Table 1 in terms of smallest feature resolution, aspect ratio, multi-material system capability, and array manufacturing capability. Furthermore, the basis for the technique in terms of lithography, injection molding, machining, printing, and laser forming is used to classify the manufacturing method. To manufacture mesoscale devices, a novel microfabrication process, the LM-RIF, has been developed based on a directed assembly, lost mold method. The LM-RIF process is listed under lithography techniques and outlined in black.

As shown in Fig. 2, multiple fabrication techniques relevant to our objective of large scale manufacturing of mesoscale compliant mechanisms were evaluated and are given in Table 1. To determine the best method of fabrication for compliant mechanisms on the mesoscale, multiple criteria were applied to the possible fabrication approaches listed in Table 1. Fabrication approaches that do not satisfy an applied criterion in the hierarchical diagram are eliminated as viable manufacturing approaches. The applied criteria include: the capability for large array fabrication, high aspect

Table 1. Manufacturing methods for micro components are summarized in terms of smallest feature resolution, aspect ratio, multi-material system capability, and array manufacturing capability. Furthermore, the basis for the technique is used to classify the various methods in terms of lithography, injection molding, machining, printing, and laser forming. The developed LM-RIF process is listed under lithography techniques and outlined in black.

Technique Basis	Micro-Fabrication Method	Smallest Feature Resolution (μm)	Aspect Ratio	2-D or 3-D	Multi-material Systems	Array Capa-bilities	Ref.
Lithography Based Techniques	Micro-fabrica MEMS EFAB	10	high	2-D/3-D	No	Yes	Microfabrica (2010)
	Casting suspensions into photo-lithographic masks on silicon	1–5	1–2	2-D	Yes	Yes	Heule et al. (2003)
	LM-RIF	1–10	1–40	2-D/3-D	Yes	Yes	Antolino et al. (2009a, b)
	Soft lithography	1–5	1–3	2-D/3-D	Yes	Yes	Xia and Whitesides (1998)
	Laminated object manufacturing	100	variable	3-D	Yes	No	Tay et al. (2003)
	Low-temperature co-fired ceramic multilayer	25–100	variable	3-D	No	No	Heule et al. (2003)
	LIGA	10–20	10	2-D	Yes	Yes	Heule et al. (2003)
Injection Molding/ Extrusion	Micro injection molding	20–100	high	3-D	Yes	No	Muller et al. (2009)
	Co-extrusion	5–16	high	3-D	Yes	No	Heule et al. (2003)
	Metal embossing	50	high	2-D	No	No	Heule et al. (2003)
Machining	Direct ceramic machining of pre-sintered bodies	50	variable	3-D	No	No	Heule et al. (2003)
	Microwire EDM	70	high	3-D	No	No	Yan et al. (2004)
	STM-tip electro-chemical etching	0.01	–	2-D	No	No	Heule et al. (2003)
	Precision grinding	50	high	3-D	Yes	No	Heule et al. (2003)
	Diamond machining (lathe)	25	5	3-D	Yes	No	Frazier et al. (1995)
	Surface micro-machining	25	variable	2-D	Yes	No	Frazier et al. (1995)
Printing	Screen printing	100	low	2-D	Yes	Yes	Heule et al. (2003)
	Ink-jet printing of suspensions	70	low	2-D	Yes	No	Heule et al. (2003)
	Freeform ink-jet printing of suspensions	170	high	3-D	Yes	No	Heule et al. (2003)
	3-DP process (printing ceramic binders)	200	variable	3-D	Yes	No	Heule et al. (2003)
	Micropen writing	250	1	3-D	Yes	No	Heule et al. (2003)
Laser Forming	Laser chemical vapor deposition	10	500	3-D	Yes	No	Wanke et al. (1997)
	Pulsed laser ablation	30–200	high	2-D	Yes	No	Heule et al. (2003)
	Micro-stereo-lithography	2	high	3-D	Yes	No	Heule et al. (2003)
	Selective laser sintering	500	high	3-D	Yes	No	Nelson et al. (1995)
	Maple direct write	10–20	low	2-D	No	No	Heule et al. (2003)

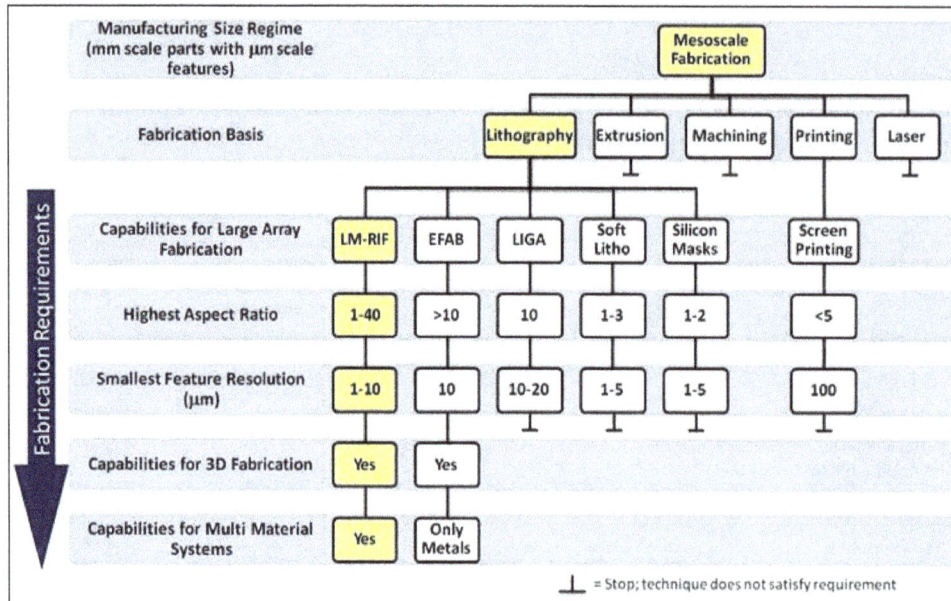

Figure 2. Hierarchy of fabrication techniques used in large array meso-scale manufacturing. Only the LM-RIF process is capable of large array fabrication of 3-D parts with high aspect ratios and small feature resolution, using multiple materials.

ratio in a free standing part, smallest feature resolution, potential for 3-D manufacturing, and capability for multiple material manufacturing. The order of the applied fabrication requirements in Fig. 2 was determined by considering the most desirable fabrication criteria first. To satisfy our objective of large arrays of mesoscale parts manufacturing, the capability for large array manufacturing is the first criterion applied. The remaining criteria were applied in order of importance, as relevant to the prototype devices discussed in Sect. 3.4. It should be noted that changes in the order of the applied requirements in Fig. 2 will result in a change in the ranking of some of the fabrication techniques. However, with all of the fabrication criteria applied, the LM-RIF process will remain the most desirable choice. Meso-scale compliant surgical instruments require components that are 1mm or less in the largest dimension while the feature resolution needs to be at the micron or even sub-micron scale. This requirement eliminates all of the possible microfabrication approaches except the EFAB and LMRIF processes. The EFAB process, invented by Cohen (2002), uses electrochemical deposition of metals combined with lithographic techniques to produce 3-D structures. However, the EFAB process is limited to metallic systems that can be electrochemically deposited. In contrast, the LM-RIF process utilizes nanometer scale ceramic or metallic particulates that in some cases are combined to form ceramic metal composites or ceramic-metal multilayers, permitting a wide range of novel and innovative design strategies. Furthermore, two materials can be combined in a hierarchical fashion, at the particulate scale, in multilayers, and in hybrid material parts in the LM-RIF process.

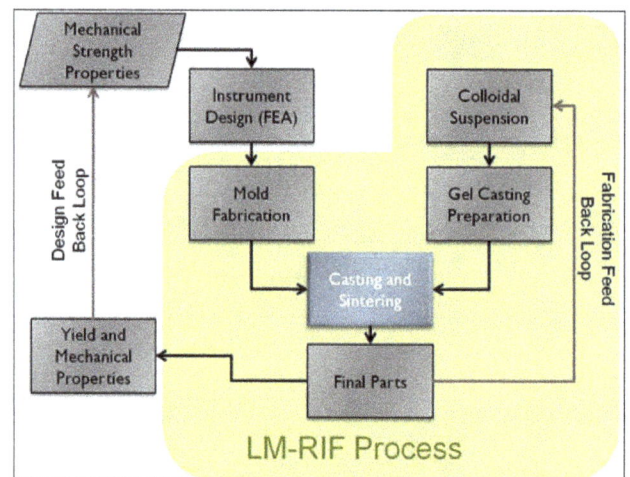

Figure 3. The LM-RIF flowchart. The process starts with a theoretically optimized mechanical design based on initial determination of mechanical properties, followed by mold fabrication, colloidal processing, and final part fabrication. Final parts are tested and characterized, with the mechanical properties used to design changes in the design and fabrication for future generations of materials and/or design components. Changes in the design geometry of the part are completed through the Design Feedback loop, while changes to the manufacturing process and material system are completed through the Fabrication Feedback loop.

3 The Lost Mold-Rapid Infiltration Forming Process (LM-RIF)

The LM-RIF process, illustrated in Fig. 3, consists of an integrated, iterative approach to both improve the mechanical design of the part being manufactured; via the design feedback loop, as well as improve the fabrication process itself; via the fabrication feedback loop that optimizes material mechanical properties. The LM-RIF process and manufacturing approach has been developed over multiple generations to improve both the basic material properties and component geometry. Antolino et al. describes the basic approach to improve material properties based on the manufacture of three mole percent yttria zirconia polycrystalline (3YTZP) mesoscale bend bars that are $15 \times 20 \times 370$ microns in dimension (Antolino et al., 2009a, b). However, in these preliminary reports, neither larger parts that can completely bridge the manufacturing gap in Fig. 1 into the 1 mm regime while maintaining micron scale features nor additional materials that expand the design space were reported. The innovations required to meet these challenges will be highlighted in this process overview section.

As illustrated in Figs. 3 and 4, the process begins with an initial compliant mechanism design based on both size and topology optimization techniques (Mehta, 2010; Aguirre and Frecker, 2007; Mehta et al., 2009; Aguirre, 2011). Once the first generation design is finalized, a lithography-based mold fabrication step is used to translate the design into a two or three dimensional mold. After molds are fabricated, a concentrated colloidal suspension (i.e., 40 to 50 volume percent solids) is formulated using the precursor particulate materials. The colloidal suspension is then cast into the mold via a screen printing squeegee, and solvent is removed by evaporation under carefully controlled conditions to minimize capillary forces to prevent part cracking. Final parts are obtained after a combined mold removal and sintering step. The finished parts are characterized, and appropriate changes can be made to the mechanical design, colloidal suspension parameters, or both to optimize components in subsequent generations. Through this process development, the ability to produce large arrays of parts from both metals and ceramics, as well as parts ranging in thickness from 10 to 400 microns has been demonstrated (Aguirre, 2011).

3.1 Mold fabrication

The optimal compliant mechanism design approach (Aguirre and Frecker, 2007; Mehta, 2010; Mehta et al., 2009, 2010), is used to generate a photomask. In this process, the prototype parts are arranged in a layout to facilitate a high volume of parts fabricated per unit area, while satisfying part proximity constraints. The separation distances among parts, also known as the proximity of parts, on the mask layout is determined via the designed mold thickness, with a 1:1 ratio of mold thickness to inter-part spacing. While each patterned part on the photomask is designed for a particular mold thickness, it is possible to have single photomasks with multiple sections designed for various part thickness.

In the mold fabrication process, polished polycrystalline, high purity (greater than 99.5 weight per cent) alumina substrates (courtesy of Kyocera Corporation or Coorstek) are used as substrates to avoid handling components between processing steps. SU8 (Microchem Corp.) photoresist molds are fabricated on the substrates using a modified UV lithography process. Initially, an antireflective coating of AZ-Barli-II 90 is spin coated onto the substrate to eliminate mold defects created by light scattering from the substrate surface. Secondly, a 10 µm under layer of SU-8 photoresist is spin coated to form the bottom layer of the mold. This under layer assures part separation from the substrate before sintering and acts as a smooth, flat bottom surface for the mold. Finally, a SU-8 layer with the targeted thickness is deposited using a calculated volume technique adapted from Lin et al. (2002) In this process, a known volume of SU-8 photoresist is deposited at 80 °C onto a substrate of specific surface area. The photoresist is prebaked at 120 °C for 4 h, with a temperature ramp of 2 °C per minute, during which the solvent is evaporated from the resist, and self leveling takes place. Next, an initial optical exposure of 3 mJ cm^{-2} micron^{-1} (thickness) is performed. The photoresist then undergoes a post exposure bake at 55 °C for 30 min with a temperature ramp of 2 °C per minute, and finally the resist is developed for 30 min with slight agitation. The mold layer is developed in propyleneglycol monomethyletheracetate (PGMEA, SU-8 Developer, Microchem Corp.). Following development, one of two additional mold manufacturing paths can be taken. For a single layer (i.e., two dimensional molds and subsequent components), a second flood exposure of approximately 4200 mJ cm^{-2} in concert with an additional heat treatment at 180 °C for 20 min fully crosslinks the resist. Alternatively following development, two or more mold layers can be laminated at 100 °C with slight pressure (0.01 MPa). The process of stacking and laminating multiple mold layers can be used to create three dimensional mold cavities and, as a consequence, more complex, three dimensional components. During the exposure steps, a UV light filter (Omega Optical, PL-360-LP) ensures vertical side walls in the final mold (del Campo and Greiner, 2007). Figure 4a shows the UV lithography layering sequence, as well as the steps going from design, to mask, to final mold cavity. In Fig. 4a, a cross sectional view of mold cavities is shown with varying length and width. It is noted that as mold thickness increases, minimum feature size increases. The minimum, stand alone, single feature size for the parts fabricated herein was taken to be approximately 1/15 of the mold thickness, while inherent features of larger part geometries can be as small as 2–3 µm (Yang and Wang, 2005).

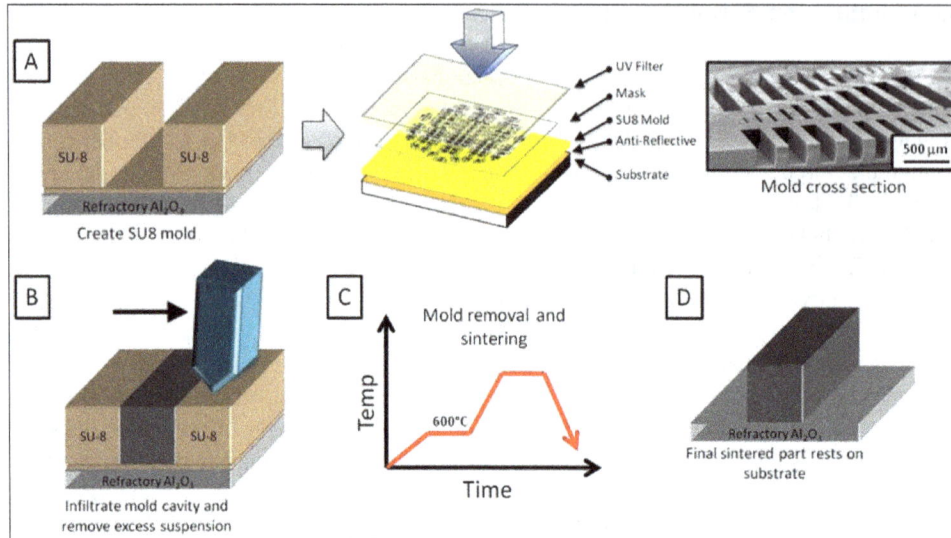

Figure 4. The LM-RIF process steps of mold fabrication, infiltration, mold and binder removal and sintering are depicted. (**A**) Molds are fabricated on high purity refractory substrates via a modified lithography technique. The lithography stack is shown, along with a cross sectional scanning electron micrograph of a sample mold. (**B**) The fabricated molds are then infiltrated with a high solids loading colloidal suspension via a screen printing squeegee and allowed to dry. (**C**) The mold and binder in the colloidal suspensions is removed at 600 °C, followed by sintering, forming a dense final part. The specific furnace cycle is as follows: In ambient atmosphere, ramp 2 °C min^{-1} to 600 °C and hold 2 h. In an atmosphere of 5 vol% H_2 and 95 vol% N_2, ramp 5 °C min^{-1} to 1300 °C, hold 2 h, and cool at 10 °C min^{-1}. (**D**) freestanding parts are left on the original substrate.

3.2   Colloidal suspension formulation

Well dispersed, high solid loading suspensions are required during particulate processing to fabricate dense parts with desired mechanical integrity. The properties of the particulate that is being processed, including particle size distribution, shape, and chemistry, can affect the processing parameters used to create a colloidal suspension. In particular, the high density from uniform particle packing of well-dispersed particulate has a positive influence on the sintering of the particulate body (Reed, 1995). Likewise, poorly packed particulates from poorly dispersed, agglomerated particulates results in poorly sintered materials with trapped porosity, grain growth and other characteristics that compromise mechanical properties (Reed, 1995). Thus, the particulate processing characteristics ultimately influence the final mechanical properties of components. If the processing parameters create colloidal suspensions with solids content too low, agglomerated particulates or other particulate created defects, the packing density in green state is compromised, and sintering to high density, to achieve mechanically strong parts, will not take place. Conversely, desired mechanical properties dictate processing parameters used to create suitable colloidal suspensions from particulates. If the desired mechanical strength is relatively low, and/or a porous final body is required for a particular application, the processing parameters can be modified to fit these requirements. The topology of the compliant mechanism design also affects the processing parameters, as intricate designs may require special pro-

cessing parameters. Finally, whether mechanical properties are dictated through design, or measured through experimentation, the design will need to be modified to fit within the given system. Principally, as well as in the context of the LM-RIF process, powder processing parameters, mechanical properties, mechanical design, and initial powder properties are interrelated aspects of powder processing. Variation in each of these attributes directly influences the others, and thus, when working to improve the system as a whole, the impact of changes in one attribute must be evaluated for all other aspects. For example, properties of the particulate material being incorporated into the LM-RIF process, such as particle size, particle shape, agglomeration characteristics, density, and chemical stability, play a direct role in the processing parameters utilized, such as solvent environment, colloidal dispersion scheme, drying method, binder removal, and sintering. Furthermore, the properties of particulates, both alone, and in conjunction with the processing parameters have direct impacts on mechanical properties and mechanical design of fabricated devices. Additionally, compliant mechanism design can dictate a specific material property, such as biocompatibility, high elastic modulus, or even desired feature size. These specifications influence the starting material powder material type and particle size, as well as desired mechanical properties, and can even constrain processing parameters. The interaction of some processing parameters, such as sintering temperature, on mechanical properties, has been described by Antolino (2010). Therefore, it

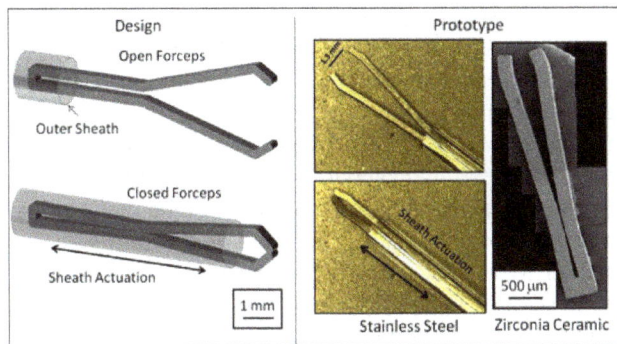

Figure 5. (Left) The design of the compliant forceps device is shown. The forceps is actuated via a sheath, a grasping action is obtained via moving the sheath forward, forcing the arms toward one another. (Right) Stainless steel prototype forceps are shown in an optical image, as fabricated by the LM-RIF process. The actuation of the forceps is shown in both the open and closed positions. Additionally, zirconia ceramic forceps are shown in a scanning electron micrograph, as fabricated by the LM-RIF process. The stainless steel and zirconia forceps demonstrate the ability of the LM-RIF process to fabricate prototype compliant mechanisms from multiple materials.

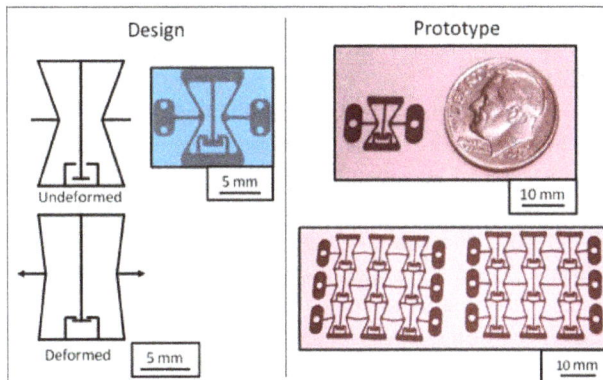

Figure 6. (Left) The design of contact-aided compliant cellular mechanisms is shown in both the undeformed and deformed cases. As the mechanism is deformed, contact occurs in the dash-pot structure, resulting in a distribution of stresses throughout the device, allowing further elastic deformation. The final design is shown in blue. (Right) Optical images of prototype designs of the C3M devices, as fabricated by the LM-RIF process are shown for both a single unit cell, and a 3×3 cellular array. A US Dime is shown for scale reference. The stainless steel parts demonstrate the ability of the LM-RIF process to fabricate mesoscale compliant mechanism devices.

is clear that the system of interactive particle processing must be considered as a whole, without eliminating the impact of one area on another.

3.3   Mold removal and sintering

During the infiltration process, an excess of suspension is placed on top of the mold and worked into the mold cavities with a squeegee while simultaneously removing any bubbles. Following infiltration and drying, the green parts are sintered in a two stage process. Initially, the mold is removed by thermolysis in the ambient, air atmosphere at 600 °C. After demolding takes place, the free standing parts are sintered in an appropriate atmosphere and temperature and time. The specific furnace cycle is as follows: mold removal in ambient atmosphere, ramp 2 °C min^{-1} to 600 °C and hold 2 h. Sintering in an atmosphere of 5 vol% H$_2$ and 95 vol% N$_2$, ramp 5 °C min^{-1} to 1300 °C, hold 2 h, and cool at 10 °C min^{-1}. Figure 4b–d depicts the casting, burnout and sintering steps leaving final parts freestanding on the original substrate.

Following fabrication, free standing parts can be evaluated for mechanical properties, as well as device functionality. As shown in Fig. 3, changes to the design, as well as changes to the colloidal processing formulations can be made through the appropriate feedback loops, permitting fine tuning of the LM-RIF process.

3.4   Fabrication of prototypes

Prototype mesoscale devices, consisting of a surgical forceps and contact aided compliant cellular mechanisms, were suc-

cessfully fabricated utilizing the LM-RIF process. Example prototype compliant forceps devices are shown in Fig. 5. The forceps are dimensionally on the mesoscale due the centimeter length scale of the forceps arms, with micrometer length scale of the gap between the forceps arms. During actuation, a sheath moves over the forceps arms, forcing them together in a grasping action. The arms come into contact with one another during grasping, distributing stresses within the arms and allowing further elastic deformation. The prototyped devices were tested in concurrent studies by Aguirre (Aguirre, 2011; Aguirre et al., 2011) and Addis (2010). Performance testing was carried out by experimentally and theoretically correlating the tip deflection of the forceps as a function of the displacement of the sheath. It was found that the finite element analysis is able to accurately predict the onset of plastic deformation of the forceps during actuation. Good agreement between the theoretical and experimental results verified the device's performance and the design and manufacturing procedure. Secondly, a pre-clinical tool assessment procedure was conducted by Addis (2010) at Penn State Hershey Medical Center (Hershey, PA, USA), which compared the performance of the prototypes against a commercially available product (Boston Scientific 1mm diameter Spybite® biopsy forceps). It was found that the prototype instrument was preferred over the standard instrument in terms of the ability to control intermediate positions between the open and closed positions of the jaws, and the prototype instrument's ability to grasp firmly is superior to that of the standard instrument. Positive feedback validated the surgical relevance of the device and provided valuable insight for improving the next generation of prototypes (Addis et al., 2011).

Example contact-aided compliant cellular mechanisms are shown in Fig. 6. These cellular mechanisms are designed to maximize horizontal elastic strain, while maintaining high stiffness. When actuated in the horizontal direction, the auxetic cell deforms, allowing for large elastic strain. Furthermore, during deformation, the dash-pot mechanism comes into contact, distributing stresses in the walls of the C3M device, allowing further elastic deformation. These C3M devices fall into the mesoscale due to the centimeter scale of the unit cell, millimeter size scale of the length of the oblique walls, and the micrometer size scale of the dash-pot contact mechanism. In a concurrent study by Mehta et al. (2009), the force and global elastic strain of the C3M devices were experimentally determined by conducting a force-deflection analysis using a force gauge actuated by a micrometer. It was found that the elastic modulus of the meso-scale stainless steel C3M parts fabricated using LM-RIF process is between 70 to 150 GPa, and that the global strain is sensitive to the size and quality of the contact gap. Good agreement was found between the theoretical and experimental global elastic strain of the C3M devices fabricated with the LM-RIF process.

Truly mesoscale parts can be fabricated using the LM-RIF process. Additionally, while just a few examples of metallic and ceramic parts are shown, these parts were manufactured in large arrays of similar parts, demonstrating the manufacturing capability of the LM-RIF process.

4 Conclusions

The collaboration between mechanical design and materials science fabrication has been described within the context of two mesoscale devices: a surgical forceps instrument, and C3M device. The possible fabrication methods for these devices have been described, listing the benefits and drawbacks of each technique. In addition to existing techniques, a new fabrication technique, the LM-RIF, was introduced and included in the comparison. Furthermore, a hierarchy was developed to easily choose the fabrication technique most applicable to the devices in question. The LM-RIF fabrication process was described that can; (1) fabricate large arrays of compliant mechanisms; and (2) be complementary to particulate based material systems. Furthermore, using this manufacturing technique for both surgical instrument and C3M device design is attractive because free standing parts are fabricated with the desired large aspect ratios while retaining good resolution on the micron scale stemming from the lithographic based molds and colloidal infiltration processes. Both zirconia ceramic and stainless steel components were manufactured with the LM-RIF process to emphasize the range of materials possible with the fabrication approach.

In concurrent studies, prototype surgical forceps devices and C3M devices were mechanically evaluated and good agreement was found between the experimental results and calculated performance.

Future extensions and improvements to the LM-RIF process, and the supporting materials science research fall into the following categories: (1) Manufacturing of multilayer, or three dimensional structures. To date, multi-layering using the described mold fabrication technique in Sect. 3.1 has been demonstrated with a 2 layer mold. Future work involves utilizing the multilayer molds to manufacture multilayer devices. (2) Improving device performance through the incorporation of two or more materials in one device layer. (3) Additional testing of the mechanical properties of the metal, ceramic, and composite components via theta-test specimens and tensile test specimens.

Acknowledgements. The work in this manuscript is partially funded under NSF STTR 0637850, NSF 0900368, and NSF 0437214, the NSF I/UCRC Ceramic and Composite Materials Center. This work was also supported by the Pennsylvania State University Materials Research Institute Nanofabrication Network and the National Science Foundation Cooperative Agreement No. 0335765, National Nanotechnology Infrastructure Network, with Cornell University. Any opinions, findings, and conclusions or recommendations expressed in this publication are those of the author(s) and do not necessarily reflect the views of Cornell University nor those of the National Science Foundation. The authors also gratefully acknowledge the partial support provided by grant number R21EB006488 from the National Institute of Biomedical Imaging And Bioengineering. The content in this paper is solely the responsibility of the authors and does not necessarily represent the official views of the National Institute of Biomedical Imaging And Bioengineering or the National Institutes of Health.

Edited by: J. Andrés Gallego Sánchez

References

Addis, M.: Evaluation of Surgical Instruments for Use in Minimally Invasive Surgery, B.S. Honor Thesis, Department of Mechanical and Nuclear Engineering, The Pennsylvania State University, University Park, PA, 2010.

Addis, M., Aguirre, M., Frecker, M., Haluck, R. S., Mathew, A., Pauli, E. M., and Gopal, J.: Development of Tasks and Evaluation of a Prototype Forceps for NOTES, JSLS-J. Soc. Laparoend., under review, 2011.

Aguirre, M.: Design and Optimization of Narrow-Gauge Contact-Aided Compliant Mechanisms for Advanced Minimally Invasive Surgery, Doctor of Philosophy Mechanical Engineering, The Pennsylvania State University, University Park, 2011.

Aguirre, M. and Frecker, M.: Size and Shape Optimization of a 1.0 mm Multifunctional Forceps-Scissors Surgical Instrument, ASME Journal of Medical Devices, 2, 015001-015001-015001-015007, 2007.

Aguirre, M. E., Hayes, G., Meirom, R., Frecker, M., Muhlstein, C., Adair, J. H., and Kerr, J. A.: The Design and Fabrication of

Narrow-Gauge (1 mm Diameter) Surgical Instruments for Natural Orifice Translumenal Endoscopic Surgery, J. Mech. Design, accepted, 2011.

Antolino, N. E.: Lost Mold-Rapid Infiltration Forming: Strength Control in Mesoscale 3Y-TZP Ceramics, Doctor of Philosophy, Materials Science and Engineering, The Pennsylvania State University, University Park, 2010.

Antolino, N. E., Hayes, G., Kirkpatrick, R., Muhlstein, C. L., Frecker, M. I., Mockensturm, E. M., and Adair, J. H.: Lost Mold Rapid Infiltration Forming of Mesoscale Ceramics: Part 1, Fabrication, J. Am. Ceram. Soc., 92, S63–S69, doi:10.1111/j.1551-2916.2008.02627.x, 2009a.

Antolino, N. E., Hayes, G., Kirkpatrick, R., Muhlstein, C. L., Frecker, M. I., Mockensturm, E. M., and Adair, J. H.: Lost Mold-Rapid Infiltration Forming of Mesoscale Ceramics: Part 2, Geometry and Strength Improvements, J. Am. Ceram. Soc., 92, S70–S78, doi:10.1111/j.1551-2916.2008.02719.x, 2009b.

Bowden, N., Terfort, A., Carbeck, J., and Whitesides, G. M.: Self-assembly of mesoscale objects into ordered two-dimensional arrays, Science, 276, 233–235, 1997.

Christian and Kenis, P. J. A.: Fabrication of ceramic microscale structures, J. Am. Ceram. Soc., 90, 2779–2783, doi:10.1111/j.1551-2916.2007.01840.x, 2007.

Clark, T. D., Tien, J., Duffy, D. C., Paul, K. E., and Whitesides, G. M.: Self-assembly of 10-mu m-sized objects into ordered three-dimensional arrays, J. Am. Chem. Soc., 123, 7677–7682, 2001.

Cohen, A. L.: Method for Electrochemical Fabrication, USA Patent, US 6 475 369 B1, 2002.

del Campo, A. and Greiner, C.: SU-8: a photoresist for high-aspect-ratio and 3D submicron lithography, J. Micromech. Microeng., 17, R81–R95, doi:10.1088/0960-1317/17/6/r01, 2007.

Frazier, A. B., Warrington, R. O., and Friedrich, C.: The Miniaturization Technologies – Past, Present, and Future, IEEE Trans. Ind. Electron., 42, 423–430, 1995.

Fu, G., Loh, N. H., Tor, S. B., Tay, B. Y., Murakoshi, Y., and Maeda, R.: Injection molding, debinding and sintering of 316L stainless steel microstructures, Appl. Phys. A-Mater., 81, 495–500, doi:10.1007/s00339-005-3273-6, 2005.

Heule, M., Vuillemin, S., and Gauckler, L. J.: Powder-based ceramic meso- and microscale fabrication processes, Adv. Mater., 15, 1237–1245, 2003.

Hotza, D. and Greil, P.: Aqueous tape casting of ceramic powders, Mater. Sci. Eng. A-Struct. Mater. Prop. Microstruct. Process., 202, 206–217, 1995.

Imbaby, M. F. and Jiang, K.: Stainless steel-titania composite micro gear fabricated by soft moulding and dispersing technique, Microelectron. Eng., 87, 1650–1654, 2009.

Janney, M. A., Omatete, O. O., Walls, C. A., Nunn, S. D., Ogle, R. J., and Westmoreland, G.: Development of low-toxicity gelcasting systems, J. Am. Ceram. Soc., 81, 581–591, 1998.

Klajn, R., Bishop, K. J. M., Fialkowski, M., Paszewski, M., Campbell, C. J., Gray, T. P., and Grzybowski, B. A.: Plastic and moldable metals by self-assembly of sticky nanoparticle aggregates, Science, 316, 261–264, 2007.

Knitter, R., Gohring, D., Risthaus, P., and Hausselt, J.: Microfabrication of ceramic microreactors, Microsyst. Technol., 7, 85–90, 2001.

Lawes, R. A.: Manufacturing costs for microsystems/MEMS using high aspect ratio microfabrication techniques, Microsyst. Technol., 13, 85–95, 2007.

Lewis, J. A., Smay, J. E., Stuecker, J., and Cesarano, J.: Direct ink writing of three-dimensional ceramic structures, J. Am. Ceram. Soc., 89, 3599–3609, doi:10.1111/j.1551-2916.2006.01382.x, 2006.

Lin, C. H., Lee, G. B., Chang, B. W., and Chang, G. L.: A new fabrication process for ultra-thick microfluidic microstructures utilizing SU-8 photoresist, J. Micromech. Microeng., 12, 590–597, 2002.

Mehta, V.: Design, Analysis, and Applications of Cellular Contact-Aided Compliant Mechanisms, Doctor of Philosophy, Mechanical Engineering, The Pennsylvania State University, University Park, 2010.

Mehta, V., Frecker, M., and Lesieutre, G. A.: Stress Relief in Contact-aided Compliant Cellular Mechanisms, ASME J. Mech. Des., 31, 1–11, 2009.

Mehta, V., Hayes, G. H., Frecker, M. I., and Adair, J. H.: Design, Fabrication, and Testing of Meso-scaled Cellular Contact-aided Compliant Mechanisms, ASME 2010 Conference on Smart Materials, Adaptive Structures and Intelligent Systems, SMASIS2010, Philadelphia, Pennsylvania, USA, 28 September–1 October 2010, 2010.

Microfabrica Inc.: http://www.microfabrica.com/, last access: December 2010.

Muller, T., Piotter, V., Plewa, K., Guttmann, M., Ritzhaupt-Kleissl, H. J., and Hausselt, J.: Ceramic micro parts produced by micro injection molding: latest developments, Microsyst. Technol., 16, 1419–1423, doi:10.1007/s00542-009-0992-1, 2009.

Nelson, J. C., Vail, N. K., Barlow, J. W., Beaman, J. J., Bourell, D. L., and Marcus, H. L.: Selective Laser Sintering of Polymer-Coated Silicon-Carbide Powders, Ind. Eng. Chem. Res., 34, 1641–1651, 1995.

Reed, J. S.: Principles of ceramics processing, Wiley & Sons, 1995.

Schonholzer, U. P., Hummel, R., and Gauckler, L. J.: Microfabrication of ceramics by filling of photoresist molds, Adv. Mater., 12, 1261–1263, 2000.

Tay, B. Y., Evans, J. R. G., and Edirisinghe, M. J.: Solid freeform fabrication of ceramics, Int. Mater. Rev., 48, 341–370, 2003.

Terfort, A., Bowden, N., and Whitesides, G. M.: Three-dimensional self-assembly of millimetre-scale components, Nature, 386, 162–164, 1997.

Wanke, M. C., Lehmann, O., Muller, K., Wen, Q. Z., and Stuke, M.: Laser rapid prototyping of photonic band-gap microstructures, Science, 275, 1284–1286, 1997.

Xia, Y. N. and Whitesides, G. M.: Soft lithography, Annu. Rev. Mater. Sci., 28, 153–184, 1998.

Yan, M. T., Huang, C. W., Fang, C. C., and Chang, C. X.: Development of a prototype Micro-Wire-EDM machine, J. Mater. Process. Tech., 149, 99–105, doi:10.1016/j.jmatprotec.2003.10.057, 2004.

Yang, R. and Wang, W. J.: A numerical and experimental study on gap compensation and wavelength selection in UV-lithography of ultra-high aspect ratio SU-8 microstructures, Sens. Actuator B-Chem., 110, 279–288, 10.1016/j.snb.2005.02.006, 2005.

16

Design, prototyping, and evaluation of a collapsible device for single-operator sheathing of ultrasound probes

S. A. Lopez[1], L. R. Hernley[1], E. N. Bearrick[1], L. M. Tanenbaum[2], M. A. C. Thomas[1], T. A. Toussaint[1], J. J. Romano[3], N. C. Hanumara[1], and A. H. Slocum[1]

[1]Mechanical Engineering Dept., Massachusetts Institute of Technology, Cambridge, MA, USA
[2]Health Sciences and Technology Dept., Massachusetts Institute of Technology, Cambridge, MA, USA
[3]Hospital Medicine, Massachusetts General Hospital, Boston, MA, USA

Correspondence to: S. A. Lopez (saulopez@mit.edu)

Abstract. During interventional ultrasound-guided procedures, sterility is maintained by covering the transducer head and cord with a sterile sheath. The current sheathing technique is cumbersome, requires an assistant to complete, and poses a risk of tangling the probe cord and breaching the sterile barrier. This paper presents the design, proof-of-concept prototyping, and evaluation of a probe holder and cartridge-style, single-use applicator that enables faster, more reliable, single-user sheathing of ultrasound probes, with a decreased risk of compromising sterility.

1 Introduction

1.1 Opportunity

The use of ultrasound guidance for interventional procedures has been shown to improve patient outcomes and decrease complications (Mercaldi et al., 2011). Catheter or needle misplacement can cause up to 4–7 excess days of hospital stay, up to USD 45 000 in extra costs, and increased morbidity rates of up to 14 % (Zhan et al., 2004). Ultrasound guidance provides real-time visualization of variations in patient anatomy during technically challenging, otherwise blind procedures, including central venous catheter placement, pleural fluid drainage, and various biopsies (Slawsky et al., 2011). In 2011, over 16 million ultrasound-guided procedures were performed in medical settings in the United States, with the estimated savings per 1000 patients totaling over USD 800 000 (Slawsky et al., 2011; Sinno and Alam, 2011).

Sterility and image quality are essential during invasive, ultrasound-guided procedures in which needle insertion into the patient occurs. Sterility is achieved through the application of a sheath to the ultrasound transducer before the start of the procedure (Moore, 2011). This technique is cumbersome, requires an assistant, and poses a risk to maintaining the sterile barrier.

The current sheath application process, detailed in Fig. 1, requires that the probe first be held upright while a thin layer of an acoustic gel is applied to the transducer head. This medium aids in transmitting ultrasound signals to tissues to preserve image quality by minimizing the difference in acoustic impedance and sound velocity between the transducer and the human tissue (Marhofer and Frickey, 2006). The sheath is then unwrapped and gloves donned by the clinician. With the assistant grasping first the probe body, inverting it, and then holding it by the cord, the sheath is pulled over the head, secured with a rubber band, and unfurled over approximately 1 m of cord. If at any point the probe touches the outside of the sheath, the sheath is scrapped and the procedure begins again. While some doctors suggest a different technique when an assistant is not available, no device exists to facilitate the process (Trotter et al., 2011).

An additional challenge comprises maintaining the layer of acoustic gel between the transducer head and sheath. Poor

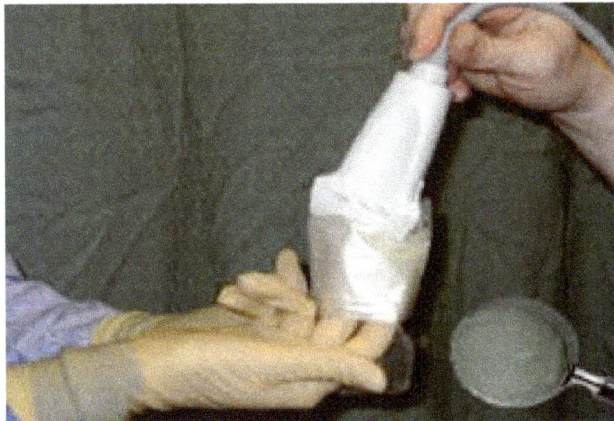

Figure 1. Ultrasound probe shown being sheathed using the current method.

application or its displacement during sheathing can result in air bubbles that reduce image quality.

Given the challenges, the goal of this work was to create a reliable sheathing device for hand-held ultrasound probes that can be operated by a single user.

1.2 Prior methods

Several existing devices and patents address sheathing an ultrasound probe; none however, fully address the single-user scenario. The Pull Up^{TM} ultrasound cover has a distinct handle-like tab that allows the user to both easily identify the sheath opening and unfurl the sheath over the length of the probe and cord (Witzky, 2012). One patent attempts to increase the ease of unfurling the sheath, often the greatest barrier to maintaining sterility, with a telescoped sheath that is easier to extend along the cord with one hand (US Patent No. 5,910,113, 1999). Additionally, the open end of the sheath is mounted to the outside of a structure that is folded flat; upon use it is opened to ease passage of the probe through the sheath. However, this patent does not adequately eliminate the difficulty of grasping the non-sterile probe while simultaneously applying a sterile sheath. One application broadly addresses covering an ultrasound probe with suggestions for integration of an acoustical couplant (gel replacement) and drawstring sheath-tightening mechanism, without specific solutions (US Patent Application No. 0139944 A1, 2008).

Condom application poses a similar challenge and US patents No. 5,316,019 from 1994 and No. 8,166,975 from 2012 address condom application by a joint packaging of condom and applicator, which extends the condom along the desired length. Sensis^{TM} tabbed condoms allow the user to unroll the condom with tabs that are freed from the condom when unrolling is complete (DDA Medical, 2009). Both of these products have methods applicable to maintaining sterility because the user holds tabs without having to touch the

product surface. The Pronto condom allows direct application of a condom straight from a break-open package (Mac-Donald, 2011). Stat Strip® adhesive bandages offer similar usability through a bandage that is opened with lateral force and can be applied directly from the packaging (Medline, 2012).

Looking outside the medical field, US patent No. 4,783,950 from 1987 covers protective wrapping of luggage by rotation along its central axis, while enveloping it with a film dispensed from a spool. US patent No. 4,827,695 from 1989 discusses the sealing of a sucker wrapper to a sucker stick through wrapping and heating.

A number of inventions attempt to discourage gel from escaping from the head of the probe. One addresses the issue of fluid frequently escaping by creating a tight fit through an inflatable sheath (US Patent No. 4,815,470, 1989). Another surrounds the ultrasound probe with a homogeneous, solid, elastic, biocompatible sheath that is conformal to each probe and eliminates the need for additional layers of gel (US Patent No. 6,039,878, 2000). Solid couplant hydrogels allow the probe to be placed directly on the patient's skin without the traditional layer of gel on the sheath exterior (US Patent No. 5,522,878, 1996).

Current devices attempt to improve the handling and application of the sheath and restricting the gel layer. Methods include adding a structure to the sheath and using novel materials for the gel. However, these do not provide a complete solution for a single user to easily grasp the probe while simultaneously applying the sterile sheath.

2 Design

A deterministic design process was used that started with determining the functional requirements for the design, progressed from identifying strategies to concepts, and concluded with detailed design and testing (Graham et al., 2007).

2.1 Functional requirements

A study of current procedural challenges and device performance constraints resulted in the following functional requirements for a new device:

– Maintain sterility – The device should enable application of a sheath that maintains the sterile barrier between the non-sterile transducer and the patient.

– Single user operation – Eliminate the need for an assistant to manipulate the non-sterile probe and cord. This will reduce procedure costs and, potentially, time.

– Maintain gel layer – Device should facilitate ~ 3 mm layer of gel application.

– Cord coverage – Sheath spans ~ 1 m of cord.

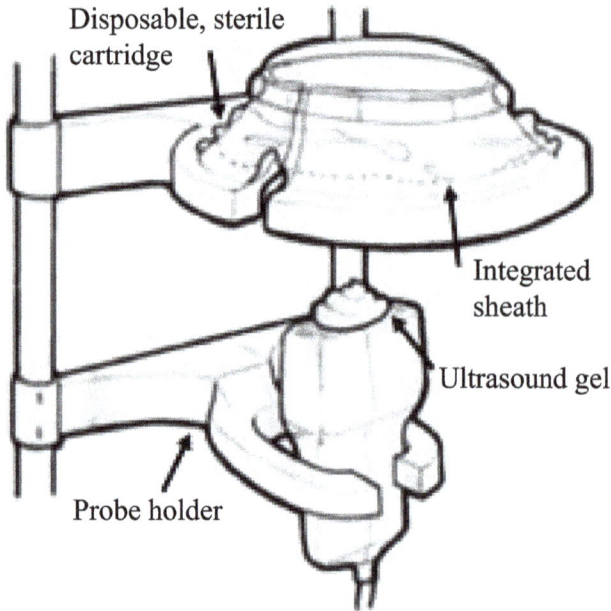

Figure 2. Design concept sketch.

- Easy to use – Reducing the complexity of the transducer sheathing process will ensure repeatable performance and minimize incidents of sterility loss.

- Multiple probe types – Device should accommodate transducers that vary in size, shape, and weight.

- Portable – Device must be portable in order to be transferred across rooms as necessary.

2.2   System overview

Shown in Fig. 2 is the two-part concept design consisting of a non-sterile probe holder and a sterile, disposable sheath cartridge that first snaps over the probe, thus protecting the outside of the sheath from contact with the probe. Handles support the rim of the cartridge so it can be manipulated with bare hands. Subsequently, while the clinician wears gloves, the probe is grasped through the sheath and pulled upwards, thus preventing contact between the gloves and the non-sterile probe as the sheath unfurls along the length of the cord. The cartridge then opens to release the now sheathed probe. A flow chart comparing the current method with the proposed method is shown in Fig. 3. The new design reduces complexity by providing a cartridge with a single correct method of deployment, which makes the procedure faster by eliminating steps that pose a risk of breaching the sterile barrier. The process can now be accomplished with a single operator at a significantly lower risk of breaking sterility. Details of the design are provided in the following section.

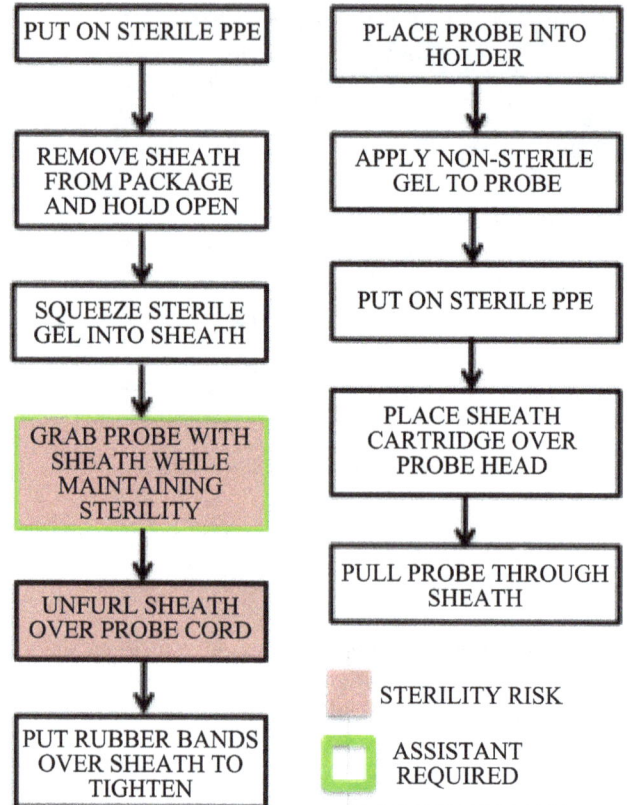

Figure 3. Process flow diagram comparing the current (left) procedure to that using the new device (right).

3   Module design and analysis

3.1   Sheath application & unfurling

Applying and unfurling the sheath is the critical part of the procedure that poses the greatest risk to sterility; therefore the most critical module of the device, i.e., sheath application, was designed to eliminate this risk. Bench level experiments and mock-ups demonstrated that fixing the ultrasound probe, rather than the sheath, provided a more ergonomic solution for a single user. Two packaging modalities for the sheath were then considered: traditional telescoping, and rolled, with the latter influencing the first design phase, taking inspiration from Sensis™ condoms in which tabs are used to directly unroll the condom from the package. Although promising, this technique proved challenging in its implementation and potential manufacturing due to the length of the sheath. Thus, it was decided that incorporating the current telescoping sheath packing would increase the likelihood of the device reaching its market.

The first iteration of this module consisted of the "sheath shell" depicted in Fig. 4. The shell would be packaged and sterilized with the sheath between the inner and outer shell layers, such that the whole assembly could be readily placed over a stationary probe without further sheath preparation. In

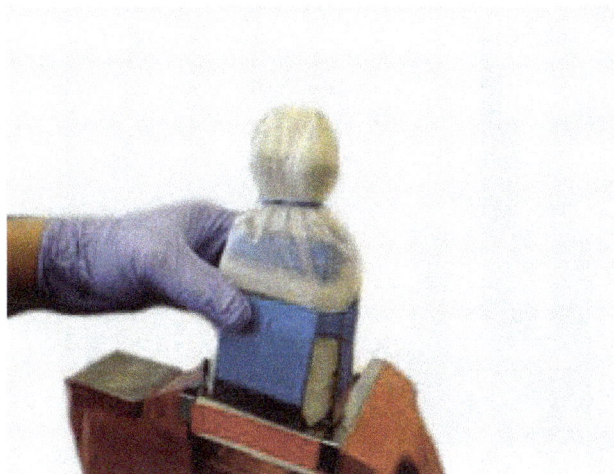

Figure 4. Initial sheath shell prototype showing its placement over a stationary probe.

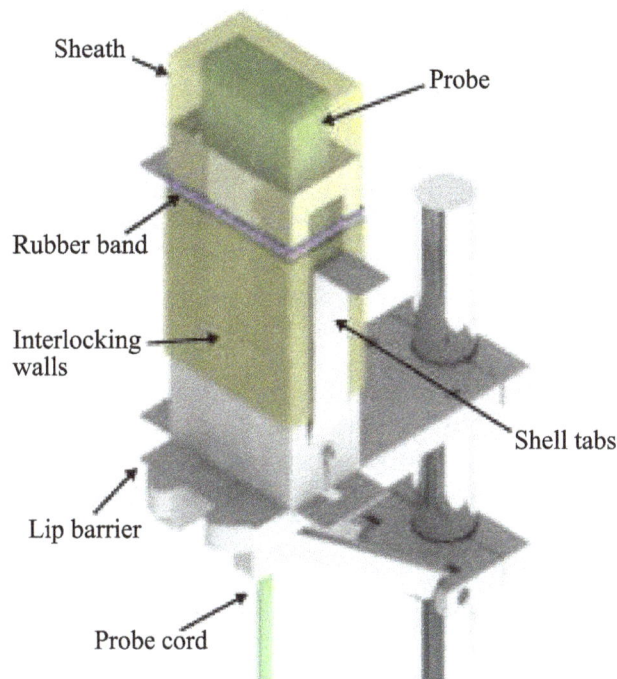

Figure 5. Rendering of the final concept showing sheath shell and integrated probe holder engaged by the sheath shell.

practice, the gel was applied to the probe head prior to employing the sheath shell, and a rubber band was loaded onto the sheath before the probe was pulled through the shell. The concept is favorable because it creates a clear barrier between non-sterile and sterile procedures. As illustrated in Fig. 3, there is a procedural break where the user can apply sterile personal protective equipment (PPE) before continuing. Additionally, validation testing demonstrated that the device allows the user to easily unfurl the sheath over the length of the cord. The next design iteration addressed cost-effective manufacturing and compact storage of the device.

3.2   Flat pack design

The current version of the prototype, illustrated in Fig. 5, addresses the shortcomings of the first sheath shell. For instance, the shell features were optimized to minimize material use, packaging space, and manufacturing complexity. Device compactness was incorporated into the design to minimize the required material and packaged volume. For ease of manufacturing and reducing costs, the sheath shell can be stamped from a sheet of material that can be folded into shape using living hinges (Fig. 6). To keep the design portable, the device folds about the living hinges to a flattened configuration when stored, allowing the use of readily available, sterilizable medical packaging, while keeping the packaged device comparable in size to the current sheathing kits. The implementation of a collapsible sheath shell introduced a number of challenges that were overcome via systematic design and bench-level experimentation. The elements to be addressed included sheath location with respect to the sheath shell, clamping of the bottom of the sheath during the unfurling process, maintenance of sterility, temporary fixation of the sheath shell to the probe holder,

cord removal following sheathing, and rubber band storage and deployment.

In the final design, the sides of the sheath shell contain left and right tabs to allow the user to grasp the shell without compromising sterility or pinching the sheath during the process. A lip was added around the base of the sheath holder to minimize user contact with the non-sterile probe holder. The rubber band is placed on one of the side handles so the user can easily remove it and place it over the probe. The front of the sheath shell has interlocking walls that allow the sheath shell to be rigid while in use yet open following sheathing. The user can pull the probe cord out of the shell while minimizing exposure of the sterile sheath to the non-sterile probe holder. Finally, a snap-fit was included at the rear of the sheath shell to allow the shell to remain fixed to the probe holder while the probe is pulled through the shell for sheathing. Afterwards, the shell is disengaged from the holder and disposed of. The device strongly considers the order of operations of the sheathing procedure and is designed to integrate with existing procedure.

3.3   Ultrasound probe holder

The probe holder was designed to mate with the sheath shell as it is placed over the probe. The sheath shell fits over the probe holder to further maintain sterility by enclosing the non-sterile probe within the sterile sheath shell. The holder also contains the complementary parts of the sheath shell snap-fits in order to temporarily fix the sheath shell to the

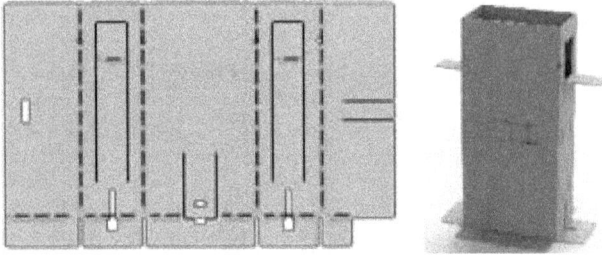

Figure 6. 2-D stamping folded into sheath shell.

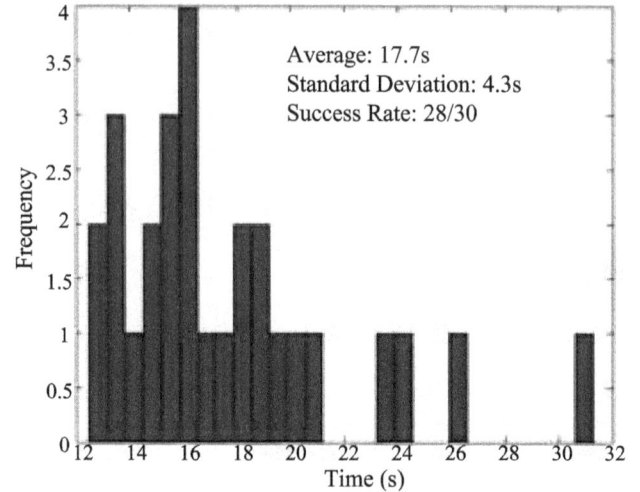

Figure 7. Histogram showing the time distribution of the 30 sheathing trials that were used to approximate the repeatability.

probe holder. The probe holder is designed to mount to IV poles, which are readily available in all hospitals and clinics, and in close proximity to portable ultrasound carts. The current prototype was CNC machined, although it would be injection molded if mass produced.

The collapsible sheath shell design made the incorporation of sterilizable packaging straightforward, as the sheath shell prototype readily fits into the currently used ethylene oxide sterilized easy-open packages, such as the Site Rite* Probe Cover Kit with Gel.

4   Testing

4.1   Repeatability of sheath shell deployment

The benefit of using a device to constrain the sheath allows the process to be streamlined across users. Thus, repeatability was an important metric of the success of the device. To approximate the reliability of the device, the probe was sheathed 30 times and the process times recorded. A histogram of the results is shown in Fig. 7. In 28 out of the 30 trials, the probe was successfully sheathed as intended. In one trial, the sheath came off the shell without telescoping and in another the snap fit did not engage. Having an adhesive hold the bottom of the sheath could eliminate the risk of the sheath being removed from the shell without telescoping. It should be noted, that the same prototype sheath shell was tested repeatedly, while the design intent is for single use. A more accurate evaluation will be performed once the final sheath shells are manufactured with true living hinges.

4.2   Sterility

The ability of the device to maintain sterility is crucial. To ensure that the sheath shell prevented user or sheath contact with the non-sterile probe, cord, or probe holder, the transfer of bacteria from the non-sterile field was simulated through coating the probe with phosphorescent paint and deploying the sheath.

The results shown in Fig. 8 show, as expected, that the transfer of paint was limited to the interior of the sheath shell, which was in direct contact with the probe and holder. The exterior of the sheath shell, the exterior of the sheath, and the

user's gloves were free of the paint. The sterility simulation was performed twice and not tested more extensively for repeatability. As a next step in the development of the device, bacterial culture testing to demonstrate that the sheath and sheath shell are sterile post sterilization will be a crucial step for regulatory approval and product deployment.

5   Future work

Although initial testing suggests that the device is reliable, future improvements will be made to ensure that the device is easy to use, maintains good image quality, and is manufactured for clinical use. Two main areas of continued investigation include: a method of automatically deploying the rubber band to secure the sheath to the transducer head, and integrating the gel into the sheath.

5.1   Rubber band deployment

Concept designs that will store a rubber band in an unstressed state, then stretch it out and deploy it onto the probe head are shown in Fig. 9.

It was envisioned that storing the rubber band around the indented part of the profile would allow the rubber band to be packaged on the sheath shell and as the probe holder pushes through the sheath shell, the rubber band would be deployed onto the probe. However, the benefits of manufacturing the device out of a single sheet outweighed the possible success of a device with an integrated rubber band.

For future considerations, the incorporation of a similar rubber band mechanism into the single-sheet design would make our device easier to use.

Eliminating the rubber band altogether by including an elastic sheath or an alternate means of tightening the sheath below the head of the probe will also be considered.

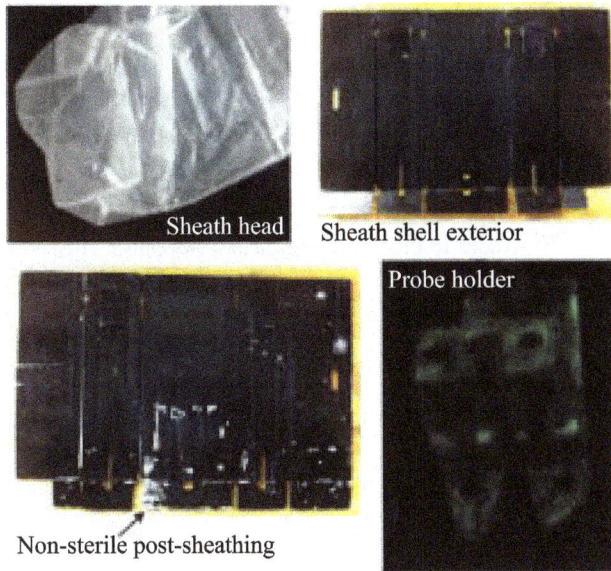

Figure 8. Results of testing with phosphorescent paint coated probe to demonstrate no probe contact outside of the sheath. Top two images show lack of sheath and outer shell contact with probe.

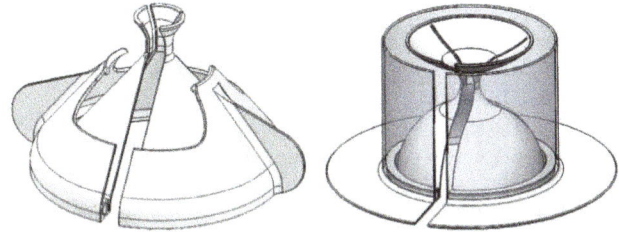

Figure 9. Conceptual designs of a shell with a narrow neck for rubber band storage. As the probe is pushed through, the rubber band is pushed off and snaps onto the probe over the sheath.

5.2 Integrated gel

Air bubbles resulting from the application of gel in the current procedure motivated significant efforts to improve ultrasound image quality through use of a semi-solid gel. Air bubbles are detrimental to image quality because the acoustic impedance and sound velocity of air are very different from that of soft tissue (Marhofer and Frickey, 2006). When considering a medium to better bridge the ultrasound probe and the surface of the body, two requirements must be satisfied: (1) minimize air bubbles within the medium and (2) match impedance and sound velocity to that of human tissues.

A medium that better adheres to the head of the transducer during the sheathing process is desirable in order to maintain image quality throughout the procedure. Ecoflex® Gel silicone was chosen for testing because the properties of silicone closely match those of human tissue. Image analysis using MATLAB® demonstrated that the image depth is comparable, but the resolution of the standard gel is better than that achieved using the silicone. However, future work to minimize bubble formation during the curing of the silicone could offer substantial improvement in resulting image quality. The use of a semi-solid gel would be beneficial to the sheathing process because it eliminates a step in the procedure as well as prevents gel migration from the head of the probe.

6 Conclusions

The proposed device satisfies the stated functional requirements. It maintains sterility during the sheathing process, can be operated by a single user, facilitates sheathing of the cord, and is portable. The design of the sheath shell and holder can also be easily modified for fabrication of various ultrasound probe-specific models and the sheath shell can be manufactured from a single sheet, allowing no additional packaging requirements.

Acknowledgements. We would like to thank the Precision Machine Design staff at MIT, as well as those who assisted with the research and testing at the Massachusetts General Hospital.

This work was supported by CIMIT under US Army Medical Research Acquisition Activity Cooperative Agreement W81XWH-09-2-0001. The information contained herein does not necessarily reflect the position or policy of the Government and no official endorsement should be inferred.

Edited by: D. Brouwer

References

Curtis, B. B. and Cermak, I. A.: Inflatable Sheath for Ultrasound Probe, USA, 4, 815, 470, 1989.

DDA Medical: Sensis Condoms, available at: http://www.sensiscondoms.com/ (last access: 9 November 2012), 2009.

Graham, M., Slocum, A., and Moreno Sanchez, R.: Teaching high school students and college freshman product development by Deterministic Design with PREP, ASME J. Mechan. Design (Special Issue on Design Engineering Education), 129, 677–681, 2007.

Hanumara, N. C., Begg, N. D., Walsh, C. J., Custer, D., Gupta, R., Osborn, L. R., and Slocum, A. H.: Classroom to Clinic: Merging Education and Research to Efficiently Prototype Medical Devices, IEEE J. Trans. Eng. Health Medic., 1, 2168–2372, 2013.

Jones, K. G.: Condom Applicator, USA, 5, 316, 019, 1994.

Larson, M. J., Rutter, J. W., and Smith, L. L.: Coupling Sheath for Ultrasound Transducers, USA, 6, 039, 694, 2000.

Logan, K. W.: Apparatus for Heat Sealing a Candy Wrapper to a Sucker Stick, USA, 4, 827, 695, 1989.

MacDonald, J.: Pronto condom aims for global distribution, available at: http://www.changehub.net/2011/10/pronto-condom-aims-for-global-distribution/ (last access: 7 November 2012), 2011.

Marhofer, P. and Frickey, N.: Ultrasonographic guidance in pediatric regional anesthesia part 1: theoretical background, Pediatric Anesthesia, 16, 1008–1018, 2006.

Medline: Stat Strip Adhesive Bandages, available at: http://www.medline.com/product/Stat-Strip-Adhesive-Bandages/Adhesive-Bandages/Z05-PF00037 (last access: 3 October 2012), 2012.

Mercaldi, C. J.: Clinical and Economic Advantage of Ultrasound Guidance Among Patients Undergoing Paracentesis, Technical Report, SonoSite, Bothell, WA, 2011.

Montecalvo, D. A. and Rolf, D.: Solid Multipurpose Ultrasonic Biomedical Couplant Gel in Sheet Form and Method, USA, 5, 522, 878, 1996.

Moore, C.: Ultrasound-guided procedures in emergency medicine, Ultrasound Clinics, 6 277–289, 2011.

Morgan, M.: Combination Prophylactic Package and Dispenser, USA, 8, 166, 975, 2012.

Pruter, R. L.: Sheath for Ultrasound Probe, USA, 5, 910, 113, 1999.

Santagati, U.: Machine for Automatic Protective Wrapping for Use with Different-Sized Baggage, USA, 4, 783, 950, 1987.

Sinno, M. and Alam, M.: Echocardiographically Detected Fibrinous Sheaths Associated with Central Venous Catheters, Echocardiography, 29, E56–E59, 2011.

Slawsky, K., McInnis, M., Goss, T. F., and Lee, D. W.: The clinical economics of ultrasound-guided procedures, Technical Report, General Electric, Wauwatosa, WI, 2011.

Trotter, M., Nomura, J. T., and Sierzenski, P.R.: Single-operator Sterile Sheathing of Ultrasound Probes for Ultrasound-guided Procedures, Academic Emergency Medicine, November 2010, 17, e153 pp., 2011.

Weymer, R. F., Cannon, M. G., Engle, L. A., Klessel, J. S., Lannutti, A. P., Randall, K. S., and Urbano, J. A.: Devices for Covering Ultrasound Probes of Ultrasound Machines, USA, 2008/0139944 A1, 2008.

Witzky, M.: Pull Up Ultrasound Probe Cover Kit, available at: http://www.coneinstruments.com/pull-up-ultrasound-probe-cover-kit/p/934436/ (last access: 3 November 2012), 2012.

Zhan, C., Smith, M., and Stryer, D.: Incidences, Outcomes and Factors Associated with Iatrogenic Pneumothorax in Hospitalized Patients, [abstract], in: AcademyHealth Annual Research Meeting, 2004, San Diego, Calif. AcademyHealth, 21, abstract no. 1862, 2004.

17

Experimental investigation of water droplets' behavior in dielectric medium: the effect of an applied D.C. electric field

H. Bararnia and D. D. Ganji

Department of Mechanical Engineering, Babol University of Technology, Babol, Iran

Correspondence to: H. Bararnia (hasan_bararnia@yahoo.com)

Abstract. In this article, the behavior of water droplets which are suspended in silicon oil is qualitatively investigated and some phenomena such as liquid's burst are reported. The movement of droplet in perpendicular line is considered while many studies have considered small droplets fluctuating between two horizontal electrodes. Additionally any deformation caused by increasing voltage was observed from start to finish until short contact occurred. As can be seen two oppositely charged drops contacting each other do not necessarily result in coalescence. Repelling can be expected occasionally. By increasing the voltage, the droplets, which spread through the domain due to frequent breakups, tend to gather in a certain line.

1 Introduction

Understanding the interaction of suspending conducting liquids with the second dielectric medium is a prominent factor in many engineering projects. One of the most important examples is emulsification. Regarding modifying the properties of the polymer, antistatic agents are added during mixing processes. Dispersion also enhances the mass transfer operations in liquid-liquid extraction. In oil industries, determining the onset and type of deformation is an important matter because the breakup of a drop into smaller droplets is a Preventive element concerning electrocoalescers. Nevertheless this separation could be utilized beneficially relating to emulsification and enhancing heat and mass transfer. Traditionally the tiny droplet gathering in continuous oil phase has attracted great attention in oil recovery technology; see Eow et al. (2001). Controlling the fusion of individual droplets in digital micro fluidic applications were investigated recently by Ahn et al. (2006). Applications, has been shown in some other aspects such as electro spraying by Bailey (1999) and nucleate boiling by Dong et al. (2006). Flows generated by electroosmosis have been used to act as a machine and enrich the mixing of substances at the micro level; see Pikal (2001). Dielectrophoresis have been used widely in separation for bio-cells and carbon nanotubes; see Aldaeus at al. (2006).

Considering the suspension of a droplet in immiscible medium, the shear and electrical stresses caused by external electric field and distinction electrical characteristic at the interface is expected. This results in burst of the droplet or elongation. Initially assuming the droplet as ideal conductor or fully insulator resulted no fluid flow because of inequality of shear stresses. Furthermore, based on electrostatic theory the consequent stresses are perpendicular to the surface and towards the fluid with lower permittivity. The deformation led the droplet to be prolate. The experiment conducted by Allan and Mason (1962) showed that conducting droplet deforms into prolate spheroid compatible with the electrostatic theory. They have considered several fluid systems and declared some droplets get oblate shape. Taylor (1966) pointed out considering drop as perfect conducting or dielectric fluid could not be proper in all situations. They stated that limited amount of permittivity or conductivity allows unbounded charge to collect at the droplet surface and this collection is responsible for inequality in shear and normal electrical stresses which causes oblate shape. This theory is named leaky-dielectric theory. As illustrated by Taylor, Fluid flow in the inner and outer part of the droplet is caused by balancing the tangential stresses by hydrodynamic tangential stresses. Taylor justified Allan and Mason results

(a)   (b)   (c)

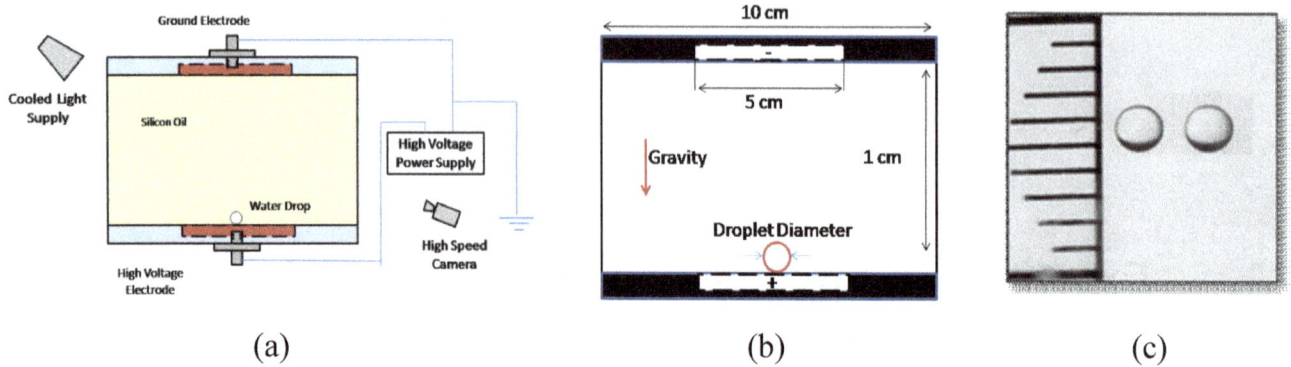

Figure 1. (a) Experimental apparatus, (b) the geometric of the cell, (c) measuring droplet's diameter.

by solving electro hydrodynamic's equation in creeping flow regime. Torza et al. (1971) revealed some discrepancies between theory and experiments. Baygents and Saville (1989) replaced leaky dielectric theory by an electrokinetic model to inspect the matters proposed by Torza et al. (1971). Feng and Scott (1966), Vizika and Saville (1992) have demonstrated that the leaky dielectric model has the capability to predict the deformation pattern while zero amounts of charges accumulated at the interface.

The research into the stabilization and breakup of an aqueous droplet is still continuing and in this section we aim to concentrate on some recent and pertinent ones. Sherwood (1988) used boundary integral method for droplets with different set of electrical properties to investigate the droplet deformation which is exposed to an electric field until burst of droplets. Ha and Yang (2000) experimented on Newtonian and non-Newtonian droplets surrounded by dielectric medium. They reported that when either the dispersed or continuous phase is non-Newtonian the droplet will be stable in low electric field. Besides, more complex behaviors are caused when both the mediums are non-Newtonian due to the zero-shear-rate viscosities ratio. Ristenpart et al. (2009) experimentally showed that there is a certain value of electric field which leads the two droplets with different sign to repel each other at the contact time while coalescence was expected. Breakup and deformation of aqueous drops in oil was investigated experimentally by Eow and Ghadiri (2003a). The results indicated that the starting point of fluctuation related to the deformation rate is about 1.9. Eow and Ghadiri (2003b) also investigated experimentally drop-drop coalescence. Their results indicated that coalescence depends on electrode geometry and orientation of the field. To achieve the maximum attractive value it is necessary to direct the center line of the droplets in line with the electric field.

Hase et al. (2006) performed experimental test to find micro-droplet's behavior which is rhythmically moved between two horizontal electrodes in silicon oil. Depending on applied electric field, three distinct zones were charac-

terized. Chiesa et al. (2005) provided the analytical formula to demonstrate the forces induced by an electric field on a falling droplet. Their results were well-matched to experimental outputs. Khorshidi et al. (2010) investigated experimentally the water droplet's motion and shape which was large in size and also calculated numerically the amounts of accumulated charges as a consequence of contacting with the electrodes. They reported that neutral water droplet tends to move toward the positive electrode and named this action as negative-electrophoresis. Recently Hokmabad et al. (2012) investigated experimentally the motion and disintegration of droplets gathering charges while leaving the nozzles. They have pointed out that more deformation is concluded by increasing the voltage and the maximum value has been depicted near the nozzles. Discharging of droplets to the negative voltage (ground) leads to two distinguished modes of drop's burst. In this report, some important aspects of water droplets' behavior in a high voltage electric field are investigated. (i) electrical charging and rhythmic motion, (ii) drop-drop coalescence and the effect of existed torque, and (iii) different breakup phenomenon and breaking the oil. All sections are examined by increasing the electric field; this is the only parameter that influences the droplets' behavior.

2 Experimental set up

The experimental cell, which is shown in Fig. 1, consisted of four Plexiglas walls (1 mm thickness) and was filled with silicone oil as the dielectric continuous phase. Table 1 gives the relevant properties of the employed fluids. The static electric field ($0 < E < 12$ kv cm^{-1}) was formed between two parallel electrodes [5×5 cm^2 stainless steel plate] and spaced 1 cm apart. The top electrode was electrically grounded, and the bottom electrode, which was connected to a D.C. power supply, acted as the high voltage electrode.

To start the experiment, two deionized water drops of 1.7 mm were released into the stationary dielectric oil using a syringe pump. Before applying the electric potential, the drops were initially 7.6 mm apart and, resting on the bottom

Table 1. Physical properties of employed liquids.

Property	Silicon oil	Water	Ratio (oil/water)
Viscosity (kg ms^{-1})	9.6e-2	1.12e-3	85.7142
Density (kg m^{-3})	963	998.2	0.9647
Conductivity (S m^{-1})	14e-13	1e-4	14e-9
Relative permittivity	2.75	80.1	3.4332e-2
Surface tension (N m^{-1})	0.020	0.072	0.2777

electrode. Then, the electric field strength was slowly raised to the critical value, above which the drops left the bottom plate and migrated towards the top (grounded) electrode. The complete history of the drop pair throughout the experiment was photographed by a CCD Camera (Casio EXILIM Pro EX-F1) at 300 fps.

3 Results

In this section, the whole process of the drops' behavior under the external electric field is discussed in detail.

3.1 Rhythmic motion of the droplets

The sequential images of the initial motion of the water drop pair in silicone oil are presented in Fig. 2. Drops are rest on the high voltage (i.e., the high voltage electrode), keeping their nearly spherical shape, while the applied voltage across the electrodes gradually increases. When the force exerted by electric field overcomes the gravity force (neglecting the surface tension between the liquid and solid surfaces), the drops are detached from the metal surface. Leaving the high voltage electrode, the drops acquire a positive charge and go against gravity towards the upper electrode. By contacting the grounded (upper) electrode, they exchange their charge and go back to the lower electrode, where they gather positive charge again and repeat this cycle. In the first few cycles, the drops travel simultaneously and reach the same electrode; eventually, the rhythmic motion becomes inverted in such a way that when the first drop touches an electrode, the other touches the opposite electrode. This change in direction causes the drops to become oppositely charged and attracted to each other while moving up and down in the surrounding dielectric oil.

It can be observed that the drops' velocity increases after contact with the electrode surface. As the drops' motion is entirely electrically driven, the increase in velocity magnitude reveals an increase in the accumulated charge on the drops' surfaces during their rhythmic motion.

The trend of drop position prior to their coalescence is plotted in Fig. 3a. As observed, the rise in the voltage yields to decline the frequency. Unlike the assumption that the reverse should be true (i.e., that a higher frequency is expected), the results have shown that the movement energy

gained from increased voltage is consumed such that the drops migrate towards each other and, consequently, become closer. Therefore, the time needed to travel between the electrodes is longer than in the case where each drop moves in a vertical line at their initial positions. The voltage value was increased up to 3.8 kv cm^{-1} when coalescence occurred. In general, anticipating the drop behavior in a high electric field is very complex. The characteristics of employed fluids, the uniformity and the orientation of the applied electric field compared to the drop, are leading factor. Under the uniform electric field, drops become polarized while bouncing up and down and develop a conical shape oriented in the orientation of the field. As the drops come closer to the electrode, the distance between droplets and electrode is reduced, leads to higher electric force and results in the elongation of the drop. The electric body forces experienced by a spherical charged droplet can be expressed as follows:

$$Fe = QE + (2\pi R)\varepsilon_1 \left(\frac{\varepsilon_2 - \varepsilon_1}{\varepsilon_2 + 2\varepsilon_1} \right) \nabla|E^2| \tag{1}$$

where Q is the free charge density. Here ε_1 and ε_2 are the continuous and dispersed phase electrical permittivity, respectively. The first term is Coulomb force (electrophoresis), and it describes the affect of the electric field on the net charge Q of the drop. In a D.C. case, Coulomb force exists only in the existence of a net charge on the droplet. The second term, referred to as the dielectrophoretic force (DEF), generated by non-uniformity in the electric field and electrical properties of materials. This force only disappears if the field is spatially homogeneous. (It is worth mentioning that the drop causes the electric field to become non-uniform in its neighboring.) Generally both of magnetism and electricity are included in the basic Maxwell's equation; however, for a wide class of problems, the magnetism is decoupled in the case of weak electric current or low magnetic field. In the vicinity of the electrode, the droplet become elongated sharper and consequently contacts the electrode by making a conical tip referred to as the "Taylor cone". Subsequently, the cone recoils, and the drops reverse their direction and move toward the opposite electrode. By increasing the voltage, the deformation rate (elongation) increases and, in some cases, will result in short contact. However at low voltages, the drop tends to keep a spherical shape at the midpoint between the electrodes due to surface tension. In the middle, the weight and drag force affect the drop and restrict the motion that originated from the electric force. However, near the electrodes drops experience excessive electric stress due to the short distance and accelerate at the time of detachment from the electrodes. Moreover, each of the polarized drops experiences a dielectrophoretic attraction force since the existence of second droplet in its vicinity, which brings them towards each other. The dipole-dipole interaction force is very small when the two drops are far, and its effect can be neglected compared to the electrophoretic force. However, when the drops are nearness in distance, the dipole-dipole attraction is

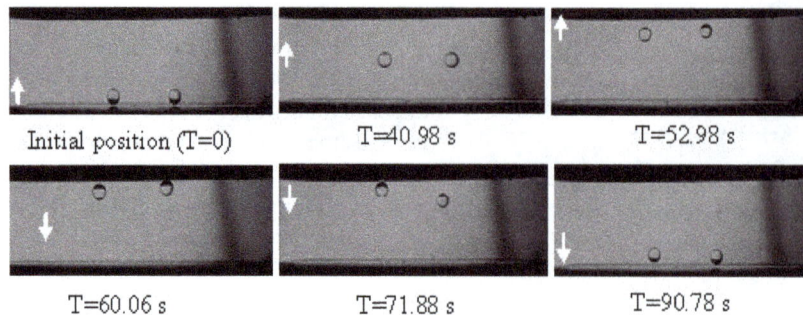

Figure 2. Sequential images showing the motion of a pair of water drops subjected to a D.C. Electric field $E_o = 1.1\,\text{kV}\,\text{cm}^{-1}$ for drop radius 0.875 mm.

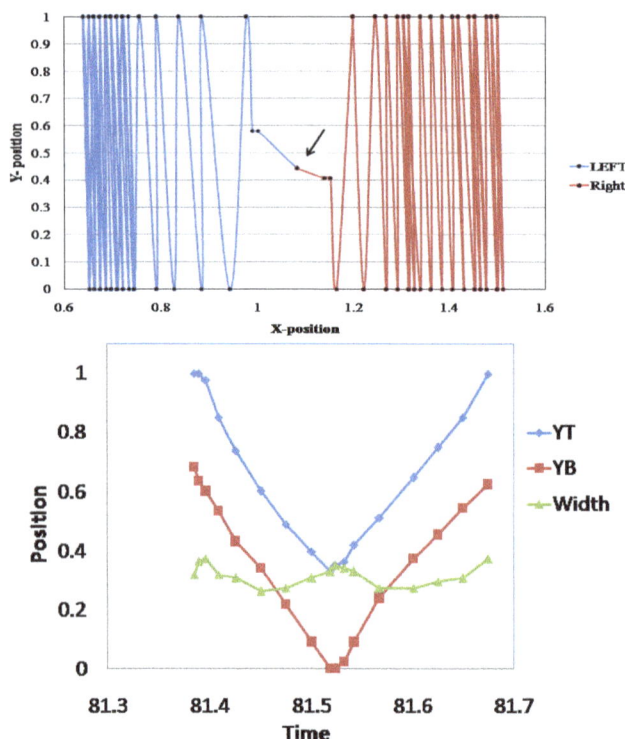

Figure 3. (a) Position of each drop versus time until coalescence, (b) Deformation of a drop in a cycle at $Ca_E = 0.018$ (Yt: the top point of droplet, Yb: the bottom point of droplet).

strong, which induces deformation on the drop pair and results in their coalescence.

3.2 Two drops deformation and coalescence

Figure 4 shows captured images where the drops move towards each other and merge into one. For two drops moving near each other, coalescence may occur due to hydrodynamical forces, however, in high electric field, electrostatic-coalescence is predominant factor. In an electro-coalescence phenomenon, when the electric force becomes high enough to push the drops towards each other, it triggers the oil film drainage between the drops and facilitates their contact, which, in most cases, results in coalescence, as illustrated in Fig. 4. When the electric field is not sufficient to develop drop-drop merging, the drops restore to their former state and keep their rhythmic motion between the electrodes.

At the time of approaching the drops, a thin layer separates the surface of the drops from the continuous phase and by rupturing the interfacial layer the drop-drop coalescence is predicted. Collectively, the roles of the electric field are; develop contact between the water drops, improve coalescence and enhance breaking the drop-interface. There are some other factors among these functions, such as the influence of electric field orientation on the coalescence of two aqueous drops, the influence of electrode geometry, which may play an important role in coalescence performance. However, it should be considered that occasionally two drops change their charges before coalescence takes place. This change can generate electrophoretic forces, which are responsible for repelling the drops from each other. One of the major factors that prevent the two dispersed drops to coalesce is drainage of the layer which separates them. By breaking this layer the chance of occurring coalescence grows rapidly. As observed, before coalescence the drops' tips face each other and deform. Furthermore, coalescence forms a dumbbell drop. There are two processes, i.e., before and after the occurring coalescence. Before coalescence, the two drops moved in a vertical direction. Considering the three dimensional behavior and distance between the two drops, it can be observed that the drops are brought closer together because of the two opposite charges, and when the distance is short enough, the torque rotates the drop while the drop pair becomes aligned in the electric field's orientation. This local rotation torque, which is due to dipole-dipole attraction, disappears when the drops become parallel and appears when they pass through each other (Fig. 5).

This torque is related to the minimum interaction energy, which depends on the distance between dipoles, permittivity of the two phases, polarization rate of the drops and direction of the induced dipole of a drop relative to the electric field orientation. Once the dipoles become far apart, the torque

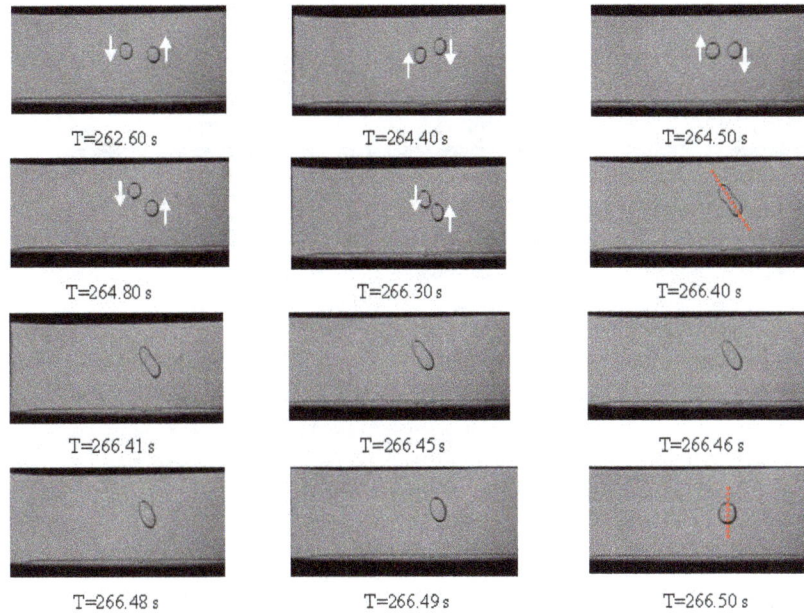

Figure 4. Interaction leading merging between two water drops at $Ca_E = 0.18$.

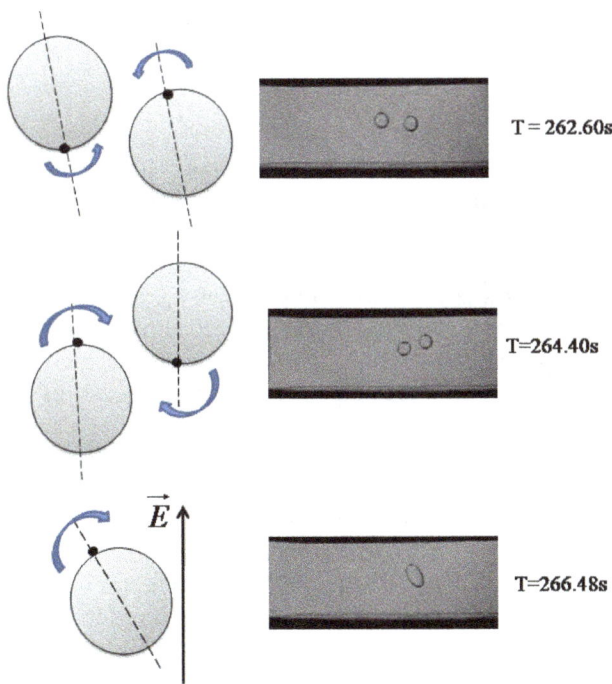

Figure 5. Schematic view of existing torque to bring drops closer together in different positions.

charges. The torque is increased and consequently the drops coalesce. The other process is after coalescence, when the two drops merge and comprise of a unique drop. This drop tends to rotate again and align itself in the electric field direction; this may occur because of the polarization that happens inside the drop which experiences a dipole moment when exposed to the electric field. This drop has a net zero charge, if neglecting the charge leakage and that equal charges are attained by the drops from the contact of the electrodes. Then the drop falls slowly due to its weight and consequently, after touching the bottom electrode, it recharges again and starts a new rhythmic motion. Similar to Fig. 2, the resultant drop fluctuates between the two electrodes, however, the drop experiences larger electrical stresses and velocity by rising the applied electric field. Therefore the deformation rate gets larger. Essentially with increasing the value of the electric field which can be defined as an voltage difference divided by the spaces of the two electrodes, the drop launches to have an acute angle at its attachment to the electrode. By increasing this value the angle increases and forms Taylor cones. In this step (Fig. 6) it can be seen higher deformation rate and a more acute angle, both at the time of contacting to the electrodes and moving between them in comparison with Fig. 2.

3.3 Break-up and bouncing

Understanding the drops' breakup under the electric field plays an important role. Since the mode of breakup, determines the distribution of aqueous drops in continuous phase in addition to the size. Breakup can be considered as a technique for the emulsification and atomization in practical

reverses and aligns the drops in the orientation of the electric field. However, the electric force tends to keep the drops in their previous direction. After a subtle rotation, the two drops return to their main direction, contact the two electrodes and acquire opposite charges. In every cycle, the drop-drop distance reduces due to the attractive force between opposite

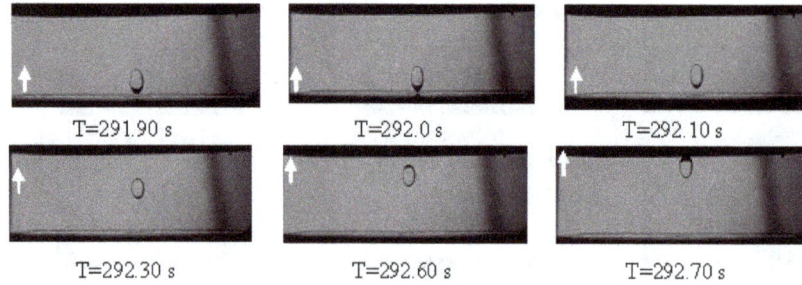

Figure 6. Rhythmic motion for drop resulted by coalescence at $Ca_E = 0.205$.

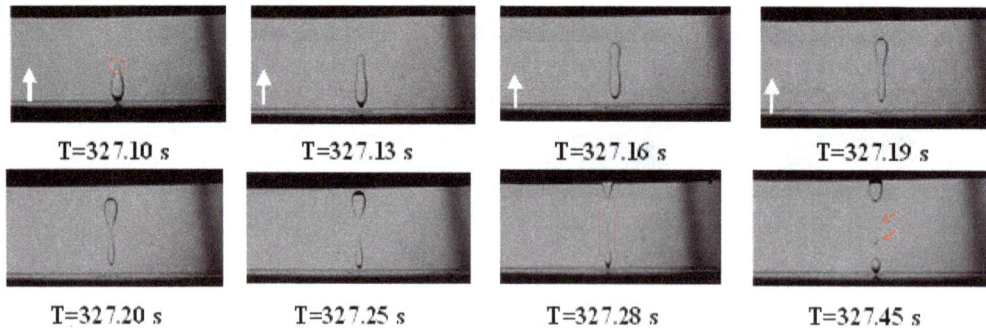

Figure 7a. The mother drop deforms greatly into an elongated shape and divides into two daughter drops along with an expanding liquid bridge between them, which eventually breaks up into more satellite drops ($Ca_E = 0.216$).

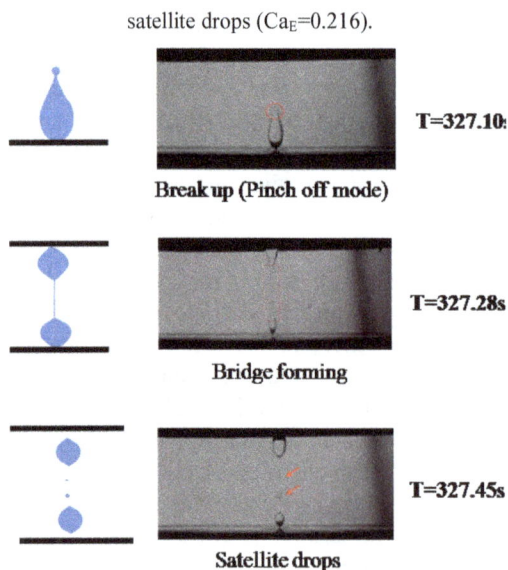

Figure 7b. Schematic view of breakup, creating thread and forming satellite drops.

projects as well as the enhancement of heat and mass transfer in droplet-based fluidic systems.

The equilibrium deformation, for a dispersed conductive drop moving through a dielectric medium of electrical permittivity ε_c and in the presence of applied electric field, can be obtained by balancing the electrical stresses ($P_{el} = \varepsilon_c E^2$) and pressure difference at the interface of droplet. If the electrical stress overcomes the restoring capillary pressure, then the drop breaks up. Two modes of drop breakup in a strong electric field have been introduced in the literature. The first type is pinch-off where the drop forms two bulbous ends, narrowing and disintegrating at a waist. The second type is tip-streaming where the tips become sharp and tiny drops are left from its ends. Figure 7 shows the single drop that moves to the lower electrode with a 4.1 kV cm^{-1} electric field strength. Shortly after touching the electrode, a pinching region develops on the top of the drop tending to move upward towards the opposite electrode. This occurrence causes the drop to be considerably elongated while leaving the metal surface and the drop is sequentially divided into two parts connected with a liquid thread. As each of the two bulbous-end drops reach the nearby electrodes, the thread expands continuously and finally breaks up into smaller satellite droplets as a result of instability caused by capillary. The ionized aqueous droplet in silicon oil with electrical resistance lower than 10^{-5}, is initially elongated into an ellipsoidal shape. By exceeding the, critical Weber number (0.21 which is in a good agreement with the numerical results of Feng and Scott (1966) and experimental results of Ha and Yang, 2000), the drop shape becomes unstable and varies with time. The deformation rate increases moderately until the main drop is broken into several tiny drops. Close to the ends, the deformation of the surface results in thinner drops. The pointed ends keep

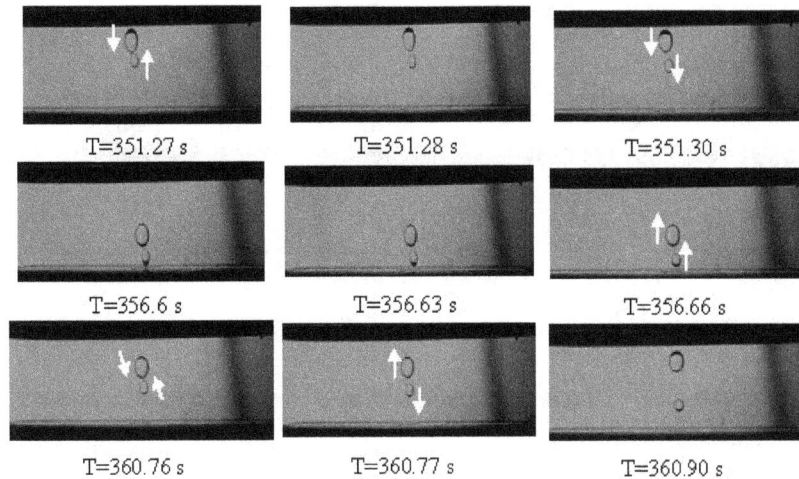

Figure 8. Repelling phenomenon of two opposite charge drops at the top, bottom and middle positions ($Ca_E = 0.226$).

the former trend and become slender until the ends which are now similar to bulb launch fragmentation into two distinct parts meanwhile the mother droplet changes steadily to become a thin cylindrical droplet as like as unstable liquid thread. As a result, satellite drops are produced. Initially a pinching zone is promoted (Fig. 7). Just after the forming of a pinching zone, the deformation of drop launches to increase quickly; meanwhile, the drop seeks to move upward due to its opposite charge. Therefore the drop stretches more and appears to break from the tip. It was expected that a daughter droplet is ejected from the pointed ends of the elongated drop, but as observed, the larger fraction of the stretched drop remains at the top and forms a bulbous end while the thinner part comprises the bottom of the main drop joined by a thread. At this time, the top and bottom drops reach the bottom and top electrodes respectively. The thread bursts into a few tiny droplets because of the capillary force and the thread becomes separated. These large drops experience collision with tiny drops. In this step, two phenomena occur when two or more drops exist in the domain (i.e., coalescence and repelling).

Tiny drops move based on gravity and coalesce with the bottom droplet. Therefore two main drops move between the electrodes. Traditionally, it was assumed that contacting the charged droplet with different sign are necessarily attracted each other. However if the electric field exceeds the critical value, repelling may take place rather than coalescence. As shown in Fig. 8, oppositely charged drops colloid with and repel each other. Figure 8 shows that the smaller drop, which is the consequence of the collision of tiny drops with the lower drop in the earlier step, moves upward because of the opposite charge and colloids with the larger drop that has been detached from the ground electrode. Electrical stresses operating on the charged surface causes the edge of drop to be elongated and consequently conical shape is formed. During rapid connection with the oppositely charged drop the

tip recoils, gathers opposite charge and change its direction. The bottom droplet after changing its charge moves downward until it reaches the bottom electrode. The top droplet after contacting the ground electrode starts to move downward due to the Columbic force (because impact occurs near the top electrode, both drops move downward with little time difference). When the bottom droplet contacts the high voltage electrode, it senses the top droplet with a negative charge at its top point, at which point the repelling phenomenon occurs, which leads the top droplet to move upward and the bottom one to move downward. This cycle repeats and is likely to occur in the center of the domain or anywhere the two oppositely charged drops have the opportunity to contact. The main question is as follows: why do the droplets bounce? The main reason for bouncing depicted in Fig. 8, is switching charge's sign because in that way drop could gain opposite charge and reverses its former direction. As a result, in non-coalescence impact, the drop cone recoils after apparent connection with the opposite sign and changes its direction.

The repelling process continues until the value of the voltage increases to ($4.5\,kV\,cm^{-1}$). The deformation rate increases and similar to the previous part, the top drop stretches to form a dumbly shape whose greater proportion is downward while the drop that is attached to the bottom electrode tends to move upward, according to its acquired charge. Figure 9, demonstrates that after fragmenting the upper drop into two parts, the tip of the bigger-detached drops aims to move upward because of the polarization, so the tip stretches until it reaches the top electrode while the bottom end reaches the bottom small drop (507.78 s). Because of the high voltage strength, the second drop, which is located on the bottom electrode, generates a Taylor cone. Near the upper drop, the two cones attach and form a thread (507.8 s). This thread cannot endure the electric field stress and burst into many tiny drops that can coalesce together. Now there are three drops that move between the two electrodes.

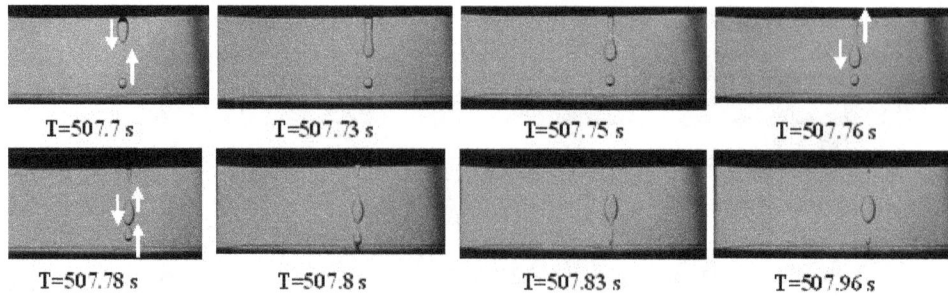

Figure 9. Making a thread due to large elongation and increasing the voltage ($Ca_E = 0.26$).

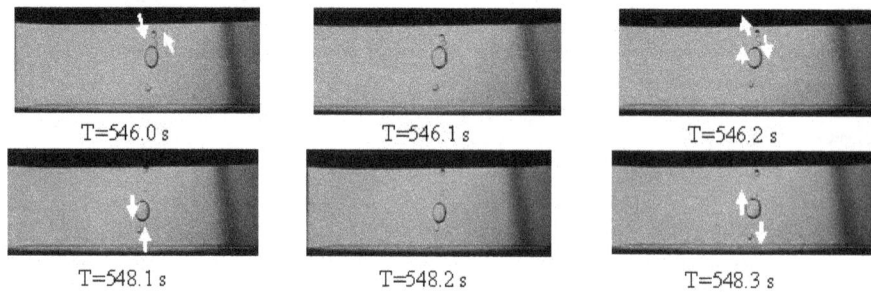

Figure 10. Bouncing droplets ($Ca_E = 0.26$).

3.4 Multi-drops motion, coalescence, repulsion and break up

By increasing the voltage, the bigger drop stretches considerably and in a certain moment, detaches a small drop from its tip. These four drops with different sizes are colliding and showing non-coalescence behavior, and during each contact they conduct, charge to each other and reverse their directions, as shown in Fig. 10.

Non-coalesce behavior is not the only feature that occurs in a high voltage electric field. In addition to the explanation stated by Ristenpart et al. (2009) which described the reasons to distinguish coalescence from repelling, and considering the presence of multiple droplets, the possibility of coalescence occurring during impact is not far-fetched. However before supporting the former argument, it is worth mentioning the reasons for repelling as fully described by Ristenpart et al. (2009). As mentioned by Khorshidi et al. (2010), the threshold field strength for droplet, hinges strongly on the amount of charges which has been transmitted during connection with the former interface. Because the charge comprises ions in the water droplet, these ions during bouncing process could transfer from one droplet to another one. Some mechanisms are suggested for charge transmitting such as conduction across the continuous phase (oil), breakdown of dielectric medium or smaller drops are sprayed by electric force (electro spray) however in the case of silicon oil, because of the aqueous thread which is formed during connection, the possibility of conduction of ions is presumably stronger than other reasons. The observation indicates that the drop takes double-cone shape near the bridge. The capillary pressure inside the thread is approximated by the Young-Laplace equation as follows:, $(\cot\theta - 1)\gamma/r_m$ while the angle is the key factor which determines the fluid flow. It means that the drops with sharper cones generate the higher pressure difference in the bridge and consequently leads the drop to be bounced. If the voltage is high enough, to break the dielectric medium (oil), then the drops prior to contact could be neutralized and the surface tension results in recoil. In other case joule heating is assumable. During connection process, equilibration of charges may heat the liquid in the thread zone and affect the neck by Maragoni force or vaporization and conductivity in this case is a leading factor. The capillary waves at connection time, expands liquid thread between them. Larger positive curvature in contrast with negative one leads to higher positive pressure in the neck than inside the droplet and breakup is developed. In other side, larger negative curvature generates higher pressure inside the droplet than in the neck and fluid is derived in to the thread and eventually coalescence occurs. During this motion, multiple drops contact each other, and the chance of coalescence and repelling happening is unavoidable. However, repelling typically occurs between them. By increasing the electric field, there is an intention to form an elongated shape again. Previously two cases are studied where the drop is placed on the top and bottom electrodes. However this is not the only moment when the drop breaks. Typically the drop can break when moving through the gap between the electrodes ($T = 629.513$ s). In Fig. 11, the top section of the drop stretches and reaches the top electrode while the bottom

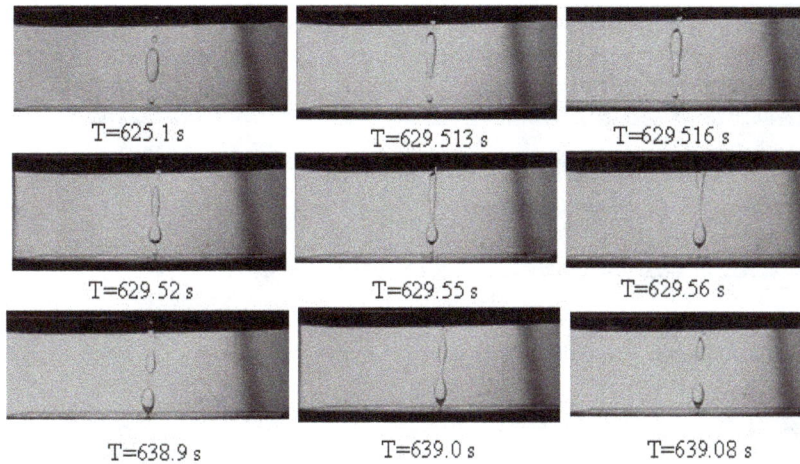

Figure 11. Pinch-off and forming thread ($Ca_E = 0.321$).

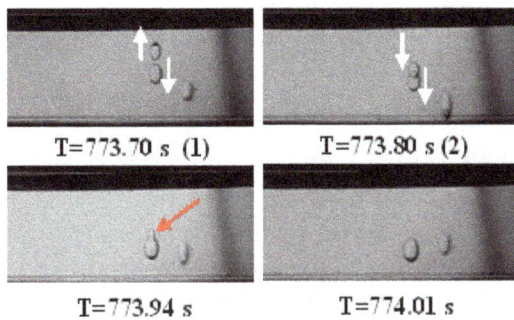

Figure 12. The second type of droplet coalescence ($Ca_E = 0.334$).

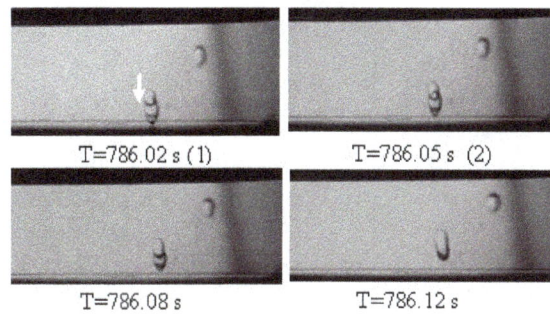

Figure 13. The first type of droplet coalescence ($Ca_E = 0.360$).

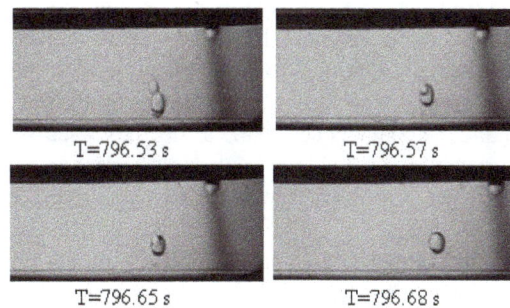

Figure 14. The first type of droplet coalescence ($Ca_E = 0.374$).

part reaches the bottom electrode. The thread appears, bursts and multiple drops form burst. This is a conventional cycle that repeats itself. These tiny drops can also generate a chain between the main drops and form a thread and short contact between the electrodes. The pinch-off appears to be a dominant mode of break up. The sizes of the main droplets and satellite drops become sensitive to the strength of electric field. An increase in the electric potential leads to a narrower size distribution of both droplets and satellites, and it eventually becomes impossible to distinguish drops from satellites.

As mentioned earlier, three phenomena can occur; contemporary break up, coalescence and repelling. Many studies revealed that coalescence happens at low strength of applied electric field. By increasing the value, the contact angle is sufficiently sharp to permit the bridge' pressure to be greater than the drops' pressure and repelling therefore occurs. As the results show ($T = 773.8$ s), the side-by side coalescence is expected in three-dimensional behaviors. It is obvious that according to the nature of the random movement of drops, there may be an instance that one drop gives a negative charge as the result of the short contact and wants to move for example upward (positive charge). The drop which transferred its opposite positive charge to its neighbor drop receives minus charge from it through the bridge (Fig. 12). The drop that attained the positive charge moves upward, contacts the ground electrode and acquires a negative charge. Because of the electrostatic force and gravity the drop moves downward. Coalescence occurs when the first drop has not yet reached the high voltage and meets the other drop.

Figure 13 shows two similarly (negative) charged drops come into contact after several prior repulsions, however they do not coalesce until the lower drop touches the metal surface, attains a positive charge and merges with the oppositely

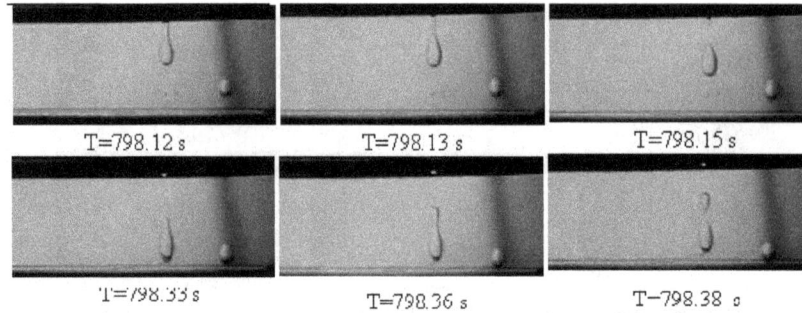

Figure 15. Pinch-off at top and bottom electrode ($Ca_E = 0.388$).

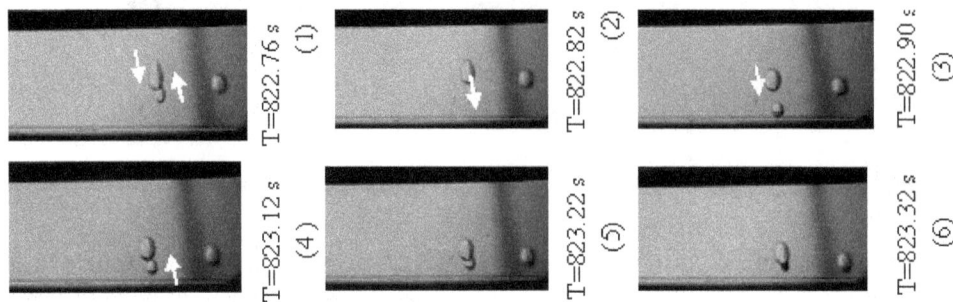

Figure 16. The third type of droplet coalescence ($Ca_E = 0.388$).

charged neighbor drop. Here, coalescence has been detected for two oppositely charged drops.

In Fig. 14 coalescence occurs again. This time, the bigger droplet attaches at the high voltage, acquires positive charge and coalesces with its upper droplet which has a negative charge. According to previous studies in the high voltage field, repelling is dominant, here, however, it seems that the direction (contact angle) at the time of contact plays an important role and leads to coalescence. The connection's line is not straight and the contact occurs side to side. As discussed above, the orientation of the drops play an important role in investigating whether coalescence or bouncing is likely to happen.

Increasing the electric field causes break up (pinch-off) to occur simultaneously. The first instance occurs, when the drop contacts the top electrode and then the bottom electrode (Fig. 15).

As mentioned above, it is possible to have coalescence in a high electric field. In Fig. 16, the top drop becomes close to the bottom one. There are two options available; coalescence and repelling. Because two opposite charges attract, it is expected that they merge and make a bigger drop, however, this has not happened and repelling has occurred instead of coalescence. So based on Ristenpart et al. (2009) charge transmission has occurred and the bottom drop obtained the negative charge and started moving downward. Because of this charge changing, it is assumed that the top drop would acquire positive charge and move upward, but the next frame

does not show this movement. In contrast, it tends to fall with its weight and the vertical distance from the bottom electrode decreases. After acquiring a positive charge from the bottom electrode it starts moving upward and contacts with the upper drop which is resting in its former position or has moved a short distance, and this is the time at which coalescence occurs. If it is justifiable that coalescence occurs with two opposite charges there would be one option left, i.e., the average amount of charge for the upper drop is negative. Therefore it may be correct to presume that the upper drop does not share its entire charge with the small droplet and that the amount of transmitted charge is not sufficient to create a positive charge within the upper droplet. This case is considered the third type of coalescence. If the last option is true, then acceleration towards the high voltage is expected for the top droplet based on electrophoretic force; however the top droplet moves very slowly similar to the droplet with zero charge. The question remains regarding the charge transmission.

4 Conclusions

In the current study, the deformation, electrical charging, coalescence and breakup of water drops moving vertically between two horizontal electrodes in an insulating liquid were investigated experimentally. Any changes that occurred during rhythmic motion have been addressed. The study confirms the recent results obtained by researchers, and the main observations are listed as follows:

- Increasing the voltage results in a Taylor cone when droplets are resting on or reaching the electrodes. By adding the voltage, the angles become smaller and the droplets leave the electrodes. The deformation rate increases with the applied voltage.

- The torque, which exists at the points of the drops, causes rotation while the electric field wants to align them in the vertical direction. This interaction continues until the distance is short enough to cause coalescence. Additionally coalescence does not necessarily occur from the tip and side by side coalescence is occasionally expected.

- After coalescence, assuming the charge transmission occurs equally between the oppositely charged droplets, the consequent droplet, which has a bigger diameter, rotates such that it aligns itself in the electric field direction (because of the polarization). Because of zero charge, the droplet falls with its weight.

- At high voltage, because of the polarization, the droplet elongates and finally disintegrates into two or more tiny droplets. These tiny droplets either coalesce or bounce against each other. Bouncing occurs because of the changing charges.

- When a thread is formed between electrodes it disintegrates immediately.

- There are three types of coalescence modes mentioned above: (i) typical coalescence, which is between oppositely charged droplets, (ii) coalescence between two droplets with the same sign maybe because of hydrodynamic coalescence and (iii) coalescence after bouncing where one of the droplets in the transmission section does not move to the opposite electrode but moves very slowly and relatively rests at its first position (the time of contact) to coalesce with the second droplet (Fig. 16).

- At very high voltage, droplets gather in a certain line, which is not in an electric field direction and deviates from the vertical line. The chain-like shape allows currents to pass through the distance between the electrodes, a spark is therefore inevitable.

Acknowledgements. We would like to express our deep appreciation to B. Khorshidi (university of Alberta) for his valuable comments during the preparation of the work.

Edited by: A. Barari

References

Ahn, K., Agresti, J., Chong, H., Marquez, M., and Weitz, D. A.: Electrocoalescence of drops synchronized by size-dependent flow in microfluidic channels, Appl. Phys. Lett., 88, 264105, doi:10.1063/1.2218058, 2006.

Aldaeus, F., Lin, Y., Amberg, G., and Roeraade, J.: Multi-step dielectrophoresis for separation of particles, J. Chromatogr. A, 1131, 261–266, 2006.

Allan, R. S. and Mason, S. G.: Particle behavior in shear and electric fields; I:deformation and burst of fluid drops, Proc. Roy. Soc. Lond. A, 267, 45–61, 1962.

Bailey, A. G.: Electrostatic spraying of liquids, Wiley, New York, 1988, dynamic processes in electrospraying, J. Aerosol Sci., 30 (Suppl. 1), 549–550, 1999.

Baygents, J. C. and Saville, D. A.: The circulation produced in a drop by an electric field: a high field strength electrokinetic model, in: Drops and Bubbles: Third Int. Colloq., edited by: Wang, T. G., American Institute of Physics, 7–17, 1989.

Chiesa, M., Melheim, J. A., Pedersen, A., Ingebrigtsen, S., and Berg, G.: Forces acting on water droplets falling in oil under the influence of an electric field: numerical predictions versus experimental observations, Eur. J. Mech. B-Fluid., 24, 717–732, 2005.

Dong, W., Li, R. Y., Yu, H. L., and Yan, Y. Y.: An investigation of behaviors of a single bubble in a uniform electric field, Exp. Therm. Fluid Sci., 30, 579–586, 2006.

Eow, J. S. and Ghadiri, M.: Motion, deformation and break-up of aqueous drops in oils under high electric field strengths, Chem. Eng. Process., 42, 259–272, 2003a.

Eow, J. S. and Ghadiri, M.: Drop-Drop coalescence in an electric field: the effects of applied electric field and electrode geometry, Colloid. Surface. A, 219, 253–279, 2003b.

Eow, J. S., Ghadiri, M., Sharif, A. O., and Williams, T. J.: Electrostatic enhancement of coalescence of water droplets in oil: a review of the current understanding, Chem. Eng. J., 84, 173–192, 2001.

Feng, J. Q. and Scott, T. C.: A computational analysis of electrohydrodynamics of a leaky dielectric drop in an electric field, J. Fluid Mech., 311, 289–326, 1966.

Ha, J. W. and Yang, S. M.: Deformation and break-up of Newtonian and non-newtonian conducting drops in an electric field, J. Fluid Mech., 405, 131–156, 2000.

Hase, M., Watanabe, S. N., and Yoshikawa, K.: Rhythmic motion of a droplet under a dc electric field, Phy. Rev. E., 74, 046301, doi:10.1103/PhysRevE.74.046301, 2006.

Hokmabad, B. V., Sadri, B., Charan, M. R., and Esmaeilzadeh, E.: An experimental investigation on hydrodynamics of charged water droplets in dielectric liquid medium in the presence of electric field, Colloid. Surface. A, 401, 17–28, 2012.

Khorshidi, B., Jalaal, M., Esmaeilzadeh, E., and Mohammadi, F.: Characteristics of deformation and electrical charging of large water drops immersed in an insulating liquid on electrode surface, Colloid Interface Sci., 352, 211–220, 2010.

Pikal, M. J.: The role of electroosmotic flow in transdermal iontophoresis, Advance Drug Delivery, Reviews, 46, 281–305, 2001.

Ristenpart, W. D., Bird, J. C., Belmonte, A., Dollar, F., and Stone, H. A.: Non-coalescence of oppositely charged drops, Nature, 461, 377–380, 2009.

Sherwood, J. D.: Breakup of fluid droplets in electric and magnetic fields, J. Fluid Mech., 188, 133–146, 1988.

Taylor, G. L.: Proc. Studies in electrohydrodynamics; I: the circulation produced in a drop by an electric field, Roy. Soc. A, 291, 159–167, 1966.

Torza, S., Cox, R. G., and Mason, S. G.: Electrohydrodynamic deformation and burst of liquid drops, Philos. Trans. R. Soc. Lond. A, 269, 295–319, 1971.

Vizika, O. and Saville, D. A.: The electrohydrodynamic deformation of drops suspended in liquids in steady and oscillatory electric fields, J. Fluid Mech., 239, 1–21, 1992.

18

Level set-based topology optimisation of a compliant mechanism design using mathematical programming

M. Otomori[1]**, T. Yamada**[2]**, K. Izui**[1]**, and S. Nishiwaki**[1]

[1]Kyoto University, Kyoto, Japan
[2]Nagoya University, Nagoya, Japan

Abstract. We propose a structural optimisation method, based on the level set method and using mathematical programming such as the method of moving asymptotes (MMA), which we apply to the design of compliant mechanisms. A compliant mechanism is a monolithic joint-free mechanism designed to be flexible to obtain a specified motion. In the design of compliant mechanisms, several requirements such as the direction of the deformation and stress concentrations must be considered to obtain the specified mechanical function. Topology optimisation, the most flexible type of structural optimisation, has been successfully used as a design optimisation method for compliant mechanisms, but the utility of topology optimisation results is often spoiled by a plethora of impractical designs such as structures containing grayscale areas. Level set-based topology optimisation methods are immune to the problem of grayscales since the boundaries of the optimal configuration are implicitly represented using the level set function. The proposed method updates the level set function using mathematical programming to facilitate the treatment of constraint functionals. To verify its capability, we apply our method to compliant mechanism design problems that include displacement constraints and stress constraints.

1 Introduction

Compliant mechanisms are gaining increasing attention as their application in myriad mechanical devices such as MEMS broadens. A compliant mechanism is a monolithic joint-free mechanism designed to be flexible to obtain a specified motion. The major advantages of compliant mechanisms are simplified manufacturing and assembly, reduced cost, lack of mechanical play, silent operation, and freedom from lubrication requirements (Howell, 2001). The first approach to compliant mechanism design was a kinematic synthesis approach in which rigid-body mechanisms were synthesized into compliant mechanisms (e.g., Her and Midha, 1987). This approach, however, is limited to lumped compliant mechanism designs. For the design of fully compliant mechanisms, topology optimisation methods using the continuum synthesis approach are used. In such methods, Sigmund (1997) formulated the objective function as the ratio between input and output forces, called the mechanical advantage. Nishiwaki et al. (1998) presented a structural topol-

ogy optimisation method for compliant mechanisms, where the concept of mutual energy was used in the formulation of flexibility.

Topology optimisation, firstly proposed by Bendsøe and Kikuchi (1988), has been successfully applied to many problems such as minimum mean compliance problems (Suzuki and Kikuchi, 1991), eigen-frequency problems (Diaz and Kikuchi, 1992), electromagnetic problems (Yoo et al., 2000) and so on. The basic concepts of topology optimisation are the extension of the design domain and replacement of the optimisation problem with a material distribution problem using the characteristic function (Murat and Tartar, 1985). Such material distribution problems are known to be ill-posed, so a relaxation technique is required, such as the homogenization design method or the SIMP method. Topology optimisation methods are the most flexible of optimisation methods since topological changes as well as shape changes are allowed during the optimisation process. However, this advantage is often offset in the optimal configurations by a plethora of impractical designs such as structures containing grayscales or excessive detail, which spoils the utility of the optimal configurations from an engineering standpoint.

Nomenclature

$\chi(x)$	characteristic function
$\chi_\phi(x)$	characteristic function using the level set function
D	fixed design domain
d	ratio of the Young's modulus of the void domains to the solid domain
∂D_D	non-design boundaries
$\partial\Omega$	solid domaim boundaries
$\mathbf{E}$	elasticity tensor
$\widehat{\varepsilon}$	global stress constraint relaxation factor
ε	strain tensor
F	objective functional
F_R	regularized objective functional
$\tilde{G}_{global}$	global stress constraint
$\phi(x)$	level set function
$f(x)$	density function of objective functional
G	constraint functional
G_{max}	upper limit of constraint functional
$g(x)$	density function of constraint functional
Γ_{in}	imposed input force boundary
Γ_{out}	output boundary
Γ_u	prescribed displacement boundary
$H_a(\phi)$	approximated Heaviside function
Ω	solid domain
μ	aggregation parameter for global stress constraint
ψ_e	stress relaxation coefficient
σ_{max}	upper limit of local stress
σ_{VM}	Von Mises stress
t_{in}	input force
t_{out}	dummy vector applied at output boundary
t'_{out}	dummy vector orthogonal to t_{out}
τ	regularization parameter
t	fictitious time
Δt	time increment
U	Sobolev space of admissible displacement
u_1	displacement when input force t_{in} is applied
V_{max}	upper limit of volume constraint
v	admissible displacement
w	transition width of the approximated Heaviside function
x	position in fixed design domain

A type of structural optimisation method using level set boundary expressions has been proposed in which the boundaries of the optimal configuration are implicitly represented using the level set function. A level set-based structural optimisation method was firstly proposed by Sethian and Wiegmann (2000) where the level set function is updated based on the Von Mises stress. Wang et al. (2003) and Allaire et al. (2004) proposed a level set-based structural optimisation method where the level set function is updated using the Hamilton-Jacobi equation, based on the shape sensitivities. Several level set-based structural optimisation methods for the design of compliant mechanisms have been proposed and applied to a multi-material problem (Wang et al., 2005), and a stress minimization problem (Allaire and Jouve, 2008). However, these particular level set-based structural optimisa-

tion methods can be categorized as a type of shape optimisation because the introduction of holes is not allowed, but the number of holes can be decreased during optimisation. As a result, the obtained optimal configurations are greatly affected by guesses concerning the initial configuration. To alleviate this problem, Allaire et al. (2005) proposed a level set-based structural optimisation method coupled with the topological gradient method (e.g., Céa et al., 2000).

Level set-based structural topology optimisation methods that do not use a level set function having the property of a signed distance function, resulting in that the introduction of holes is allowed, have also been proposed, such as by Wei and Wang (2009), in which a piecewise constant level set function is used. Their method formulates the objective function as the sum of a primary objective functional and the perimeter of the structure, and a constraint is applied so that the level set function becomes piecewise constant. However, the magnitude of the constraint parameter greatly affects the optimal configuration so that, again, initial configuration settings often determine the utility, or lack thereof, of the obtained optimal configurations. Luo and Tong (2008) proposed a level set-based topology optimisation method incorporating a radial basis function for the design of compliant mechanisms. However, some experience is required when choosing appropriate values for the radial basis function parameters. Yamada et al. (2010) proposed a level set-based topology optimisation method incorporating a fictitious interface energy derived from the phase field concept. In this method, the objective functional is the sum of a primary objective functional and a fictitious interface energy. A piecewise constant level set function is used in this method and the updating scheme uses a reaction-diffusion equation.

The aim of this paper is to extend Yamada's method to the design of compliant mechanisms so that displacement and stress constraints can be easily included. In our proposed method, the level set function is updated using the MMA (Svanberg, 1987) here, to facilitate the treatment of constraint functionals. In Sect. 2, the formulation of the level set-based topology optimisation procedure and optimisation problems are discussed. In Sect. 3, the numerical implementation is discussed, and numerical examples considering displacement and stress constraints are presented in Sect. 4, to confirm the validity and utility of the proposed method.

2 Formulation

2.1 Level set-based topology optimisation

Here, we briefly discuss the level set-based topology optimisation incorporating a fictitious energy. Topology optimisation is formulated in a fixed design domain D that consists of a domain filled with solid material Ω and a domain filled with void material. Using the characteristic function $\chi(\boldsymbol{x})$ defined as

$$
\chi(\boldsymbol{x}) = \begin{cases} 1 \text{ if } \boldsymbol{x} \in \Omega \\ 0 \text{ if } \boldsymbol{x} \in D \backslash \Omega \end{cases} \tag{1}
$$

the optimisation problem can be replaced by a material distribution problem. The optimisation problem is formulated as

$$
\begin{aligned}
&\inf_{\chi} F(\chi) = \int_D f(\boldsymbol{x}) \chi \mathrm{d}\Omega \\
&\text{subject to } G(\chi) = \int_D g(\boldsymbol{x}) \chi \mathrm{d}\Omega - G_{\max} \le 0
\end{aligned} \tag{2}
$$

where F is the objective functional, G is a constraint functional and $G_{\max}$ is the upper limit of the constraint functional.

The following level set function $\phi(\boldsymbol{x})$ is used to represent the boundaries of the structure, where positive values represent the solid domain, negative values represent the void domain, and zero represents the boundary surfaces.

$$
\begin{cases} 1 \ge \phi(\boldsymbol{x}) > 0 & \text{for } \boldsymbol{x} \in \Omega \backslash \partial\Omega \\ \phi(\boldsymbol{x}) = 0 & \text{for } \boldsymbol{x} \in \partial\Omega \\ 0 > \phi(\boldsymbol{x}) \ge -1 & \text{for } \boldsymbol{x} \in D \backslash \Omega \end{cases} \tag{3}
$$

As a result, the characteristic function $\chi(\boldsymbol{x})$ is replaced by following function, $\chi_\phi(\boldsymbol{x})$.

$$
\chi_\phi(\boldsymbol{x}) = \begin{cases} 1 & \text{if } 1 \ge \phi(\boldsymbol{x}) \ge 0 \\ 0 & \text{if } 0 > \phi(\boldsymbol{x}) \ge -1 \end{cases}
$$

The above optimisation problem is an ill-posed problem since the optimal configuration expressed by the level set function is not required to be continuous. To regularize the optimisation problem, Yamada et al. (2010) proposed a regularization technique based on Tikhonov regularization. Thus, the above optimisation problem is replaced with the following:

$$
\begin{aligned}
&\inf_{\chi} F_R(\chi_\phi(\phi)) = \int_D f(\boldsymbol{x}) \chi_\phi \mathrm{d}\Omega + \int_D \tfrac{1}{2} \tau |\nabla\phi|^2 \mathrm{d}\Omega \\
&\text{subject to } G(\chi_\phi(\phi)) = \int_D g(\boldsymbol{x}) \chi_\phi \mathrm{d}\Omega - G_{\max} \le 0
\end{aligned} \tag{4}
$$

The proposed method updates the level set function $\phi(\boldsymbol{x})$ using mathematical programming, the MMA.

2.2 Optimal synthesis of compliant mechanisms

Here, the optimum design of compliant mechanisms is briefly discussed. The main goal of the present optimum design process is to maximize the output displacement in a desired direction. Consider the design domain D where the displacement is fixed at boundary Γ_u, an input force $\boldsymbol{t}_{in}$ is applied at boundary Γ_{in}, and a dummy vector $\boldsymbol{t}_{out}$ is introduced at the output port, boundary Γ_{out}, along the desired output direction.

Two functions are required for the design of compliant mechanisms. One is to provide sufficient flexibility for deformation along a desired direction specified by a dummy

vector $\boldsymbol{t}_{out}$ when an input force is applied. The mutual mean compliance between Γ_{in} and Γ_{out} is used in this paper to formulate the flexibility of the target structure. By maximizing the mutual mean compliance, the output displacement is maximized along the direction of the dummy vector $\boldsymbol{t}_{out}$. The second requirement is for sufficient stiffness to maintain the integrity of structural shapes when workpiece reaction forces and the input force are applied. Dummy springs are imposed at the input and output ports to represent the input and reaction forces. The optimisation problem under a volume constraint is then formulated as follows.

$$
\begin{aligned}
&\inf_{\chi} F(\chi_\phi) = -l_2(\boldsymbol{u}_1) \\
&\text{subject to } a(\boldsymbol{u}_1, \boldsymbol{v}) = l_1(\boldsymbol{v}) \text{ for } \forall \boldsymbol{v} \in U, \forall \boldsymbol{u}_1 \in U
\end{aligned} \tag{5}
$$

$$
G(\chi_\phi) = \int_\Omega \mathrm{d}\Omega - V_{\max} \le 0
$$

where $\boldsymbol{u}_1$ is the displacement when $\boldsymbol{t}_{in}$ is applied at Γ_{in}, $V_{\max}$ is the upper limit of the volume constraint, and the other notations are defined as follows.

$$
a(\boldsymbol{u}, \boldsymbol{v}, \chi_\phi) = \int_D \boldsymbol{\varepsilon}(\boldsymbol{u}) : \mathbf{E} : \boldsymbol{\varepsilon}(\boldsymbol{v}) \chi_\phi \mathrm{d}\Omega \tag{6}
$$

$$
l_1(\boldsymbol{v}) = \int_{\Gamma_{in}} \boldsymbol{t}_{in} \cdot \boldsymbol{v} \mathrm{d}\Gamma \tag{7}
$$

$$
l_2(\boldsymbol{v}) = \int_{\Gamma_{out}} \boldsymbol{t}_{out} \cdot \boldsymbol{v} \mathrm{d}\Gamma \tag{8}
$$

$$
U = \left\{ \boldsymbol{v} = v_i \boldsymbol{e}_i : v_i \in H^1(\Omega) \text{ with } \boldsymbol{v} = 0 \text{ on } \Gamma_u \right\} \tag{9}
$$

The sensitivity of the objective functional is simply obtained as follows using the adjoint variable method.

$$
\begin{aligned}
&\left\langle \frac{\mathrm{d}F(\chi_\phi)}{\mathrm{d}\phi}, \delta\phi \right\rangle = -\left\langle \frac{\partial l_2(\boldsymbol{u}_1)}{\partial \boldsymbol{u}_1}, \delta\boldsymbol{u}_1 \right\rangle \left\langle \frac{\partial \boldsymbol{u}_1}{\partial \phi}, \delta\phi \right\rangle + \left\langle \frac{\partial a(\boldsymbol{u}_1, \boldsymbol{v}, \chi_\phi)}{\partial \boldsymbol{u}_1}, \delta\boldsymbol{u}_1 \right\rangle \\
&\left\langle \frac{\partial \boldsymbol{u}_1}{\partial \phi}, \delta\phi \right\rangle + \left\langle \frac{\partial a(\boldsymbol{u}_1, \boldsymbol{v}, \chi_\phi)}{\partial \phi}, \delta\phi \right\rangle = \left\langle \frac{\partial a(\boldsymbol{u}_1, \boldsymbol{v}, \chi_\phi)}{\partial \phi}, \delta\phi \right\rangle
\end{aligned} \tag{10}
$$

where the adjoint field is defined as follows.

$$
a(\boldsymbol{v}, \boldsymbol{u}_1) = l_2(\boldsymbol{u}_1) \text{ for } \forall \boldsymbol{u}_1 \in U, \boldsymbol{v} \in U \tag{11}
$$

2.3 Optimum design problem with mutual mean compliance constraint

The formulation of the optimum design problem is now extended to a problem with a constraint so that the displacement in the direction orthogonal to the desired output direction will be constrained.

$$
\begin{aligned}
&\inf_{\chi} F(\chi_\phi) = -l_2(\boldsymbol{u}_1) \\
&\text{subject to } a(\boldsymbol{u}_1, \boldsymbol{v}) = l_1(\boldsymbol{v}) \text{ for } \forall \boldsymbol{v} \in U, \forall \boldsymbol{u}_1 \in U
\end{aligned} \tag{12}
$$

$$
G_1(\chi_\phi) = \int_\Omega \mathrm{d}\Omega - V_{\max} \le 0
$$

$$
G_2(\chi_\phi) = l_3(\boldsymbol{u}_1) = 0
$$

where $l_3(\boldsymbol{u}_1)$ is the mutual mean compliance when the dummy load $\boldsymbol{t}'_{\text{out}}$ which is orthogonal to $\boldsymbol{t}_{\text{out}}$ is applied, as follows.

$$l_3(\boldsymbol{v}) = \int_{\Gamma_{\text{out}}} \boldsymbol{t}'_{\text{out}} \cdot \boldsymbol{v} \, \mathrm{d}\Gamma \tag{13}$$

The sensitivity of $G_2(\chi_\phi)$ is also simply obtained using the adjoint variable method.

2.4 Optimum design problem with stress constraint

Here, an optimum design problem with a stress constraint is discussed. Since the utility of compliant mechanisms depends on their structural flexibility, stress concentrations easily occur at thin locations that are subject to repeated flexing. Therefore, the implementation of a stress constraint in the optimisation method can be advantageous for the design of reliable compliant mechanisms that avoid structural failures over their projected lifetime.

Several different stress constraint formulations have been studied, such as local stress constraints (e.g. Duysinx and Bendsøe, 1998), global stress constraints (e.g. Martins and Poon, 2005; París et al., 2009) and the block aggregated approach (e.g. París et al., 2010a) which is a hybrid approach combining local and global stress constraints. These formulations are compared in the literature (París et al., 2009; París et al., 2010b). The implementation of local stress constraints uses a straightforward approach, with stress constraints imposed at predefined points such as the centre of finite elements, however this often increases computational demands to the point of intractability. On the other hand, global stress constraints impose a single global constraint that aggregates the effect of all local stress constraints. Although local stress constraints may not be strictly satisfied, the use of a global stress constraint greatly reduces computational demands, and we use a global constraint in this research.

The global stress constraint is formulated as follows.

$$G_{\text{global}}(\boldsymbol{u}_1, \phi) = \tilde{G}_{\text{global}}(\sigma_{\text{VM}}(\boldsymbol{u}_1), \phi) \le 0 \tag{14}$$

where

$$\tilde{G}_{\text{global}}(\sigma_{\text{VM}}, \phi) = \frac{1}{\mu} \ln\left[\int_\Omega \exp^{\mu(\sigma^*-1)} d\Omega \right] - \frac{1}{\mu} \ln\left(\int_\Omega d\Omega \right) \tag{15}$$

where μ is a parameter called the "aggregation parameter". Higher magnitudes of parameter μ impose higher penalties for violated local constraints. σ^* in the above equation is defined as follows.

$$\sigma^* = \frac{\sigma_{\text{VM}}}{\sigma_{\text{max}} \psi_e} \tag{16}$$

where σ_{VM} represents the Von Mises stress, σ_{max} is an applied stress constraint and ψ_e is a parameter called the "stress relaxation coefficient" which is introduced to avoid singularity phenomena and is formulated as follows.

$$\psi_e = 1 - \widehat{\varepsilon} + \frac{\widehat{\varepsilon}}{H_a(\phi)} \tag{17}$$

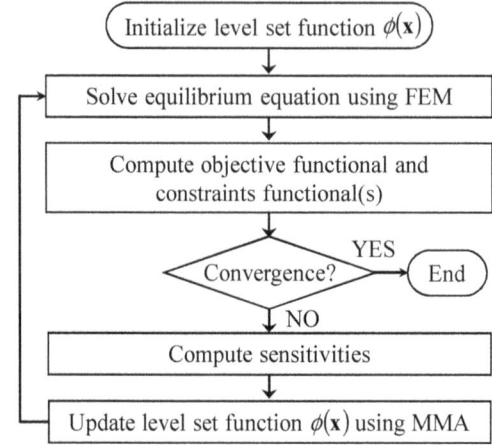

Figure 1. Optimization flowchart.

where $\widehat{\varepsilon}$ is a parameter called the "relaxation factor" which adjusts the magnitude of relaxation and $H_a(\phi)$ is a Heaviside function to approximate the equilibrium function, which will be described in Sect. 3.3.

The sensitivity of the global stress constraints is formulated as follows.

$$\left\langle \frac{\partial G_{\text{global}}(\boldsymbol{u}_1, \phi)}{\partial \phi}, \delta\phi \right\rangle = \left\langle \frac{\partial \tilde{G}_{\text{global}}(\sigma_{\text{VM}}, \phi)}{\partial \sigma_{\text{VM}}}, \delta\sigma_{\text{VM}} \right\rangle \left\langle \frac{\partial \sigma_{\text{VM}}}{\partial \boldsymbol{u}_1}, \delta\boldsymbol{u}_1 \right\rangle \left\langle \frac{\partial \boldsymbol{u}_1}{\partial \phi}, \delta\phi \right\rangle$$
$$+ \left\langle \frac{\partial \tilde{G}_{\text{global}}(\sigma_{\text{VM}}, \phi)}{\partial \phi}, \delta\phi \right\rangle - \left\langle \frac{\partial a(\boldsymbol{u}_1, \boldsymbol{v}, \phi)}{\partial \boldsymbol{u}_1}, \delta\boldsymbol{u}_1 \right\rangle \left\langle \frac{\partial \boldsymbol{u}_1}{\partial \phi}, \delta\phi \right\rangle - \left\langle \frac{\partial a(\boldsymbol{u}_1, \boldsymbol{v}, \phi)}{\partial \phi}, \delta\phi \right\rangle$$
$$= \left\langle \frac{\partial \tilde{G}_{\text{global}}(\sigma_{\text{VM}}, \phi)}{\partial \phi}, \delta\phi \right\rangle + \left\langle \frac{\partial a(\boldsymbol{u}_1, \boldsymbol{v}, \phi)}{\partial \phi}, \delta\phi \right\rangle \tag{18}$$

where the adjoint variable is obtained by solving the following equation.

$$\left\langle \frac{\partial \tilde{G}_{\text{global}}(\sigma_{\text{VM}}, \phi)}{\partial \sigma_{\text{VM}}}, \delta\sigma_{\text{VM}} \right\rangle \left\langle \frac{\partial \sigma_{\text{VM}}}{\partial \boldsymbol{u}_1}, \delta\boldsymbol{u}_1 \right\rangle - \left\langle \frac{\partial a(\boldsymbol{u}_1, \boldsymbol{v}, \phi)}{\partial \boldsymbol{u}_1}, \delta\boldsymbol{u}_1 \right\rangle = 0$$

$$\text{for } \forall \boldsymbol{u}_1 \in U, \boldsymbol{v} \in U \tag{19}$$

3 Numerical implementation

3.1 Optimisation algorithm

The optimisation flowchart is shown in Fig. 1. First, the level set function is initialized. Second, the equilibrium equation is solved using the Finite Element Method (FEM) and the objective functional and constraint functionals are then calculated, also using the FEM. If the objective functional has converged, the optimisation procedure is terminated. If not, the sensitivities of objective and constraint functionals, derived as a continuous expression in the previous section, are computed at Gaussian points of the finite elements. The sensitivities are mapped to the nodes of the finite elements, the level set function is then updated using the MMA, and the process returns to the second step.

Note that the method cited earlier (Yamada et al., 2010), the level set function is updated using a reaction-diffusion equation that is derived based on the Lagrange multiplier method. In the case of multi-constraint problems, the derivation of the Lagrange multiplier used in reaction-diffusion equations becomes complicated, so the proposed method uses the MMA to update the level set function, which facilitates the treatment of constraints. The objective and constraint functionals are approximated using a convex function, and the approximated subproblem is solved at each iteration.

3.2 Approximated $\nabla^2 \phi$

In the proposed method, $\frac{1}{2}\tau|\nabla\phi|^2$ is used as a regularization term. Therefore, $-\tau\nabla^2\phi$ must be calculated to compute the sensitivity, and we use the following approximation technique. Note that the variation of the regularization term is formulated in Gurtin (1996). Here, a Dirichlet boundary condition is applied on the non-design boundaries, and Neumann boundary condition is applied on the other boundaries to represent the level set function independently of the outside of the fixed design domain (Yamada et al., 2010). First, we introduce the following time evolutionary equation.

$$\begin{cases} \frac{\partial\phi}{\partial t} = -\tau\nabla^2\phi & \text{in } D \\ \frac{\partial\phi}{\partial n} = 0 & \text{on } \partial D\backslash\partial D_D \\ \phi = 1 & \text{on } \partial D_D \end{cases} \quad (20)$$

∂D_D represents non-design boundaries where the Dirichlet boundary condition is applied. Next, the above equation is discretized in the time domain using the Finite Difference Method, which leads to the following equation.

$$\begin{cases} \frac{\phi(t+\Delta t)-\phi(t)}{\Delta t} = -\tau\nabla^2\phi & \text{in } D \\ \frac{\partial\phi}{\partial n} = 0 & \text{on } \partial D\backslash\partial D_D \\ \phi = 1 & \text{on } \partial D \end{cases} \quad (21)$$

The above equation is then expressed in weak form as follows.

$$\begin{cases} \int_D \frac{\phi(t+\Delta t)}{\Delta t}\tilde{\phi}dD + \int_D \nabla^T\phi(t+\Delta t)\tau\nabla\tilde{\phi}dD = \int_D \frac{\phi(t)}{\Delta t}\tilde{\phi}dD \\ \phi = 1 \text{ on } \partial D_D \end{cases} \quad (22)$$

The above equation can be solved using the FEM. $\nabla^2\phi$ is approximately computed using following equation.

$$\nabla^2\phi \cong -\frac{\phi(t+\Delta t)-\phi(t)}{\tau\Delta t} \quad (23)$$

3.3 Approximated equilibrium equation

In this paper, the equilibrium equation is approximated using the ersatz material approach, following the literature (e.g. Allaire et al., 2004; Yamada et al., 2010). The equilibrium equation, Eq. (6), is approximated using following equation.

$$\int_D \varepsilon(u):\mathbf{E}:\varepsilon(v)H_a(\phi)d\Omega = \int_{\Gamma_{in}} t_{in}\cdot v d\Gamma \quad (24)$$

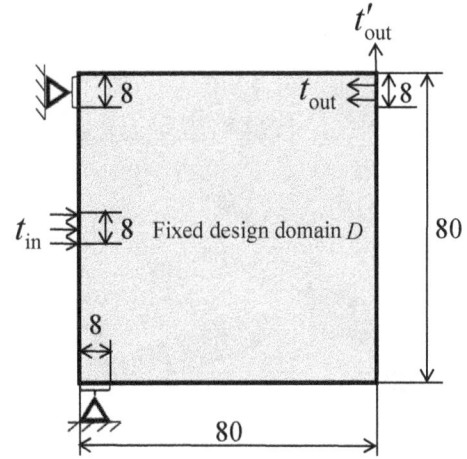

Figure 2. Fixed design domain and boundary conditions of model A.

where $H_a(\phi)$ is the Heaviside function defined as follows.

$$H_a(\phi) = \begin{cases} d & (\phi < -w) \\ \left(\frac{1}{2} + \frac{\phi}{w}\left(\frac{15}{16} - \frac{\phi^2}{w^2}\left(\frac{5}{8} - \frac{3}{16}\frac{\phi^2}{w^2}\right)\right)\right)(1-d)+d & (-w < \phi < w) \\ 1 & (w < \phi) \end{cases} \quad (25)$$

where d is the ratio of the Young's modulus of the void domains to the solid domain and w is the transition width of the Heaviside function.

4 Numerical examples

In this section, two numerical examples are illustrated to verify the utility and validity of our method. In the following examples, the Young's modulus of the elastic material is 210 GPa, Poisson's ratio is 0.33, and the upper limit of the volume constraint is set to 30% of the design domain. The initial configuration is filled with material in the fixed design domain.

4.1 Compliant mechanism design problem with mutual mean compliance constraint

Figure 2 shows the fixed design domain and boundary conditions for model A. The load is applied at the centre of the left edge, with fixed segments at the top of the left edge and extreme left of the bottom edge. A dummy vector t_{out} is applied at the top right of the domain in the horizontal direction. The fixed design domain is descretized using an 80×80 mesh of quadrilateral finite elements. The regularization parameter τ is set to 7.0×10^{-5}. The transition width of the Heaviside function w is set to 0.2 to stabilize the optimisation procedure.

Figure 3a shows the optimal configuration without an applied constraint, with Fig. 3c showing its deformed shape. Figure 3b shows the optimal configuration with a constraint

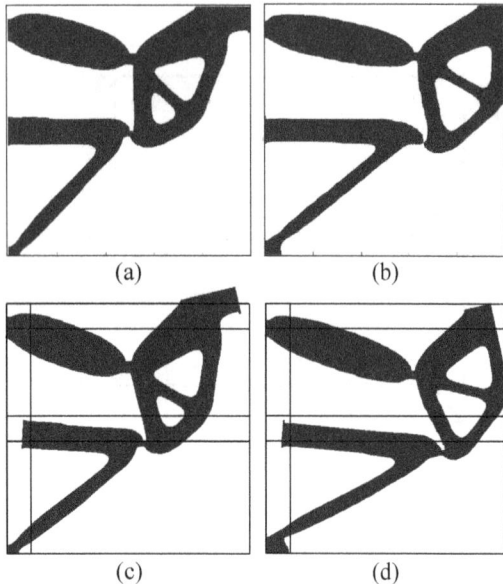

Figure 3. Compliant mechanism optimal configurations: (**a**) without displacement constraints; (**c**) deformed shape of (**a**); (**b**) with displacement constraint; (**d**) deformed shape of (**b**).

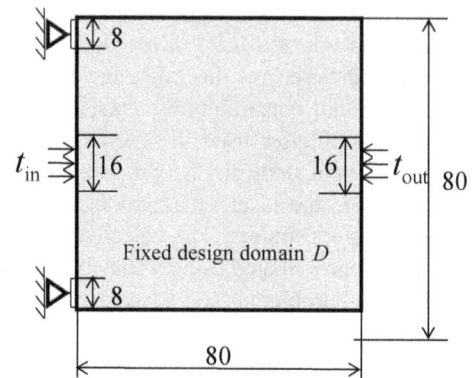

Figure 4. Fixed design domain and boundary conditions of model B.

set so that the structure only deforms in the desired horizontal direction, and Fig. 3d shows its deformed shape. As can be seen in the Fig. 3d illustration, the constraint effectively prevents deformation in the orthogonal direction at the output port. The mutual mean compliance represented by dummy vector t'_{out} is -2.43×10^{-9} without a constraint and -1.42×10^{-10} with a constraint, indicating that displacement in the vertical direction is significantly reduced. The mutual mean compliance represented by dummy vector t_{out}, however, shows little differece in value without and with a constraint, 3.65×10^{-9} and 3.60×10^{-9}, respectively.

Since the transition width of the Heaviside function w is set to 0.2, the optimal configuration contains grayscale areas to some extent. The grayscale areas in the optimal configuration are removed by setting $w = 1.0 \times 10^{-3}$. The mutual mean compliance values represented by dummy vector t'_{out} and t_{out} are then -1.42×10^{-10} and 3.60×10^{-9}, respectively, which are essentially the same as before removing the grayscale areas.

From the numerical results, we can confirm that the proposed method successfully imposed a displacement constraint for the design of compliant mechanisms, which the existing method, where the level set function is updated using a reaction diffusion equation, cannot easily accomplish.

4.2 Compliant mechanism design problem with a stress constraint

Figure 4 shows the fixed design domain and boundary conditions for model B. Small segments at the top and bottom of the left side are fixed, and a load is applied at the centre

of the left side. Since the design domain is symmetric, only the top half is analyzed in the optimisation process. The applied stress constraint σ_{max} is 5.0×10^3. The design domain is discretized using an 80×80 mesh of quadrilateral finite elements and regularization parameter τ is set to 1.0×10^{-4}. The transition width of the Heaviside function w is set to 0.8 and the global stress constraint relaxation factor $\widehat{\varepsilon}$ is set to 0.1.

Figure 5a shows the optimal configuration without an applied stress constraint and Fig. 5c shows the Von Mises stress distribution of this configuration. Figure 5b shows the optimal configuration with the stress constraint applied, and Fig. 5d shows the corresponding Von Mises stress distribution. The Von Mises stresses at the centre of finite elements are considered. The maximum value of the Von Mises stress is 7.18×10^3 without the stress constraint and 5.83×10^3 with the stress constraint. Although the local stress constraints are not strictly satisfied, the maximum value of the Von Mises stress is reduced and the obtained mutual mean compliance values show little difference, namely, 7.62×10^{-10} without the stress constraint and 7.14×10^{-10} with the stress constraint.

Figure 6 shows the density distribution before and after removing grayscale areas of optimal configuration, by setting $w = 1.0 \times 10^{-3}$. The maximum value of the Von Mises stress after removing grayscale areas is 5.86×10^3, while the obtained mutual mean compliance is 7.25×10^{-10}. Although the maximum value of the Von Mises stress is slightly increased after removing grayscale areas, it is sufficiently small compared with the value obtained without an applied stress constraint. Thus, useful optimal configurations can be qualitatively obtained using the proposed method.

5 Conclusions

This paper proposed a level set-based topology optimisation using mathematical programming. In the proposed method, the level set function is updated using mathematical programming, the MMA, to facilitate the treatment of the constraint functional. This is more difficult with the existing

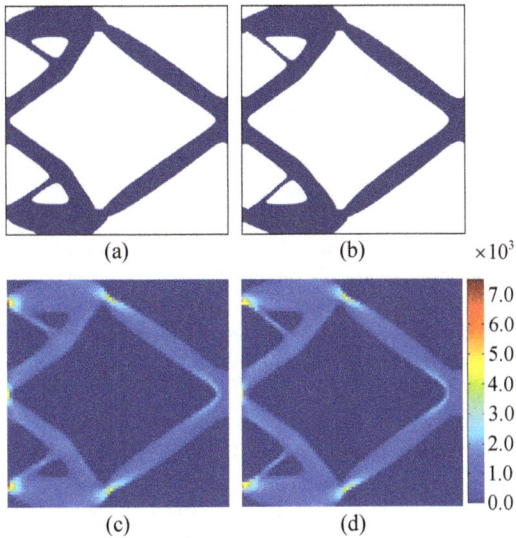

Figure 5. Optimal configurations: **(a)** without stress constraint; **(c)** Von Mises stress distribution of **(a)**; **(b)** with stress constraint; **(d)** Von Mises stress distribution of **(b)**.

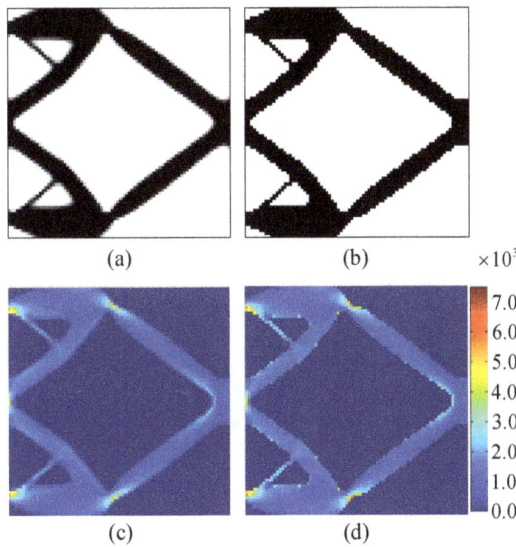

Figure 6. **(a)** Optimal configuration density distribution; **(b)** Von Mises stress distribution of **(a)**; **(b)** Configuration after removing grayscale areas; **(d)** Von Mises stress distribution of **(b)**.

method, in which the level set function is updated using a reaction diffusion equation, where the derivation of the Lagrange multiplier becomes complicated in problems with multiple constraints. A topology optimisation method for compliant mechanisms considering a mutual mean compliance constraint and a stress constraint was presented and optimisation problems were formulated. Although the proposed approach can not explicitly prevent the creation of lumped compliant mechanisms, applying the stress constraint strongly inhibits this, since small areas subject to flex-

ure, and notch hinges, tend to be locations where stress is high. The proposed method was applied to compliant mechanism design problems considering a mutual mean compliance constraint and a stress constraint. A global stress constraint is applied but because it does not require that the stress constraint at every point in design domain be satisfied, the optimal configuration does not strictly satisfy all local stress constraints, even though the global stress constraint is satisfied. We confirmed that the maximum stress was effectively reduced in the obtained optimal configuration. Although the optimal configurations contained grayscale areas to some extent, it was confirmed that useful optimal configurations can be qualitatively obtained using the proposed method.

Acknowledgements. The authors would like to gratefully acknowledge Krister Svanberg's help in providing MMA code. The first author is partially supported by AISIN AW CO., LTD., and sincerely appreciates this assistance.

Edited by: N. Tolou

References

Allaire, G. and Jouve, F: Minimum stress optimal design with the level set method, Eng. Anal. Bound. Elem., 32, 909–918, 2008.

Allaire, G., Jouve, F., and Toader, A. M.: Structural optimization using sensitivity analysis and a level-set method, J. Comput. Phys., 194, 363–393, 2004.

Allaire, G., Gournay, F., Jouve, F., and Toader, A. M.: Structural optimization using topological and shape sensitivity via a level set method, Control Cybern., 34, 59–80, 2005.

Bendsøe, M. P. and Kikuchi, N.: Generating optimal topologies in structural design using a homogenization method, Comput. Method. Appl. M., 71, 197–224, 1988.

Céa, J., Garreau, S., Guillaume, P., and Masmoudi, M.: The shape and topological optimizations connection, Comput. Method. Appl. M., 188, 713–726, 2000.

Diaz, A. R. and Kikuchi, N.: Solutions to shape and topology eigenvalue optimization problems using a homogenization method, Int. J. Numer. Meth. Eng., 35, 1487–1502, 1992.

Duysinx, P. and Bendsøe, M. P.: Topology optimization of continuum structures with local stress constraints, Int. J. Numer. Meth. Eng., 43, 1453–1478, 1998.

Gurtin, M. E.: Generalized Ginzburg-Landau and Cahn-Hilliard equations based on a microforce balance, Physica D, 92, 178–192, 1996.

Her, I. and Midha, A.: A compliance number concept for compliant mechanisms, and type synthesis, J. Mech. Transm-T. ASME, 109, 348–355, 1987.

Howell, L.: Compliant Mechanisms, John Wiley & Sons, Inc., 2001.

Luo, Z. and Tong, L.: A level set method for shape and topology optimization of large-displacement compliant mechanisms, Int. J. Numer. Meth. Eng., 76, 862–892, 2008.

Martins, J. R. R. A. and Poon, N. M. K.: On structural optimization using constraint aggregation, in: (WCSMO6) Proceedings, 6th

World Congress on Structural and Multidisciplinary Optimization, Rio de Janeiro, 2005.

Murat, F. and Tartar, L.: Optimality conditions and homogenization, in: Nonlinear variational problems, edited by: Marino, A., Modica, L., Spagnolo, S., and Degiovannni, M., Pitman Publishing Program, Boston, 1–8, 1985.

Nishiwaki, S., Frecker, M. I., Min, S., and Kikuchi, N.: Topology optimization of compliant mechanisms using the homogenization method, Int. J. Numer. Meth. Eng., 42, 535–559, 1998.

París, J., Navarrina, F., Colominas, I., and Casteleiro, M.: Topology optimization of continuum structures with local and global stress constraints, Struct. Multidiscip. O., 39, 419–437, 2009.

París, J., Navarrina, F., Colominas, I., and Casteleiro, M.: Improvements in the treatment of stress constraints in structural topology optimization problems, J. Comput. Appl. Math., 234, 2231–2238, 2010a.

París, J., Navarrina, F., Colominas, I., and Casteleiro, M.: Stress constraints sensitivity analysis in structural topology optimization, Comput. Method. Appl. M., 199, 2110–2122, 2010b.

Sethian, J. A. and Wiegmann, A.: Structural boundary design via level set and immersed interface methods, J. Comput. Phys., 163, 489–528, 2000.

Sigmund, O.: On the design of compliant mechanisms using topology optimization, Mech. Struct. Mach., 25, 493–524, 1997.

Suzuki, K. and Kikuchi, N.: A homogenization method for shape and topology optimization, Comput. Method. Appl. M., 93, 291–318, 1991.

Svanberg, K.: The method of moving asymptotes – a new method for structural optimization, Int. J. Numer. Meth. Eng., 24, 359–373, 1987.

Wang, M. Y., Wang, X., and Guo, D.: A level set method for structural topology optimization, Comput. Method. Appl. M., 192, 227–246, 2003.

Wang, M. Y., Chen, S., Wang, X., and Mei, Y.: Design of multimaterial compliant mechanisms using level-set methods, J. Mech. Design, 127, 941–956, 2005.

Wei, P. and Wang, M. Y.: Piecewise constant level set method for structural topology optimization, Int. J. Numer. Meth. Eng., 78, 379–402, 2009.

Yamada, T., Izui, K., Nishiwaki, S., and Takezawa, A.: A topology optimization method based on the level set method incorporating a fictitious interface energy, Comput. Method. Appl. M., 199, 2876–2891, 2010.

Yoo, J., Kikuchi, N., and Volakis, J. L.: Structural optimization in magnetic devices by the homogenization design method, IEEE T. Magn., 36, 574–580, 2000.

19

Design and optimization of a XY compliant mechanical displacement amplifier

A. Eskandari and P. R. Ouyang

Department of Aerospace Engineering, Ryerson University, Toronto, Canada

Correspondence to: P. R. Ouyang (pouyang@ryerson.ca)

Abstract. Piezoelectric actuators are increasingly becoming popular for the use in various industrial, pharmaceutical, and engineering applications. However, their short motion range limits their wide applications. This shortcoming can be overcome by coupling the piezoelectric actuators with a mechanical displacement amplifier. In this paper, a new design for a XY planar motion compliant mechanical displacement amplifier (CMDA) based on the design of a symmetric five-bar compliant mechanical amplifier is introduced. Detailed analysis with Finite Element Method (FEM) of static and dynamic characteristics of the proposed XY CMDA design is also provided. Finally, the optimization process and results to increase the Amplification Ratio (AR) of the proposed XY compliant mechanism with minimal compromise in Natural Frequency (NF) is discussed.

1 Introduction

High precision manipulation systems have a variety of uses in many industrial applications, especially where the positioning of components with high accuracy (i.e. in micrometer or nanometer scales) is required (Yong and Lu, 2009). Examples of these applications may include the alignment of fibre-optics and lasers, the positioning of specimens in a scanning-electron-microscope, the positioning of masks in lithography, cells manipulation in micro-biology, and assembly and manipulation of micro-scale components in microassembly applications (Yong and Lu, 2009).

Piezoelectric (PZT) actuators are micro motion generators capable of producing a high displacement resolution and low strain with high force outputs (Ouyang et al., 2008a). However, due to their relatively short motion ranges, the functions of PZT actuators become limited or infeasible for many of the above mentioned applications. One technique to overcome the mentioned shortcoming is to integrate a PZT actuator with a mechanical displacement amplifier (Ouyang et al., 2008b). Such an amplification mechanism can be based on a compliant mechanical displacement amplifier (CMDA) (Timoshenko and Gere, 1961; Howell, 2001; Hull and Canfield, 2006; Lu and Kota, 2006; Su and McCathy, 2007). A CMDA has many advantages such as no friction losses,

no need for lubrication, no tolerance, and et al. over conventional rotating pin-joint mechanisms (Lobontiu, 2002). Hence, the primary goal of a CMDA is to achieve a large output displacement in desired direction(s) for a given input displacement generated by a PZT actuator, and to keep a high positioning resolution at the same time. This, however, may cause a reduction in the structure's generated output force and the natural frequency (NF) of the mechanism (Ouyang et al., 2008b). Nevertheless, the consequent reduction of the output force due to a CMDA can be tolerated since PZT actuators are capable of generating large amount of force, and a relatively high NF can be achieved through properly structural design.

The topic of design, characteristics, and application of CMDAs is not a new one for one direction motion amplification. Numerous literatures regarding this topic can be found in Ananthasuresh and Saxena (2000); Bharti and Frecker (2004); Furukawa et al. (1995); Kota et al. (1999); Pokines and Garcia (1998); Tian et al. (2009); Yang et al. (1996). In general, the performance of a CMDA is a function of some important parameters such as the material properties and the flexure hinge profile of the mechanical displacement amplifier. Xu and King (1996) introduced and compared the performance (in terms of flexibility and accuracy) of three topologies of flexure hinge structures that

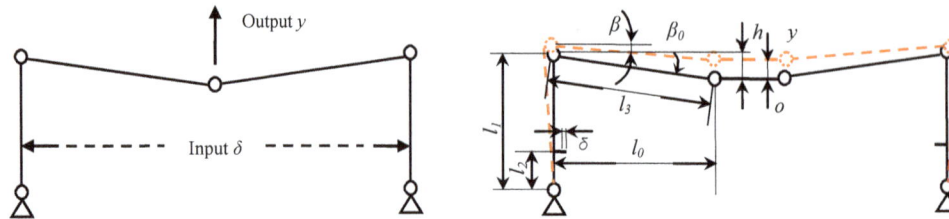

Figure 1. Symmetric five-bar topology.

comprise the elemental components of the majority of compound amplifiers. Shuib et al. (2007) also conducted a review of existing compliant technologies, their applications, and their design limitations. Ma et al. (2006) suggested that increasing the thickness of the flexure hinges will increase the NF of the CMDA. However, this will be accompanied by a significant decrease in the total output displacement. The design of compliant mechanisms with multiple optimally placed and sized PZT actuators for obtaining maximized output deflection and force was also discussed by Bharti and Frecker (2004).

Ouyang et al. (2008b) proposed a new design of CMDA based on a symmetric five-bar topology and compared its performance characteristics to other existing topologies for displacement amplification purpose. But the compliant mechanism developed in Ouyang et al. (2008b) can only produce one direction motion and has its limitations for real applications. It is nature to expand the previous design from one direction motion to two direction motions that suits most micro precision applications. The performed optimization analysis in this paper is the extension of the previous research to XY planar motions and also leads to the introduction of the parameters which have the most effect on the performance of the device.

There are some other advantages of using piezoelectric actuators with mechanical displacement amplifiers with the redundancy in design. That is to say, such a system has a potential of redundancy in providing actuations in robotics. Such redundancy may be utilized for improving the resilience of the system and for improving dynamic performance of the system (Ouyang, 2011). It should be mentioned that the designed CDMA can be used as the micro motion part for a Hybrid macro-micro mechanism (Sun et al., 2011). Therefore, the designed CMDA has its potential in different applications.

2 Topology of a symmetric five-bar structure

The design and optimization of the planar motion generator CMDA, discussed in this paper, is based on the symmetric five-bar structure proposed by Ouyang et al. (2008b). This particular CMDA was designed to be symmetrical in its configuration. As shown in Fig. 1a a PZT actuator is used to produce the input displacement required for simultaneous ro-

tations of two driving links in opposite directions. This then generates an output that can be constrained in one direction only. This topology is actually a combination of a symmetric four-bar topology and a lever arm topology (Yong and Lu, 2009). In another word the vertical bars are of lever arm structures, and between the two lever arms is a symmetric four-bar structure, see Fig. 1a. The advantage of this configuration over the symmetric four-bar topology is its high NF and Amplification Ratio (AR) in a compact size. Figure 1b also shows a pseudo-rigid body model (PRBM) of this model. The PRBM model can be used to predict the output displacement and overall stiffness of the CMDA. A detailed PRBM analysis on the symmetric five-bar structure CMDA is provided by Ouyang et al. (2008b).

It should be mentioned that, from the PRBM in Fig. 1b, the AR is directly related to the ratio of l_1/l_2, h and l_0. Also it is certain that the profile of the flexible hinge has significant contribution to the AR and the NF. Please note that all the design parameters and their importance on the performance of the 5 bar CMDA has been discovered and explained in details. The purpose of this paper was mainly design a planar motion generator based on the design of the 5 bar CMDA and optimize the performance of the device base on the proposed design parameters mentioned (Ouyang et al., 2008b).

3 Design requirements and constraints

As it is mentioned earlier, the symmetric five-bar structure CMDA is only capable of generating an output displacement constrained in only one direction. However, many industrial applications of micro motion devices require manipulators be capable of generating an XY planar motion with high NF. The goal of this research is to exploit the advantages of the symmetric five-bar structure CMDA and propose a novel XY CMDA capable of generating large displacements in both X and Y directions.

The proposed planar motion CMDA comprises of two identical legs located in X and Y directions as shown in Fig. 2. Each leg then is connected with a pair of arms to the output of the device. Leg A of the device translates the horizontal driving force and motion of the PZT to the Y direction motion, while the Leg B does that in the X direction motion. Please note that both legs A and B are each constrained in all degree of freedoms from the base.

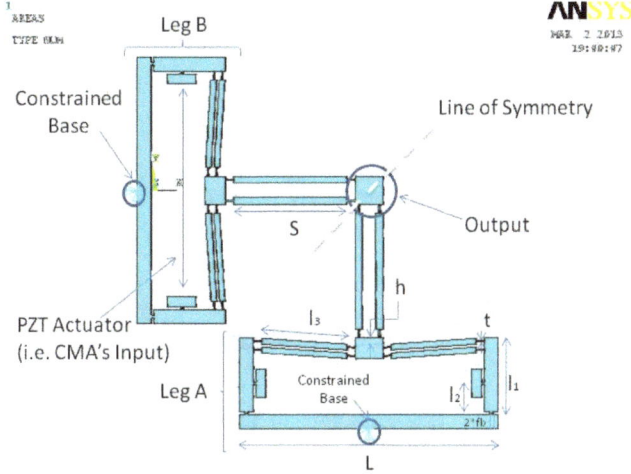

Figure 2. Planar motion generator CMDA based on symmetric five-bar topology.

According to Fig. 2, the proposed planar motion generator CMDA has the following characteristics:

1. The mechanism is a symmetric compliant mechanism with two degree of freedoms (DOF) due to two identical legs in X and Y directions.

2. A CMDA is applied to amplify the stroke of a PZT actuator for each leg.

3. The designed mechanism can be used to generate both planar and linear micro motions.

4. The moving output or end-effector of the device is design to be a cube.

5. Each leg of the planar motion generator CMDA consists of two pairs of the inclined bars of the length l_3 to increase the NF of the designed mechanism.

Some of the key parameters that have significant effect on the performance of the CMDA are also shown in Fig. 2. Please note that the parameters shown on the Leg A can be used to construct the rest of the device since the mechanism is symmetric about the illustrated line of symmetry.

Table 1 below also lists the initial values of the most important parameters used to design the planar motion generator CMDA. Please note that the minimum dimensions that parameters l_2 and L can adopt are related to the dimensions of *NPA50SG* actuators that will be integrated within each leg. S is determined to make two directional motions possible in a compact size. Parameter h also stipulates the inclined angle of the double parallel bars, referring to Fig. 1b. Finally parameters t and fb represent the thickness of the rectangular and circular flexure hinges, respectively. The thickness of the CMDA is 10 mm.

Table 1. Design parameters of the planar motion generator CMDA (unit: mm).

Parameters	L	l_1	S	t	l_2	h	fb	l_3
Dimension	93	26.696	40.5	0.35	11	2.5	0.35	30.57

Table 2. Stainless steel material properties.

Modulus of Elasticity [GPa]	200
Density [kg m^{-3}]	8000
Poisson's Ratio	0.285
Yield Strength [MPa]	703

4 Initial FEM static and dynamic analysis

A 3-D model of the planar motion generator CMDA was constructed using ANSYS Mechanical APDL (ANSYS, 2009) with the primary parameters listed in Table 1. Then by assigning the stainless steel material properties (Table 2) to the model, the CMDA was meshed using SOLID 186 element type. SOLID186 is a higher order 3-D 20-node solid element that exhibits quadratic displacement behavior. The element is characterized by 20 nodes having three translational degrees of freedom per node in the nodal X, Y, and Z directions. This element supports plasticity, hyperelasticity, creep, stress stiffening, large deflection, and large strain capabilities (Ouyang, 2011).

The initial static and dynamic FEM analyses for the following four cases are performed:

– Case 1: input displacements of 10 [μm] were applied to each of the device's four inputs to obtain the mechanism's overall AR and NF.

– Case 2: input displacements of 10 [μm] were applied to each of the two inputs of the Leg A of the device only to get the AR of the mechanism in vertical direction.

– Case 3: input displacements of 10 [μm] were applied to each of the two inputs of the Leg B of the device only to get the AR of the mechanism in horizontal direction.

– Case 4: two forces of 5 [N] each were applied to the devices output (as shown in Fig. 3) while an input displacement of 10 [μm] were applied to each of the device's four inputs. Then the mechanism's overall AR and NF were obtained by performing FEM analysis.

Please note that the AR is the ratio of the maximum output displacement (y) of the device to the input displacement (δ) created by PZT actuator:

$$AR = y/\delta \tag{1}$$

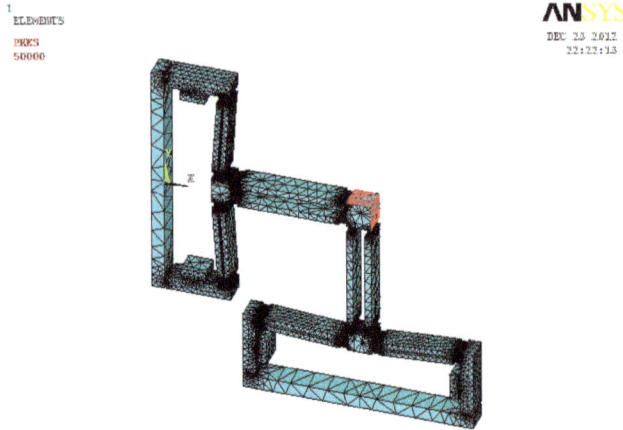

Figure 3. FEM model of the planar motion generator CMDA with the locations of the applied forces shown in red.

Table 3. Initial dynamic and static analysis results.

Case Number	NF [Hz]	AR	Cross-Talk
1	293.09	33.3	–
2	289.27	24.3	19.5
3	289.27	24.3	19.5
4	293.09	27.6	–

The cross-talk of the CDMA was calculated using the following two formulas (Loberto et al., 2004):

$$A_{yx} = \frac{\Delta y}{\Delta x_s} \qquad (2)$$

$$A_{xy} = \frac{\Delta x}{\Delta y_s} \qquad (3)$$

where Δy is the desired displacement in the Y direction and Δx_s is the spurious displacement found in X direction when attempt to displace in Y, while in the case of A_{xy}, Δx is the desired displacement in the X direction and Δy_s is the spurious displacement found in Y direction when attempt to displace in X.

The results obtained from the above analysis are listed in Table 3.

As it can be seen from Table 3 the mechanism has an overall AR of 33.3 and a NF of 293.09 [Hz]. The NF of the system remains the same for Case 4. However; the AR reduces from 33.3 for Case 1 to 27.6 for Case 4 due to device being under loading for this scenario, as the external force will act to resist the output motion of the CMDA. Table 3 also shows that the mechanism possess a NF of 289.286 [Hz] and an AR of 24.3 in each vertical and horizontal directions when only one direction motion is produced using only one PZT actuator. The cross-talk level between Y and X displacement as well as the X and Y displacement for both cases 2 and 3 were also found to be 19.51.

Figure 4. Contour plot of the Von-Mises stress of the initial CMDA.

Figure 4 illustrates the contour plot of the Von-Mises stress of the CMDA when displacements of 10 [µm] were applied to all of its inputs. From Fig. 4, it can be observed that the CMDA experiences a maximum stress of 124 [MPa] at the rectangular flexure hinge attached to the left vertical arm that connects the Leg A of the device to cubic output. This high stress area is marked on the figure by MX. Certainly, the maximum stress is much less than the Yield Strength of the material.

It should be mentioned that this initial design is a sub-optimal design based on previous one dimensional CMDA presented in Ouyang et al. (2008b), and it will be used as the first iteration in the following optimal design process.

5   Optimal design

5.1   Design parameters

The goal of an optimization process for the planar motion generator CMDA is to obtain the values of some significant design parameters that maximize the AR of the device with taking into account the maximum stress occurred in the flexural hinges and the maximum force available by PZT actuators to create a desired input displacement. These significant parameters were discovered in Ouyang et al. (2008b) and were regarded as having the most effects on the AR of the symmetric five-bar CMDA. Table 4 lists the design parameters used for the optimization of the planar motion generator CMDA with their corresponding allowable ranges. It should be noted that selected ranges of the design parameters are based on some sub-optimal design results.

5.2   Optimization process

As mentioned earlier, one of the optimization constraints is the maximum stress that flexural hinges can undergo. This

Table 4. Design parameters and their range used for optimization (unit: mm).

Design Parameter	Range
l_1	26.696–28.700
t	0.300–0.500
l_2	10.500–12.500
h	2.000–3.000
fb	0.250–0.400

Table 5. Design parameters obtained after optimization (unit: mm).

Design Parameter	Value
l_1	26.696
t	0.303
l_2	10.526
h	2.090
fb	0.347

Table 6. Optimized dynamic and static analysis results.

Case Number	NF [Hz]	AR	Cross-Talk
1	243.91	39.2	–
2	239.86	28.5	15.52
3	239.86	28.5	15.52
4	243.91	32.8	–

Table 5 design parameters l_1, l_2, and h have the most effect on the AR of the device. On the other hand, parameters fb and t maintained relatively constant values throughout the optimization process. The thickness t of the rectangular flexure hinge is very close to the minimum value of the allowable range, while the thickness fb of the circular flexure hinge is determined mainly by the allowable Please note that smaller values of these two parameters yield a higher AR. However, thinner flexure hinges can sustain less stress for a given input displacement and will result in the CMDA failure.

5.3 Static and dynamic FEM analysis on the optimized design

After obtaining the optimized design parameters, static and dynamic FEM analysis were performed on the proposed optimized CMDA design for four different cases. These analyses were done for the same four cases as of the initial design. Table 6 lists the dynamic and static results of the proposed optimized design for each case.

Figure 6 illustrates the contour plot of the Von-Mises stress of the optimized CMDA when displacements of 10 [µm] were applied to all of its inputs. From Fig. 6, it can be observed that the location of the maximum stress experienced by the CMDA has changed to the location of rectangular flexure hinge attached to the top horizontal arm S that connects the Leg B of the device to cubic output. This high stress area is shown on the figure by MX and the maximum stress experienced by the hinge is 132 [MPa]. Such an increase in the amount of the stress experienced by the rectangular flexure hinge is due to the fact that the thickness of the hinge has reduced from 0.35 [mm] for the un-optimized CMDA to 0.303 [mm] for the optimized one. The same reason has also led to the decrease of the cross-talk level between Y and X displacement as well as the X and Y displacement from 19.51 for the un-optimized case to 15.52 for the optimized one.

From the optimized results, it can be observed that the overall AR of the device increased from 32.8 to 39.2 while the NF decreased from 293.09 [Hz] for the initial design to 243.91 [Hz] for the optimized design. Such a result is expected as the goal of the optimization is to achieve high AR value with a small sacrifice of NR. Figure 7 shows the corresponding mode shape to the found NF for the first case. Similar results have also been obtained for cases 2 to 4. Comparing Table 6 with Table 3, one can see that the ARs are

was defined by stainless steel yield strength listed in Table 2. In another word during optimization process, the maximum stress (σ) experienced within flexural hinges was not allowed to exceed stainless steel yield strength (σ_y) with a safety factor (SF). Another constraint considered for the optimization was the force required to create a given amount of input displacement. This force also was set to not exceed *NPA50SG* actuators push load capacity of 1000 [N] during the optimization process.

Therefore, the optimization problem of the XY motion CMDA can be described as:

$$\text{Maximum}: \text{AR} = y/\delta \tag{4}$$

Subject to:

$$\sigma < \sigma_y/SF \tag{5}$$

$$F + (20E6) \cdot (\delta) < 1000 \tag{6}$$

where F is the force required to create a displacement at each of the device's input, and the constant $20E6$ in Eq. (5) is the actuator inherent stiffness.

As there is no explicit formula to build the connections between the design parameters and the objective function and constraints, a numerical optimization analysis using ANSYS is approached. ANSYS first order optimization method in conjunction with random design method were used to find the design parameters (within the given range) that yields the best AR for the proposed design for a constant 10 [µm] input displacement, while two constraints are satisfied at the same time. The optimized parameters under this specific condition are obtained through ANSYS and listed in Table 5.

Figure 5 is the plot of the design parameters versus the design objective (DMAX). As it can be seen from the graph and

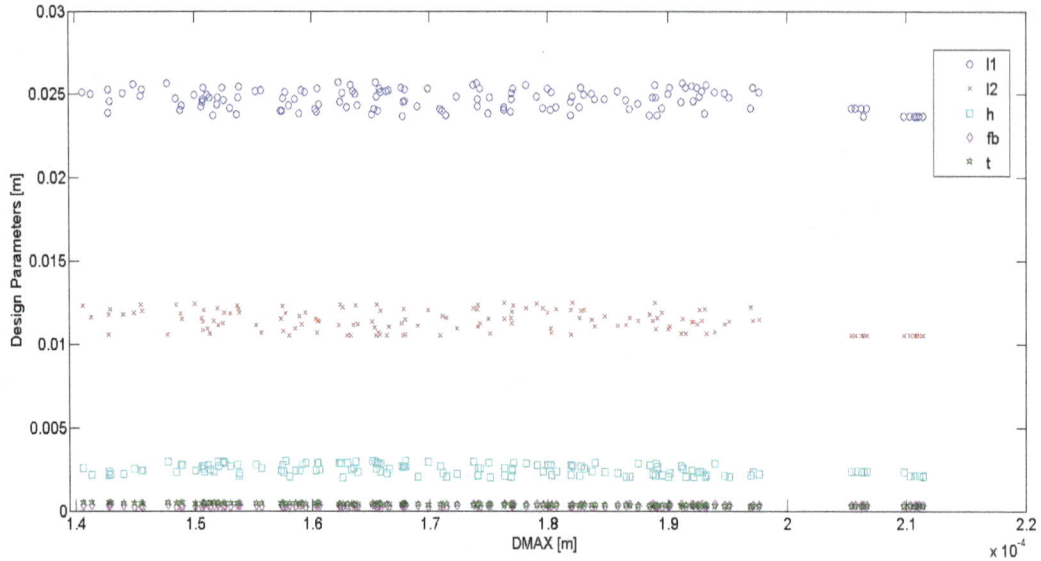

Figure 5. Plot of design parameters versus design objective (unit: m).

Figure 6. Contour plot of the Von-Mises stress of the optimized CMDA.

larger and the NRs are less for the optimized designs. All these results are expected as the objective function for the optimization is the maximum amplification ratio in the design. Figure 8. also shows the mode shape corresponding to the second case where displacements of 10 [μm] in the X direction were applied to the Leg A of the device.

6 Conclusions

In this paper, the design, modeling, and optimization of a planar motion generator CMDA based on the design of a symmetric five-bar CMDA are presented. The goal of optimization process is to achieve high amplification ratio for the device and have a relative high natural frequency. The planar motion generator CMDA is capable of converting and amplifying the linear motion of PZT actuator to a large rang planar (2-D) output motion. Some important parameters are identified and the optimized values are obtained for the purpose of the maximum output displacement of the designed CMDA for a selected specified PZT actuator. Detailed analysis with finite element method of static and dynamic characteristics of the proposed XY CMDA design is also provided. Through the optimization process, a XY compliant mechanical displacement amplifier with high AR compared to the initial sub-optimal design is provided. In the future, a full scale model of the proposed device will be manufactured and tested to compare the obtained FEM results to that obtained by real experiments.

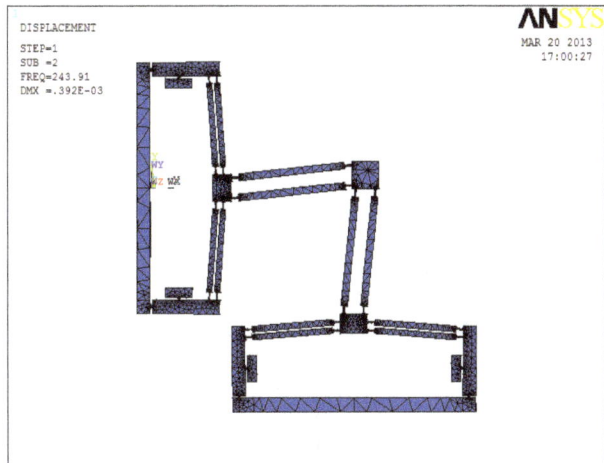

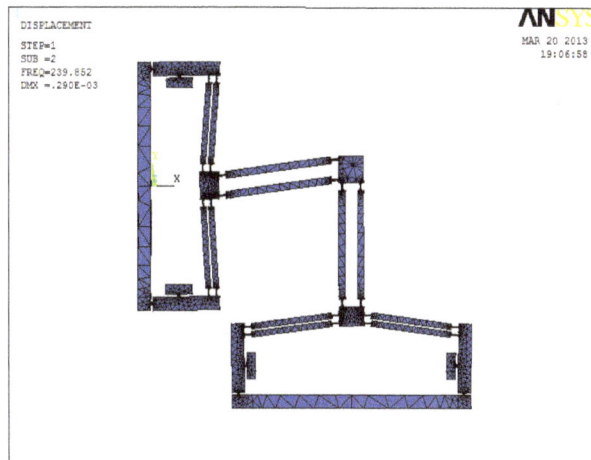

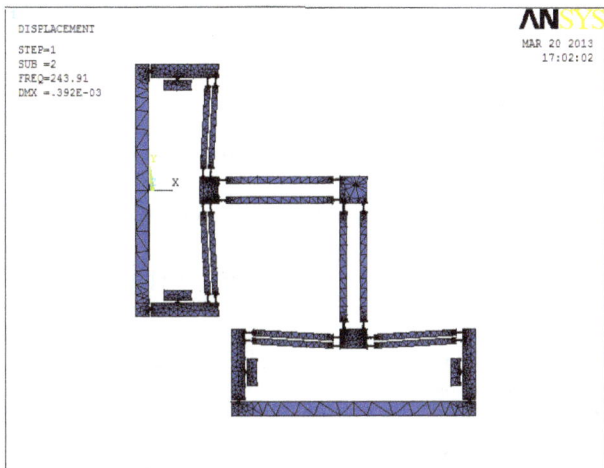

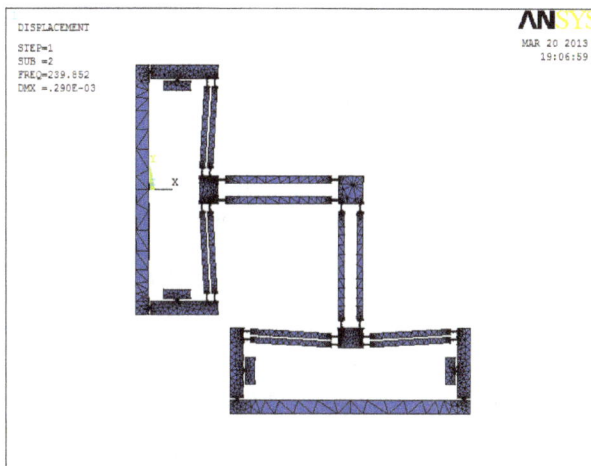

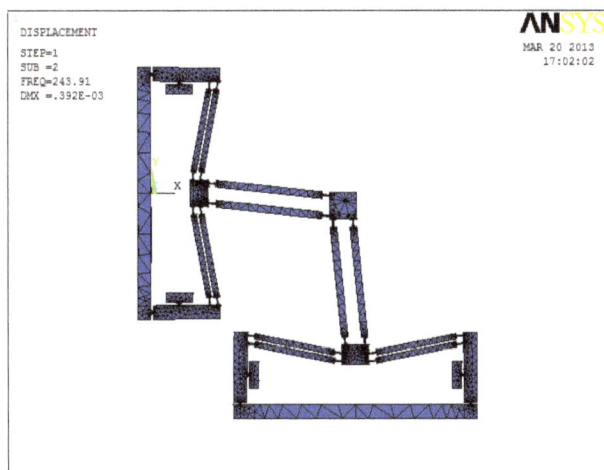

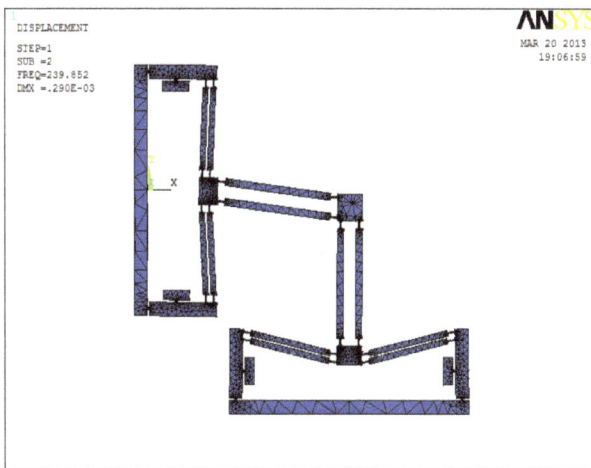

Figure 7. Mode shape of the device corresponding to the first case Fig. 8. Mode shape of the device corresponding to the second case.

Figure 8. Mode shape of the device corresponding to the second case.

Acknowledgements. This research is supported by the Natural Sciences and Engineering Research Council of Canada (NSERC) through a Discovery Grant.

Edited by: G. Hao

References

Ananthasuresh, G. K. and Saxena, A.: On an optimal property of compliant topologies, Structural Multidisciplinary Optimization, 19, 36–49, 2000.

ANSYS 12.0: Help Manual, 2009.

Bharti, S. and Frecker, M. I.: Optimal design and experimental characterization of a compliant mechanism piezoelectric actuator for inertially stabilized rifle, J. Intel. Mat. Syst. Str., 15, 93–106, 2004.

Furukawa, E., Mizuno, M., and Doi, T.: Development of a flexure-hinged translation mechanism driven by two piezoelectric stacks, Int. J., Ser. C, 38, 743–748, 1995.

Howell, L. L.: Compliant Mechanisms, New York, Wiley and Sons, 2001.

Hull, P. V. and Canfield, S.: Optimal synthesis of compliant mechanisms using subdivision and commercial FEA, Mech. Des., 128, 337–348, 2006.

Kota, S., Hetrick, J., Li, Z., and Saggere, L.: Tailoring unconventional actuators using compliant transmissions: design methods and applications, IEEE/ASME Trans. Mechatronics, 4, 396–408, 1999.

Loberto, G. A., Mechhl, A., Nader, G., and Silva, E. C. N.: Development of a XY piezoelectric nanopositioner, ABCM Symposium Series In Mechatronics, 1, 662–671, 2004.

Lobontiu, N.: Compliant Mechanisms, Dsign of Flexural Hinges, in: N. Lobontiu, Compliant Mechanisms, Dsign of Flexural Hinges, Newyork, CRC Press LLC, 2002.

Lu, K. J. and Kota, S.: Topology and dimensional synthesis of compliant mechanisms using discrete optimization, Mech. Des., 128, 1080–1091, 2006.

Ma, H. W., Yao, S. m., Wang, L. Q., and Zhang, Z. L.: Analysis of the displacement amplification ratio of bridge type flexure hinge, Sensor. Actuator., 132, 730–736, 2006.

Ouyang, P. R.: A spatial hybrid motion mechanism: design and optimization, Mechatronics, 21, 479–489, 2011.

Ouyang, P. R., Tjiptoprodjo, R. C., Zhang, W. J., and Yang, G. S.: Micro-motion devices technology: The state of arts review, Int. J. Adv. Manuf. Technol., 38, 463–478, 2008a.

Ouyang, P. R., Zhang, W. J., and Gupta, M. M.: A New compliant mechanical amplifier based on a symmetric five-bar topology, J. Mech. Design, 130, 104501–104505, 2008b.

Pokines, B. J. and Garcia, E.: A Smart material microamplification mechanism fabricated using LIGA, Smart Mater. Struct., 7, 105–112, 1998.

Shuib, S., Ridzwan, M. I. Z., and Kadarman, A. H.: Methodology of compliant mechanisms and its current developments in applications: A Review, Am. J. Appl. Sci., 4, 160–167, 2007.

Su, H. J. and McCathy, J. M.: Synthesis of bistable compliant four-bar mechanisms using polynomial homotopy, Mech. Des., 129, 1094–1098, 2007.

Sun, Z. H., Zhang, B., Cheng, L., and Zhang, W. J.: Application of the redundant servomotor approach to design of path generator with dynamic performance improvement, Mech. Mach. Theory, 46, 1784–1795, 2011.

Tian, Y., Shirinzadeha, B., and Zhang, D.: A flexure-based five-bar mechanism for micro/nano manipulation, Sensor. Actuator., 153, 96–104, 2009.

Timoshenko, S. P. and Gere, J. M.: Theory of Elastic Stability, New York, McGraw-Hill, 1961.

Xu, W. and King, T.: Piezoactuator displacement amplifiers: flexibility, accuracy, and stress considerations, Precis. Eng., 19, 4–10, 1996.

Yang, R. Y., Jouaneh, M., and Scheweizer, R.: Design and characterization of a low-profile micropositioning stage, Precis. Eng., 18, 20–29, 1996.

Yong, Y. K. and Lu, T. F.: Kinetostatic modeling of 3-RRR compliant micro-motion stages with flexure hinges, Mech. Mach. Theory, 38, 1156–1175, 2009.

20

Experimental tests on operation performance of a LARM leg mechanism with 3-DOF parallel architecture

M. F. Wang, M. Ceccarelli, and G. Carbone

LARM: Laboratory of Robotics and Mechatronics, DICeM-University of Cassino and South Latium, Cassino (Fr), Italy

Correspondence to: M. F. Wang (wang@unicas.it)

Abstract. In this paper, a prototype of a LARM leg mechanism is proposed by using a tripod manipulator and its operation performance is investigated through lab experimental tests. In particular, an experimental layout is presented for investigating operational performance. A prescribed motion with an isosceles trapezoid trajectory is used for characterizing the system behavior. Experiment results are analyzed for the purpose of operation evaluation and architecture design characterization of the tripod manipulator and its proposed prototype.

1 Introduction

Legged locomotion has a number of advantages as compared with conventional wheeled and crawler-type locomotion, such as higher mobility, better obstacle overcoming ability, energy efficiency, active suspension and achievable speed, especially when it operates in rough or unconstructed environment (Pfeiffer, 2004; Carbone and Ceccarelli, 2005; Siciliano and Khatib, 2008).

Biped locomotors, as a significant hot topic, have attracted interests of many research communities in the past decades, and a lot of prototypes have been built in the laboratories and even for specific application tasks (Carbone and Ceccarelli, 2005; Siciliano and Khatib, 2008). In addition, parallel manipulators are well known for higher payload capability, stiffness, accuracy and dynamic performance in contrast to traditional serial manipulators, and have been widely studied both in industry and academia (Ceccarelli, 2004; Merlet, 2006). However, most of the existing biped locomotors are based on leg designs with human-like architectures by using serial chain solutions, such as ASIMO (Sakagami et al., 2002), HUBO (Park et al., 2005), HRP (Kaneko et al., 2002) and so on. WL-16 (Waseda Leg-No.16) is a design that achieved world first dynamic biped walking as based on leg designs with Gough-Stewart parallel manipulators (Hashimoto et al., 2009). Ota et al. (1998) and Sugahara et al. (2002) have also proposed to use Gough-Stewart parallel manipulators for leg modules in other systems. Neverthe-

less, the potentiality of parallel manipulators for leg mechanisms has not been fully investigated, since the typical six degrees of freedom (DOFs) manipulators also suffer from some disadvantages, e.g., limited workspace, difficult mechanical design, complex direct kinematics, and control algorithms. To overcome the above disadvantages, parallel manipulators with fewer than six DOFs, namely reduced DOF manipulators, have been widely studied both in industry and academia (Tsai, 1999; Merlet, 2006). In the field of leg designs for biped locomotors, Ceccarelli and Carbone (2009) have investigated the possibility of using parallel manipulator mechanisms with less than six DOFs for leg designs as inspired from the human leg muscular system. It is worthy to note that several architectures of 3-DOF purely translational parallel mechanisms (TPMs) like Delta in Clavel (1988), Orthoglide in Chablat and Wenger (2003), or others like those in Tsai and Joshi (2001) can be used for leg designs, since turning capability of the biped locomotors can be solved by the waist rotation.

In general, the cost features can be related to the mechanical design mainly concerning with joints and actuators, and to the control equipment both in software and hardware, with numerical evaluations that according to authors' knowledge are not reported in the literature. Hence, from the viewpoints of low-cost and easy-operation, although both TPMs in Clavel (1988) and Chablat and Wenger (2003) show regular workspace and proper dynamic performance, the rela-

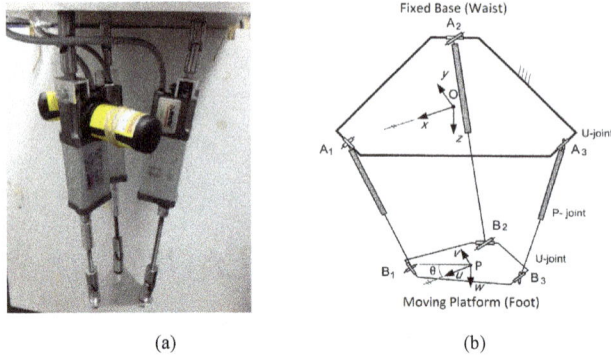

(a)   (b)

Figure 1. A design of LARM tripod leg mechanism: **(a)** a prototype at LARM; **(b)** a kinematic scheme of 3-UPU parallel manipulator.

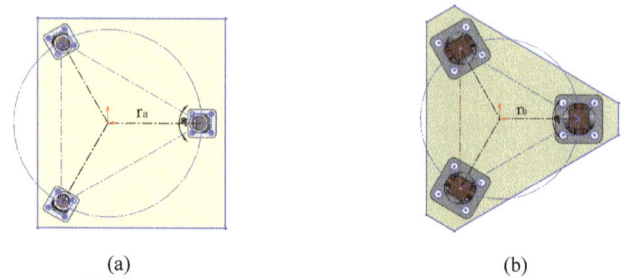

(a)   (b)

Figure 2. A scheme of arrangement of U-joints: **(a)** in the upper plate (waist); **(b)** in the lower plate (foot).

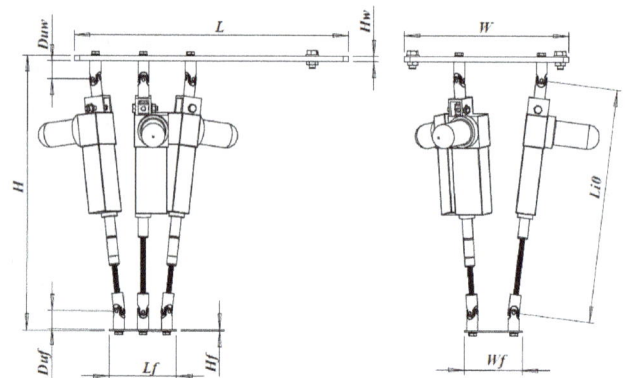

Figure 3. Dimension parameters of the prototype in Fig. 1a.

tively large required operation space and high cost of parallelogram pairs containing S- or U-joints (where S and U stand for spherical and universal pairs, respectively) cannot be considered suitable for leg designs in biped locomotors. In addition, by comparing the four 3-DOF TMPs in the Tsai and Joshi (2001), the rail guides of linear actuators in 3-PUU TMPs (where P stands for a prismatic pair) make the fixed platform too large to be a waist for biped locomotors, while 3-RUU and UPU TMPs (where R stands for a revolute pair) can be expected to be useful for leg designs. Furthermore, Bhutani and Dwarakanath (2014) have presented a high-precision prototype as based on 3-UPU TMP and they validated the practical feasibility of this design in terms of repeatability and trajectory following accuracy for various payloads.

Although few existing prototypes of biped locomotors have been built as based on leg designs with 3-DOF TMPs, there are some other legged locomotors with 3-DOF leg mechanisms. Zhang and Li (2011) have presented a walking locomotor as based on a 3-RPC parallel mechanism by a specific operation mode with two moving platforms. Wang et al. (2009) have proposed a quadruped/biped reconfigurable walking locomotor as based on four 3-UPU parallel leg mechanisms for quadruped walking and by converting them into two 6-SPU parallel mechanisms for biped working. Pan and Gao (2013) have presented a hexapod walking locomotor for situations when a nuclear disaster happens and it is based on 3-DOF parallel mechanisms. In addition, parallel manipulators with 3-DOF, have been widely investigated for relevant applications and they have simpler structure and kinematics, larger workspace, and more convenient control with respect to hexapods (Merlet, 2006). Indeed, 3-DOF manipulators could be enough for reducing the total cost and operations of leg mechanisms for biped locomotors.

At the Laboratory of Robotics and Mechatronics (LARM) of University of Cassino and South Latium, a research line is devoted to design and analysis of parallel mechanisms with reduced DOFs for multiple purposes, and several prototypes of parallel manipulators have been built with low-cost and

easy-operation characteristics (Ceccarelli, 2012). Research activities have been carried out on both theoretical study and application aspects. A 3-DOF parallel manipulator has been proposed, studied and used as a novel design solution of a LARM tripod leg mechanism for a biped locomotor (Wang and Ceccarelli, 2013).

In this paper, a prototype of a LARM leg mechanism is presented as based on a 3-DOF parallel manipulator. The experimental layout of the foot motion control system is presented for experimental experiences. Operational performance of the proposed LARM tripod leg mechanism has been investigated by experimental tests. A fairly simple motion with an isosceles trapezoid trajectory is prescribed for characterizing the system behavior. Experimental results are analyzed for the purpose of operation performance evaluation and architecture design characterization of the proposed LARM tripod leg mechanism.

2   A LARM tripod leg mechanism

A prototype of the proposed LARM tripod leg mechanism and its kinematic scheme of a 3-UPU parallel manipulator are shown in Fig. 1a and b, respectively. The tripod leg mechanism is a parallel manipulator consisting of a waist plate, a moving foot plate, and three identical limbs of linear actuators with U-joints at each end. Hence, in the tripod leg

Table 1. Main specifications of LARM tripod leg mechanism

DOF	Weight (kg)	Dimension ($L \times W \times H$ mm)	Step size ($S_L \times S_H$ mm)	Step cycle (s/step)
3	4.5	$223 \times 200 \times 463$	200×50	6

mechanism, there are eight links that are connected by six U-joints and three P-joints, and the DOFs of the mechanism can be calculated as 3 from the expression of Grübler–Kutzbach criteria

$$F = 6(n - j - 1) + \sum_{i=1}^{j} f_i \qquad (1)$$

Since the joint DOFs of each limb are equal to five, each limb provides one constraint to the moving platform. As shown in Fig. 1b, U-joints in each limb are arranged with two outer revolute joint axes that are parallel to each other and the two inner revolute joint axes that are parallel to one another, so that each limb provides one rotational constraint to the moving platform. A combination of three limbs in each leg completely constrains the moving platform from any instantaneous rotation. Hence, since the limb constraints are independent from each other, the moving platform possesses purely translational motion, as indicated in (Tsai and Joshi, 2000).

Additionally, the upper and lower three U-joints are installed in equilateral triangle arrangement with one ahead and the other two rear, as shown in Fig. 2, where each three inner revolute axes are installed pointing to the corresponding circumcenter of the triangle and the circumradiuses are r_a and r_b, respectively.

The main specifications and details of mechanical design parameters of the LARM tripod leg mechanism are listed in Tables 1 and 2, as referring to dimension parameters in Fig. 3, where L, W, and H are length, width and height of the mechanism, respectively; S_L and S_H are step length and height of the foot; H_w and H_f are the thicknesses of waist and foot plates; L_f and W_f are length and width of the foot, respectively. In each limb, the distance between the rotation center of upper U-joint and waist plate is equal to the distance between the rotation center of lower U-joint and foot plate, i.e. the half length of the universal joint, given by $D_{uw} = D_{uf}$. The distance between the two rotation centers of U-joints, is the length of each limb that is indicated as l_i ($i = 1, 2, 3$), with the initial value given as L_{i0}, which determines the initial height of the proposed mechanism. Finally, the stroke of linear actuators is indicated as L_s.

Since the LARM tripod leg mechanism is developed for biped locomotors, which will be capable of moving with flexibility and versatility in practical applications, during the activity for mechanical design, particular attention has been paid to characteristics for low cost solution, load capacity, easy operation, lightweight and compact design. The proposed solution is worked out by choosing proper commercial

Table 2. Mechanical design parameters of LARM tripod leg mechanism (mm).

H_w	H_f	L_f	W_f	r_a	r_b	$D_{uw} = D_{uf}$	L_{i0}	L_s
10	3	122.4	106	100	50	25	403	100

products, which have been also used in design modelling, and by adopting aluminium alloy as the material for the plates of waist and foot for its stiffness, mass density, and cheap cost.

For kinematic analysis, a static coordinate frame A: O-xyz and a moving coordinate frame B: P-uvw are assumed on the fixed base and moving platforms, and points O and P are the centers of platforms, respectively, as shown in Fig. 1b. A position vector $\boldsymbol{p} = [p_x, p_y, p_z]^T$ of a reference point P in the center of moving platform is given for indexing walking performance. The ith ($i = 1, \ldots, 3$) actuated limb is connected to the moving platform at point B_i and to the fixed base at point A_i, where A_i and B_i are the rotation center of corresponding U-joint. Furthermore, points A_i ($i = 1, \ldots, 3$) lie on an equilateral triangle in the O-xy plane at a radial distance of r_a from point O, and B_i ($i = 1, \ldots, 3$) lie on an equilateral triangle in the P-uv plane at a radial distance of r_b from point P. Hence, the position of the moving platform and length of each limb can be respectively obtained in closed-form through the expressions as

$$
\begin{aligned}
p_x &= \left(-2l_1^2 + l_2^2 + l_3^2\right)/6w \\
p_y &= \left(l_3^2 - l_2^2\right)/2\sqrt{3}w \\
p_z &= \sqrt{l_1^2 - (p_x - w)^2 - p_y^2}
\end{aligned}
\qquad (2)
$$

$$
\begin{aligned}
l_1 &= \sqrt{(p_x - w)^2 + p_y^2 + p_z^2} \\
l_2 &= \sqrt{(p_x + w/2)^2 + \left(p_y - \sqrt{3}w/2\right)^2 + p_z^2} \\
l_3 &= \sqrt{(p_x + w/2)^2 + \left(p_y + \sqrt{3}w/2\right)^2 + p_z^2}
\end{aligned}
\qquad (3)
$$

where $w = r_a$-r_b, and the negative values for limb length and root of p_z cannot be considered for this mechanism. Hence, the displacement L_i of each linear actuator can be expressed as

$$L_i = l_i - L_{i0} \qquad (4)$$

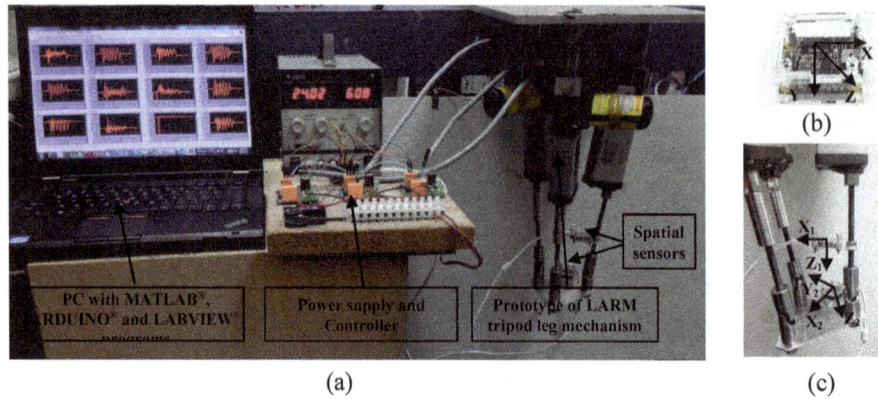

Figure 4. Layout for a foot motion control system of the proposed LARM tripod leg mechanism for experimental tests: (**a**) a layout of whole system; (**b**) a Phidgets® spatial 3/3/3 1044_0 sensor; (**c**) Location of two sensors.

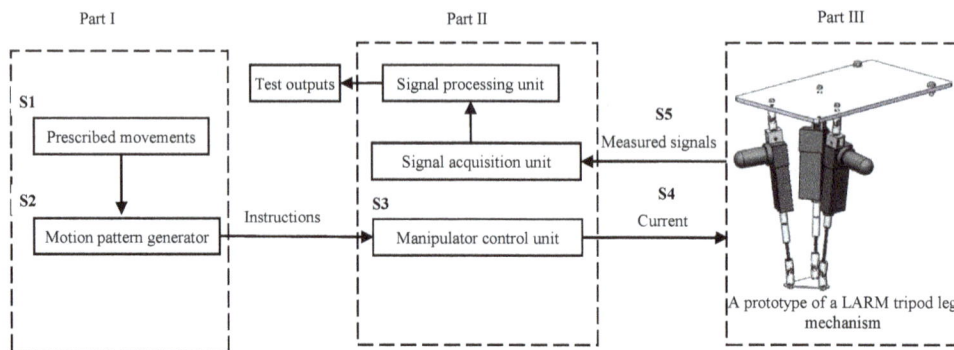

Figure 5. A schematic diagram of the experimental layout in Fig. 4 for an operation procedure.

3   An experimental layout and test modes

Figure 4 shows the foot motion control system of the proposed prototype of the LARM tripod leg mechanism for experimental tests. In Fig. 4a, the leg mechanism is actuated by three linear actuators with 24V DC motors which are controlled by three AQMD2410NS DC motor drive cards with instructions from an ARDUINO card. Figure 4b shows a spatial Phidgets® sensor which has been used to measure acceleration, angular rate, and magnetic field strength in the Cartesian space. In Fig. 4c, two similar spatial sensors that are fixed on the linear actuator and the foot platform, respectively, are used for measuring the acceleration of linear actuators along z axis and the angle information of the foot platform along three axes.

Figure 5 shows a scheme of the experimental operation procedure for the prototype of LARM tripod leg mechanism during experimental tests. The whole system consists of three parts: Part I is a motion pattern generator with suitable program running in MATLAB® environment; Part II consists of the manipulator control unit, signal acquisition unit, and PC with LABVIEW® software; Part III is the built prototype of the LARM tripod leg mechanism.

A controlled operation can be performed by following five steps (S1 to S5), which can be described as

– S1: it gives inputs of the motion generator for the leg mechanism of the prototype in Fig. 1a;

– S2: displacements for the three linear actuators are computed by using Eqs. (3) and (4) in MATLAB® environment;

– S3: the computed motion trajectories in the motion pattern generator are transformed to control instructions in ARDUINO® environment, as the inputs of the manipulator controllers;

– S4: each limb follows the prescribed motion trajectory by driving DC motor of linear actuator under an open-loop control;

– S5: for each period of sampling time, angle positions of the moving platform and accelerations of each linear actuator are measured in LABVIEW® environment, and they are stored in the PC for data analysis and characterization purposes of operation performance.

Table 3. Prescribed motion parameters of LARM tripod leg mechanism (mm).

Axis	P_0 ($t = 0$ s)	P_1 ($t = 1$ s)	P_2 ($t = 2$ s)	P_3 ($t = 3$ s)	P_4 ($t = 4$ s)	P_5 ($t = 5$ s)	P_6 ($t = 6$ s)
x	0	50	100	0	−100	−50	0
y	0	0	0	0	0	0	0
z	400	400	450	450	450	400	400

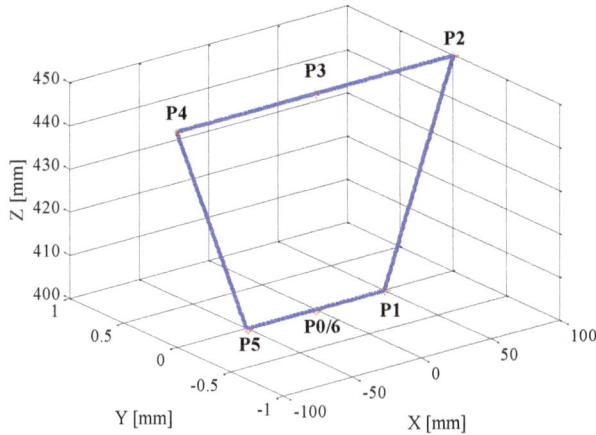

Figure 6. A prescribed input motion for experimental test.

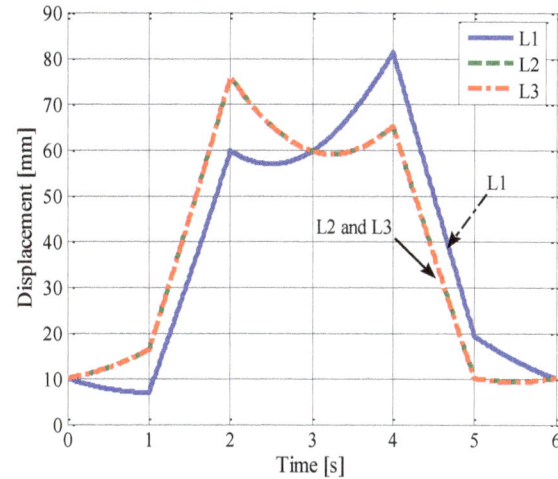

Figure 7. Input displacements of the three linear actuators.

4   Experimental test results

For human normal walking, the motion of a leg can be divided into two phases, i.e. a swing phase and a supporting phase (Carbone and Ceccarelli, 2005), and the human foot can be considered as the end-effector for a leg since it is moved to achieve proper motions and actions during the leg movements.

In general, the trajectory of a human-liked foot step is an ovoid curve where the straight line segment represents the supporting phase and the curve segment represents the swinging phase (Rose and Gamble, 2006). In this section, a prescribed input motion of foot platform has been considered for experimental experiences, as shown in Fig. 6, where the curve segment is simplified by linear segments in O-XZ plane (Zielinska, 2004). Each step of the motion, whose trajectory actually is an isosceles trapezoid, is divided into six segments that are identified by seven prescribed positions, as shown in Table 3, where the start and end values P_0 and P_6 are equally given as the initial position of the foot motion. Since the three linear actuators in the prototype are driven under a position and velocity open-loop control, the prescribed displacements are computed in MATLAB® environment and then they are transformed into ARDUINO® programs. The displacements of three linear actuators can be obtained accordingly as shown in Fig. 7, where L2 and L3 are coincident because of the structure symmetry and no offset of the foot in y axis, which can be easily obtained from Eq. (3).

Figure 8 shows the measured rotation angles of the foot platform about three axes. It takes 6 s for the tested prototype to finish a prescribed movement. The foot platform rotates from −4.0 to 4.5° about x_2 axis and from −3.2 to 3.2° about z_2 axis, while it rotates from −14.1 to 10.9° about y_2 axis. In addition, the rotation angles about y_2 axis are normally between from −7.5 to 3°, and the two peak values only happen at about $t = 2.2$ s and $t = 4.1$ s, i.e., close to points P_2 and P_4. From the test results with micro rotation angles of the foot platform, it can be considered that the foot platform always maintains relatively translational motion but with certain pitch at the extreme positions as points P_2 and P_4.

Figure 9 shows the measured linear accelerations of the three linear actuators. According to the prescribed input motions of the three linear actuators, as shown in Fig. 7, the motions of L2 and L3 should be the same, so that only linear accelerations of L1 and L2 are measured and plotted. Since the prescribed velocity of linear actuators in each segment is approximately constant, the accelerations should be approximately equal to zero, but at the beginning of each segment, i.e. at points Pi ($i = 0, \ldots, 5$) in Fig. 6, a sudden variation of acceleration occurs because of the velocity changes, which can be deduced by the input motion in Fig. 7. In addition, maximum and minimum acceleration values of L1 and L2 are measured as $A_{L1max} = 0.96$ m s^{-2}, $A_{L1min} = -1.34$ m s^{-2} and $A_{L2max} = 0.98$ m s^{-2}, $A_{L2min} = -1.28$ m s^{-2}, respectively.

(a)

(b)

(c)

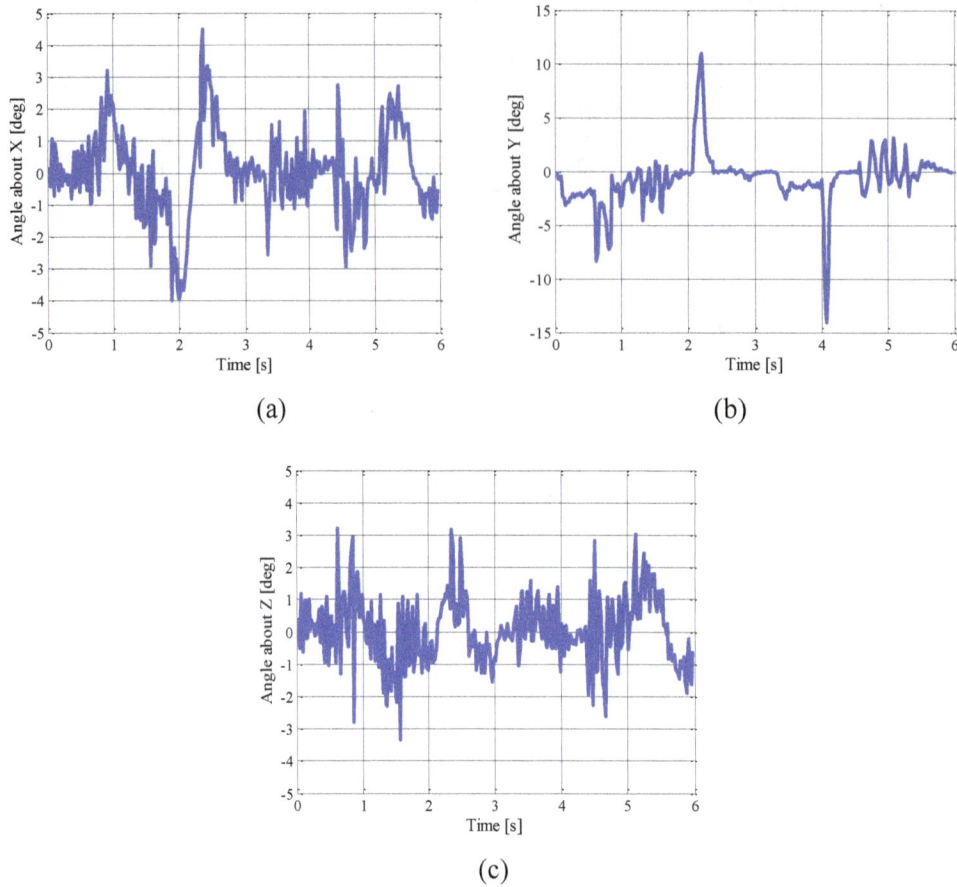

Figure 8. Measured rotation angles of the foot platform: (**a**) about x_2 axis; (**b**) about y_2 axis; (**c**) about z_2 axis.

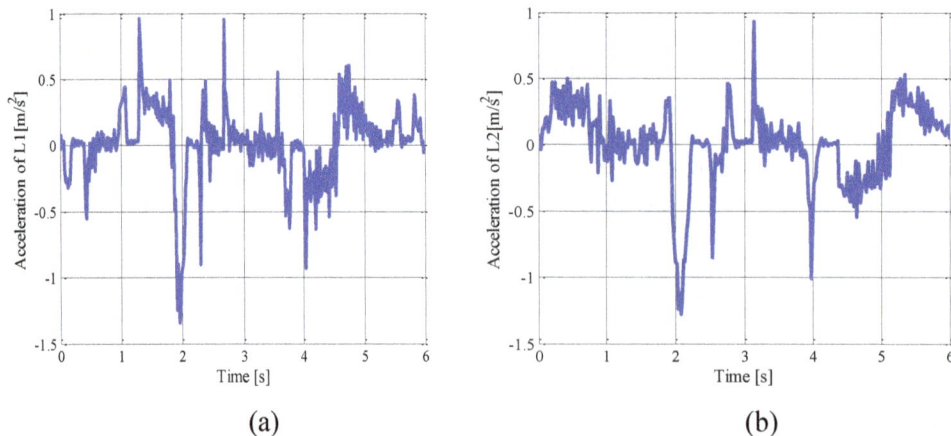

(a)   (b)

Figure 9. Measured axial linear acceleration along z axis for two linear actuators: (**a**) actuator L1; (**b**) actuator L2.

Figure 10 shows a sequence of leg configurations during an experimental test. Experimental test time has been indicated for each snapshot, which is coincident with the prescribed motion parameters in Table 3. It can be noted that the prototype starts from P0 and follows the prescribed trajec-

tory by six reference points, which are marked by red labels in Fig. 10.

From the experimental test results, it can be noted that the built prototype of the LARM tripod leg mechanism in Fig. 4 can perform a predefined movement with the aim to follow an isosceles trapezoid trajectory. Since it has been experienced

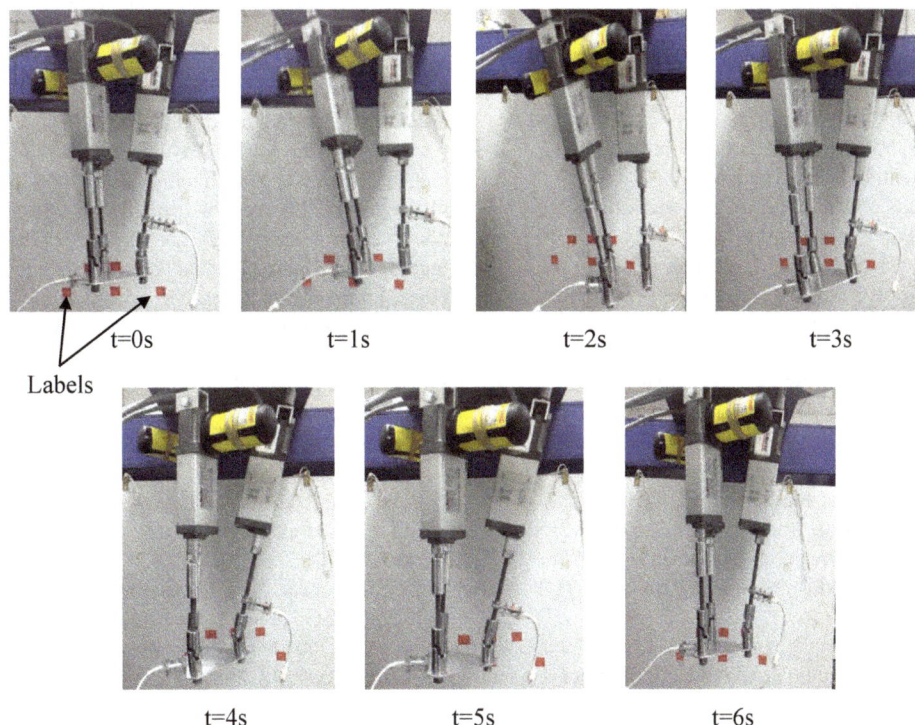

Figure 10. A sequence of snapshots of the prototype during a test following the prescribed motion for the test.

difficultly to maintain the relationship between U-joints in each limb during the operation, the motion of the foot platform has shown purely translational movements but with certain pitch at the extreme positions for points P2 and P4 in Fig. 6. In addition, the precision of the proposed tripod leg mechanism is function of errors in mechanical design and control equipment. In particular, each of the used linear actuator consists of a 24 V permanent magnet motor that is coupled to an ACME lead screw whose lead is 2 mm and actuation error can be estimated in 0.2 mm. The motor drive card is based on current control whose resolution is 0.1 A and estimated error is about 0.02 A. By considering the stroke and current range of the linear actuators, which are 100 mm and 1 to 7 A, respectively, the proposed arrangement provides relatively low control resolution along the prismatic motion to the limbs of the tripod leg mechanism. Nevertheless, since the tripod leg mechanism is built for a biped locomotor, small errors of rotation angles of the foot platform can be considered acceptable during the practical application also because they give an operation that is comparable with human walking. The experimental test results are also quite useful for identifying the operation performance of the prototype, validating the mechanical design, and looking for enhancements.

5   Conclusions

A prototype of a LARM tripod leg mechanism has been built by using a 3-DOF parallel manipulator. The built prototype is a mechanical design solution with low-cost easy-operation features that can be useful for a tripod leg mechanism in a biped locomotor. Experimental tests have been performed successfully with the prototype to follow a prescribed step movement of the foot platform. Test results have validated the feasibility of the proposed design and have characterized its operation with suitable motion characteristics as a tripod leg mechanism for biped locomotors.

Acknowledgements. The first author would like to acknowledge China Scholarship Council (CSC) for supporting his PhD study and research at the Laboratory of Robotics and Mechatronics (LARM) in the University of Cassino and South Latium, Italy, for the years 2013-2015.

Edited by: A. Barari

References

Bhutani, G. and Dwarakanath, T. A.: Practical Feasibility of A High Precision 3-UPU Parallel Mechanism, Robotica, 32, 341–353, 2014.

Carbone, G. and Ceccarelli, M.: Legged Robotic Systems, Cutting Edge Robotics, ARS Scientific Book, Vienna, 553–576, 2005.

Ceccarelli, M.: Fundamentals of Mechanics of Robotic Manipulation, Kluwer Academic Publishers, Dordrecht, 2004.

Ceccarelli, M.: An Illustrated History of LARM in Cassino, in: Proc. of Int. Workshop on Robotics in Alpe-Adria-Danube Region (RAAD), Naples, Edizioni Scientifiche e Artistiche, 85–92, 2012.

Ceccarelli, M. and Carbone, G.: A New Leg Design with Parallel Mechanism Architecture, in: Proc. of Int. Conf. on Advanced Intelligent Mechatronics (AIM), Singapore, 1447–1452, 2009.

Chablat, D. and Wenger, P.: Architecture Optimization of A 3-DOF Translational Parallel Mechanism for Machining Applications, the Orthoglide, IEEE Trans. Robot. Autom., 19, 403–410, 2003.

Clavel, R.: Delta, A Fast Robot with Parallel Geometry, in: Proc. of the 18th Int. Symposium on Industrial Robots, Lausanne, 91–100, 1988.

Hashimoto, K., Sugahara, Y., Lim, H. O., and Takanishi, A.: Biped Landing Pattern Modification Method and Walking Experiments in Outdoor Environment, J. Rob. Mechat., 20, 775–784, 2009.

Kaneko, K., Kanehiro, F., Kajita, S., Yokoyama, K., Akachi, K., Kawasaki, T., and Isozumi, T.: Design of Prototype Humanoid Robotics Platform for HRP, in: Proc. of IEEE/RSJ Int. Conf. on Intelligent Robots and Systems (IROS), Lausanne, 3, 2431–2436, 2002.

Merlet, J. P.: Parallel Robots (2nd Edn.), Springer, Dordrecht, 2006.

Ota, Y., Inagaki, Y., Yoneda, K., and Hirose, S.: Research on a Six-Legged Walking Robot with Parallel Mechanism, in: Proc. of Int. Conf. on Intelligent Robots and Systems (IROS), Victoria, 241–248, 1998.

Pan, Y., and Gao, F.: Payload Capability Analysis of A New Kind of Parallel Leg Hexapod Walking Robot, in: Proc. of Intl. Conf. on Advanced Mechatronic Systems (ICAMechS), Luoyang, 541–544, 2013.

Park, I. W., Kim, J. Y., Lee, J., and Oh, J. H.: Mechanical Design of Humanoid Robot Platform KHR-3 (KAIST Humanoid Robot-3: HUBO). In: Proc. of IEEE/RAS Int. Conf. on Humanoid Robots (HUMANOIDS), Tsukuba, 321–326, 2005.

Pfeiffer, F.: Technological Aspects of Walking, in: Walking: Biological and Technological Aspects, edited by: Pfeiffer, F. and Zielińska, T., Springer, Wien, 119–154, 2004.

Rose, J. and Gamble, J. G.: Human walking (3rd Edn.), Lippincott Williams & Wilkins, Philadelphia, 2006.

Sakagami, Y., Watanabe, R., Aoyama, C., Matsunaga, S., Higaki, N., and Fujimura, K.: The Intelligent ASIMO: System Overview and Integration, in: Proc. of IEEE/RSJ Int. Conf. on Intelligent Robots and Systems (IROS), Lausanne, 3, 2478–2483, 2002.

Siciliano, B. and Khatib O.: Handbook of robotics, Part G, Legged Robots, Springer, Heidelberg, 361–390, 2008.

Sugahara, Y., Sugahara, Y., Endo, T., Lim, H. O., and Takanishi, A.: Design of a Battery-powered Multi-purpose Bipedal Locomotor with Parallel Mechanism, n: Proc. Int. Conf. on Intelligent Robots and Systems (IROS), Lausanne, 3, 2658–2663, 2002.

Tsai, L. W.: Robot Analysis: The Mechanics of Serial and Parallel Manipulators, John Wiley & Sons, New York, 1999.

Tsai, L. W. and Joshi, S.A.: Kinematics and optimization of a spatial 3-UPU parallel manipulator, ASME J. Mechan. Design, 122, 439–446, 2000.

Tsai, L. W. and Joshi, S.: Comparison Study of Architectures of Four 3 Degree-Of-Freedom Translational Parallel Manipulators, n: Proc. of the 2001 IEEE Int. Conf. on Robotics and Automation (ICRA), Seoul, 1283–1288, 2001.

Wang, H. B., Qi, Z. Y., Hu, Z. W., and Huang Z.: Application of Parallel Leg Mechanisms in Quadruped/Biped Reconfigurable Walking Robot, J. Mechan. Eng., 45, 24–30, 2009.

Wang, M. F. and Ceccarelli, M.: Design and Simulation for Kinematic Characteristics of a Tripod Mechanism for Biped Locomotors, in: Proc. of Intl. Workshop on Robotics in Alpe-Adria-Danube Region (RAAD), Portorož, 124–131, 2013.

Zhang, C. J. and Li, Y. W.: A New Walking Robot Based on 3-RPC Parallel Mechanism, Chinese J. Mechan. Eng., 47, 25–30, 2011 (in Chinese).

Zielinska, T.: Motion Synthesis, in: Walking: Biological and Technological Aspects, edited by: Pfeiffer, F. and Zielińska, T., Springer, Wien, 155–191, 2004.

21

Modelling and experimental investigation of process parameters in WEDM of WC-5.3 % Co using response surface methodology

K. Jangra and S. Grover

Department of Mechanical Engineering, YMCA University of Science and Technology, Faridabad 121006, India

Correspondence to: K. Jangra (kamaljangra84@gmail.com)

Abstract. Tungsten carbide-cobalt (WC-Co) composite is a difficult-to-machine material owing to its excellent strength and hardness at elevated temperature. Wire electrical discharge machining (WEDM) is a best alternative for machining of WC-Co composite into intricate and complex shapes. Efficient machining of WC-Co composite on WEDM is a challenging task since it involves large numbers of parameters. Therefore, in present work, experimental investigation has been carried out to determine the influence of important WEDM parameters on machining performance of WC-Co composite. Response surface methodology, which is a collection of mathematical and experimental techniques, was utilised to obtain the experimental data. Using face-centered central composite design, experiments were conducted to investigate and correlate the four input parameters: pulse-on time, pulse-off time, servo voltage and wire feed for three output performance characteristics – cutting speed (CS), surface roughness (SR) and radial overcut (RoC). Using analysis of variance on experimental data, quadratic vs. two-factor interaction (2FI) models have been suggested for CS and RoC while two-factor interaction (2FI) has been proposed for SR. Using these mathematical models, optimal parameters can be determined easily for desired performance characteristics, and hence a trade-off can be made among different performance characteristics.

1 Introduction

Demand for tungsten carbide-cobalt (WC-Co) composite has been growing rapidly due to its excellent mechanical properties like high wear and corrosion resistance, which makes it most suitable for cutting tools, dies and other special wear-resisting applications. Machining of WC-Co composite is very difficult with conventional machining processes like turning, milling and grinding (Liu and Li, 2001; Liu et al., 2003; Engqvist et al., 1999; Jia and Fischer, 1997) because of its high hardness and high melting temperature. Due to the low material removal rate and difficulty in machining of complex and intricate profiles in WC-Co composite, cost associated with conventional machining processes is very high.

Wire electrical discharge machining (WEDM) is a specialised form of electrical discharge machining (EDM) process which is potentially used to generate intricate and com-plex geometries in hard conductive materials without making any mechanical contact (Jangra et al., 2010). In WEDM, material is eroded due to the melting/evaporation of work surface which is mainly due to the localised high temperature generation in plasma channel between the work material and downward moving wire electrode (as shown in Fig. 1). In order to achieve high productivity within specified tolerance on WEDM, selection of optimal process parameters plays a key role which depends on large numbers of variables such as composition and grain size of work material, machining parameters, wire electrode material, wire diameter, cutting conditions and work geometry etc. (Jangra et al., 2011).

Several attempts have been made to determine optimal machining conditions for WC-Co composite on EDM and WEDM. Lee and Li (2001) investigated the influence of EDM parameters such as the electrode materials, electrode polarity, open circuit voltage, peak current, pulse duration,

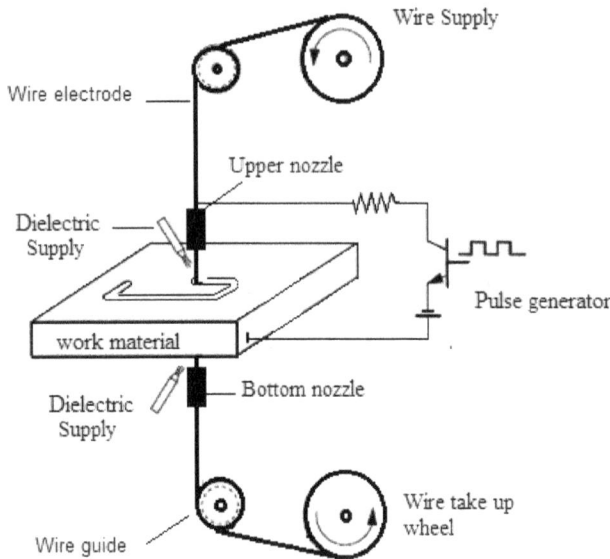

Figure 1. Representation of WEDM process.

pulse interval and flushing on the machining characteristics, such as material removal rate (MRR), surface finish and relative tool wear in EDM of tungsten carbide. Lee and Li (2003) elucidated the effect of discharge energy on integrity of EDMed surface of tungsten carbide. It was found that the surface roughness is a function of two main parameters: peak current and pulse duration. At high peak current and/or long pulse duration, rough surface and abundance of micro-cracks were observed. Mahdavinejad and Mahdavinejad (2005) studied the instability in EDM of WC-Co composites. This machining instability was mainly due to open circuit, short circuit and arcing pulses. Increase in pulse duration results in more melting and recasting of material, which causes arcing and rougher surface.

The grain size of WC and cobalt composition also shows noticeable influence on machining performance of WC-Co composite with WEDM (Kim and Kruth, 2001; Lauwers et al., 2006). Varying the cobalt concentration and grain size of WC alters the thermal conductivity of the material, which affects the machining performance. Chen et al. (2010) utilised Taguchi method, back propagation neural network and genetic algorithm to optimize the WEDM parameters in machining of pure tungsten. Pulse-on time was the most significant parameter that influences both the cutting speed and surface roughness. Jangra et al. (2012) optimized the multi-machining characteristics in WEDM of WC-5.3% Co composite using an integrated approach of Taguchi, GRA and entropy method. Six input parameters – taper angle, discharge current, pulse-on time, pulse-off time, wire tension and dielectric flow rate – were investigated for four output machining characteristics: MRR, SR, RoC and angular error.

Although WEDM can easily machine complex profiles in WC-Co composite but damaged work surface consisting

of recast layer, large micro-cracks are a major concern in WEDM (Çaydaş et al., 2009; Jangra, 2012). In order to obtain fine surface finish and low surface defects, low discharge parameters with high dielectric flushing rate are required. but they lower the cutting rate in WEDM. This implies that high cutting rate with low surface roughness and minimum surface defects is difficult to obtain in a single setting of process parameters. In order to achieve an efficient machining of WC-Co composite into desired shape, inter-relationship (mathematical modelling) between input WEDM parameters and output performance characteristics should be available to the manufacturers, so that a trade-off can be made among various performance characteristics.

Two kinds of approaches, theoretical and empirical, have been commonly used in modelling of WEDM process (Patil and Brahmankar, 2010). Owing to the simplified and unavoidable assumptions, the theoretical models yield large errors between predicted and experimental results. On the other hand empirical models are limited to specific experimental conditions. Response surface methodology (RSM) is the most used statistical technique for determining the relationship between various input parameters and output responses (Myers and Montgomery, 1995; Hewidy et al., 2005; Kansal et al., 2005). RSM is a collection of mathematical and experimental techniques that requires sufficient number of experimental data to analyse the problem and to develop mathematical models for several variables and output performance characteristics.

Because the WEDM involves multi-performance characteristics, the objective of the present study is to investigate the influence of process parameters and to develop the mathematical models for three performance characteristics namely cutting speed (CS), surface roughness (SR) and radial overcut (RoC) in WEDM of WC-5.3% Co composite. Response surface methodology with face-centered central composite design has been utilised to conduct the experimentation which helps to investigate and to correlate the input parameters with output performance characteristics. Using these mathematical models, optimal combination of WEDM parameters can be selected for desired performance characteristics for WC-Co composite.

2 Experimental procedure

2.1 Work material and machining parameters

Tungsten carbide composite with low cobalt concentration (5.3%) has been taken as a work material in the form of a rectangular block of thickness of 13 mm. The density and hardness of WC-5.3% Co composite were measured as 14.95 g cm^{-3} and 77 HRC, respectively.

Experiments were performed on 5-axis sprint cut (epulse-40) WEDM, most widely used in Indian industries, manufactured by Electronica Machine Tools Ltd., India. In present machine tool, range of the important parameters is as

Table 1. Process parameters and their levels.

Symbol	Parameters	Levels		
		(−1)	(0)	(+1)
A	Pulse-on Time (T_{on})	108	115	122
B	Pulse-off Time (T_{off})	30	40	50
C	Servo Voltage (SV)	20	30	40
D	Wire Feed Rate (WF)	4	6	8

follows: discharge current, 10–230 amp; pulse-on time, 101–131 μs; pulse-off time, 10–63 μs; servo voltage 0–90 V; dielectric flow rate, 0–12 litre per minute ($l\,min^{-1}$); wire feed, 1–15 m min^{-1}; wire tension, 1–15 N.

In present investigation, four important WEDM parameters, namely pulse-on time (T_{on}), pulse-off time (T_{off}), servo voltage (SV) and wire feed (WF) have been considered with three levels each (Table 1). Discharge current is kept at optimum value of 90 amp which is taken on the basis of earlier investigations (Jangra et al., 2011, 2012). Similarly, other parameters were kept constant at their optimal values. As the workpiece thickness is low (13 mm), wire tension was kept fixed at 10 N. Zinc-coated brass wire of diameter 0.25 mm was used as an electrode because of its good capability to sustain high discharge energy. High flow rate results in quick and complete flushing of melted debris out of the spark gap. Therefore, dielectric flow rate is kept at maximum value of $12\,l\,min^{-1}$. Vertical cutting was performed at zero wire offset.

2.2 Performance characteristics

CS was measured as surface area removed per minute ($mm^2\,min^{-1}$). It was obtained by multiplying the workpiece thickness (13 mm) with linear cutting speed (mm min^{-1}) displaying on machine tool monitor screen. Surface roughness is a good indicator of the surface quality which can be easily measured by a tool manufacturer. In present work, SR value (in μm) was measured in terms of mean absolute deviation (Ra) using the digital surface tester Mitutoyo 201P. RoC (in μm) is the gap between machined work surface and wire periphery (as shown in Fig. 2). "D" represents the wire diameter. RoC helps to predict accurate wire offset for precise dimensional tolerance. Neglecting the wire lag compensation for straight cutting, wire offset is given by the sum of wire radius and RoC (Sarkar et al., 2008). RoC was measured using optimal microscope.

2.3 Experimental design using RSM

For developing an adequate relationship between input WEDM parameters and output performance characteristics, response surface methodology (RSM) (Myers and Montgomery, 1995) has been employed. By using the design of experiments and applying regression analysis, the modelling of the desired response (Y) to several independent input vari-

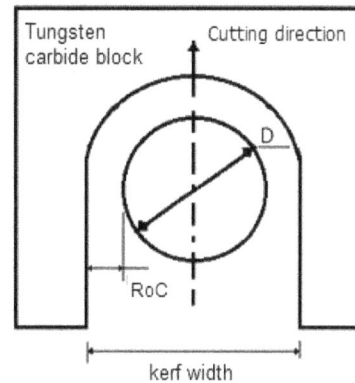

Figure 2. Representation of RoC in WEDM.

ables (x_i) can be gained. In the RSM, the quantitative form of relationship between desired response and independent input variables could be represented as

$$Y = \Phi(x_1, x_2, \ldots\ldots\ldots, x_k) \pm e_r. \tag{1}$$

The function Φ is called response surface or response function. The residual e_r measures the experimental errors (Cochran and Cox, 1962).

In applying the RSM, the dependent variable is viewed as a surface to which a mathematical model is fitted. For the development of regression equations related to various performance characteristics of WEDM process, the second-order response surface has been assumed as

$$Y = b_0 + \sum_{i=1}^{k} b_i X_i + \sum_{i=1}^{k} b_{ii} X_i^2 + \sum_{i<j=2}^{2} b_{ij} X_i X_j \pm e_r. \tag{2}$$

This assumed surface Y contains linear, squared and cross product terms of variables X_i. The model parameters can be estimated most effectively if proper experimental designs are used to collect the data. In present case, a standard second-order experimental design called face-centered central composite design (CCD) has been adopted for analysing and modelling four input parameters. This design consists of factorial portion with all parameters at three levels, eight star points and six central points. The star points are at the face of the cube portion on the design which corresponds to α-value of 1, the distance from design centre. The centre points, as implied by the name, are points with all levels set to coded level 0, the midpoint of each parameter range. Table 2 shows the experimental conditions.

3 Results and discussion

The 30 experiments were conducted and CS, SR and RoC were obtained for each experimental run (as listed in Table 2).

Table 2. Test conditions in face-centered central composite design.

Trial No.	A (T_{on}) coded	A (T_{on}) actual	B (T_{off}) coded	B (T_{off}) actual	C (SV) coded	C (SV) actual	D (WF) coded	D (WF) actual	CS (mm min^{-1})	SR (μm)	RoC (μm)
1	−1	108	−1	30	−1	20	−1	4	16.691	1.51	16.97
2	1	122	−1	30	−1	20	−1	4	27.97	2.2	19.78
3	−1	108	1	50	−1	20	−1	4	6.52	1.104	19.03
4	1	122	1	50	−1	20	−1	4	14.22	1.78	21.35
5	−1	108	−1	30	1	40	−1	4	16.59	1.2	16.87
6	1	122	−1	30	1	40	−1	4	27.88	1.9	19.45
7	−1	108	1	50	1	40	−1	4	6.29	0.84	18.68
8	1	122	1	50	1	40	−1	4	14.12	1.52	21.05
9	−1	108	−1	30	−1	20	1	8	18.57	1.4	17.87
10	1	122	−1	30	−1	20	1	8	32.58	2.02	21.61
11	−1	108	1	50	−1	20	1	8	8.4	1.01	19.46
12	1	122	1	50	−1	20	1	8	16.36	1.7	22.08
13	−1	108	−1	30	1	40	1	8	16.53	1.17	16.98
14	1	122	−1	30	1	40	1	8	30.54	1.86	20.56
15	−1	108	1	50	1	40	1	8	5.12	0.81	18.77
16	1	122	1	50	1	40	1	8	14.32	1.5	22.09
17	−1	108	0	40	0	30	0	6	12.01	1.14	17.58
18	1	122	0	40	0	30	0	6	22.25	1.82	21.22
19	0	115	−1	30	0	30	0	6	24.8	1.66	18.83
20	0	115	1	50	0	30	0	6	12.22	1.3	20.23
21	0	115	0	40	−1	20	0	6	19.04	1.61	19.96
22	0	115	0	40	1	40	0	6	17.97	1.35	19.6
23	0	115	0	40	0	30	−1	4	17.99	1.55	19.56
24	0	115	0	40	0	30	1	8	19.03	1.39	20.32
25	0	115	0	40	0	30	0	6	18.35	1.48	19.82
26	0	115	0	40	0	30	0	6	18.69	1.5	19.76
27	0	115	0	40	0	30	0	6	18.87	1.46	19.89
28	0	115	0	40	0	30	0	6	18.48	1.48	20.01
29	0	115	0	40	0	30	0	6	18.54	1.47	19.8
30	0	115	0	40	0	30	0	6	18.08	1.51	19.58

3.1 Mathematical model for CS, SR and RoC

Using the experimental data, regression equations have been developed for correlating the output performance characteristics and input WEDM parameters. Analysis of variance (ANOVA) has been applied on the experimental data to select the adequate model. Design expert (DX8), a statistical tool, has been utilised to analyse the experimental data. Quadratic vs. two-factor interaction (2FI) models have been suggested for CS and RoC, while two-factor interaction (2FI) has been proposed for SR. Table 3 shows the fit summary for proposed models.

In Table 3, a p-value for the model terms that are less than 0.05 (i.e. $\alpha = 0.05$, or 95 % confidence level) indicates that the obtained models are considered to be statistically significant (Kanagarajan et al., 2008). It shows that the terms in the model have significant effect on responses. From Table 3, coefficient of determination (R^2) and R^2 (adj.) for all three responses are closer to unity, which is desirable for the re-

sponse model to fit the actual data. The lack of fit is a measure of the failure of the model to predict the data in the upper and lower limits of the parameters. The "lack of fit" for all three characteristics is insignificant because of high probability value (F-value), which is desirable for selecting the models. Figures 3 to 5 show that the residuals are normally distributed about a straight line and there is no problem with the observed results. Consequently, the proposed models for three responses can be considered as significant for fitting and predicting the experimental results within the specified experimental domain.

Table 4 shows the "F-value" and "p-value" for each term in performance characteristics CS, SR and RoC, respectively. The terms having p-value less than 0.05 are considered to be significant while insignificant terms can be eliminated from the final predicted models. In case of CS, the model terms A, B, C, D, AB, AD, BD, CD and A^2 are significant. Similarly, A, B, C, D and CD for SR, and A, B, C, D, AB, AD and A^2 for RoC are significant model terms.

Table 3. The ANOVA table for fitted models.

Source	Sum of Squares	Degree of freedom	Mean Square	F-value	p-value	
(a) For CS						
Model	1277.4	14	91.24	771.23	< 0.0001	Significant
Residual	1.774	15	0.118			
Lack of fit	1.401	10	0.140	1.873	0.2532	Not significant
Pure error	0.374	5	0.074			
Cor. total	1279.17	29				
Standard deviation = 0.34396; R^2 = 0.9986; R^2 (Adj.) = 0.9973						
(b) For SR						
Model	3.0092	10	0.3009	680.53	< 0.0001	Significant
Residual	0.0084	19	0.0004			
Lack of fit	0.0066	14	0.00047	1.374	0.3859	Not significant
Pure error	0.0017	5	0.00034			
Cor. Total	3.0175	29				
Standard deviation = 0.02229; R^2 = 0.9972; R^2 (Adj.) = 0.9957						
(c) For RoC						
Model	56.653	14	4.0466	89.536	< 0.0001	Significant
Residual	0.678	15	0.0452			
Lack of fit	0.576	10	0.0576	2.823	0.1318	Not significant
Pure error	0.102	5	0.0204			
Cor. Total	57.33	29				
Standard deviation = 0.2126; R^2 = 0.9881; R^2 (Adj.) = 0.9771						

Table 4. F-value and p-value for each model term for CS, SR and RoC.

Source/ Symbol	DOF	CS		SR		RoC	
		F-value	p-value	F-value	p-value	F-value	p-value
A	1	4106.921	<0.0001	4699.663	<0.0001	894.779	<0.0001
B	1	6165.125	<0.0001	1415.063	<0.0001	234.773	<0.0001
C	1	56.727	<0.0001	599.29	<0.0001	20.262	0.0004
D	1	81.560	<0.0001	69.546	<0.0001	60.232	<0.0001
AB	1	169.249	<0.0001	0.1832	0.6735	5.983	0.0273
AC	1	1.0075	0.3314	0.9973	0.3305	0.179	0.6781
AD	1	26.488	0.0001	0.4433	0.5136	13.984	0.0020
BC	1	1.0046	0.3321	0.8164	0.3775	1.4957	0.2402
BD	1	19.266	0.0005	2.6143	0.1224	3.8106	0.0699
CD	1	41.648	<0.0001	16.7263	0.0006	3.2796	0.0902
A^2	1	43.014	<0.0001	–	–	5.9025	0.0282
B^2	1	0.0101	0.9213	–	–	2.0886	0.1690
C^2	1	0.0153	0.9030	–	–	0.2004	0.6608
D^2	1	0.0101	0.9213	–	–	2.7525	0.1179

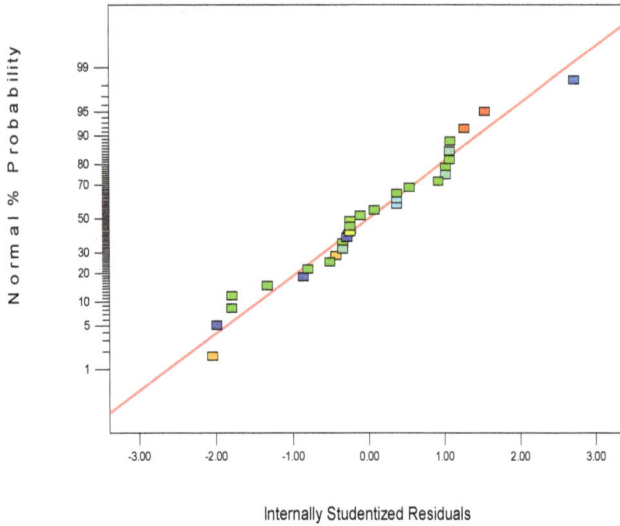

Figure 3. Residuals plot for CS.

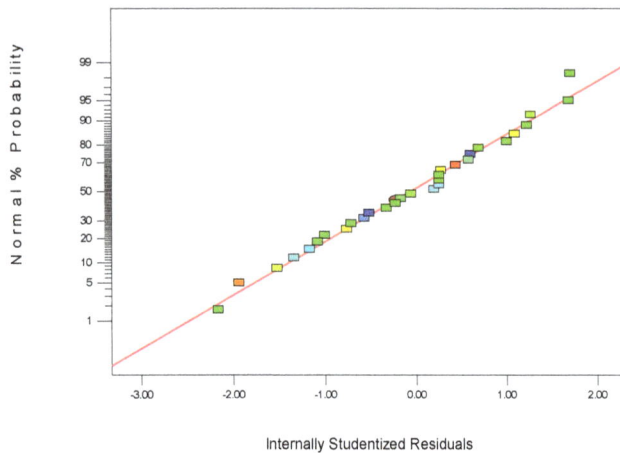

Figure 4. Residuals plot for SR.

Using backward elimination method, non-significant terms were eliminated. Final regression equations for performance characteristics have been obtained as follows.

3.1.1 Cutting speed (CS)

Regression equation is in terms of coded parameters:

$$CS = 18.51 + 5.20A - 6.37B - 0.61C + 0.73D$$
$$-1.12AB + 0.44AD - 0.38BD - 0.55CD - 1.45A^2 \quad (3)$$

Regression equation is in terms of actual parameters:

$$CS = -494.76 + 8.0055A + 1.314B + 0.1054C - 1.682D$$
$$-0.01598AB + 0.031612AD - 0.01887BD$$
$$-0.02774CD - 0.0296A^2 \quad (4)$$

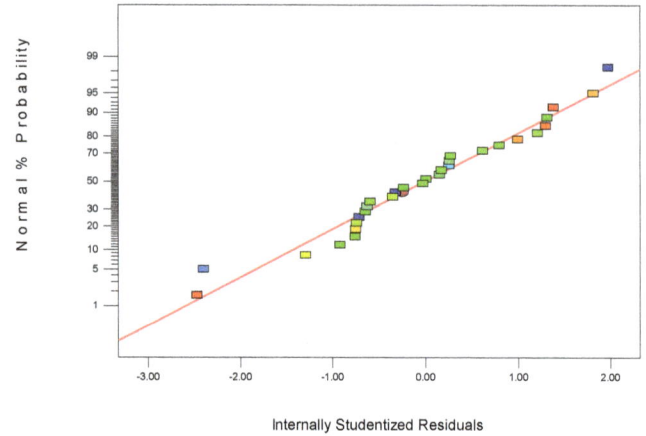

Figure 5. Residuals plot for RoC.

3.1.2 Surface roughness (SR)

Regression equation is in terms of coded parameters:

$$SR = 1.47 + 0.34A - 0.19B - 0.12C - 0.041D + 0.022CD \quad (5)$$

Regression equation is in terms of actual parameters:

$$SR = -2.679 + 0.04854A - 0.01864B - 0.01858C$$
$$-0.05292D + 0.001075CD \quad (6)$$

3.1.3 Radial overcut (RoC)

Regression equation is in terms of coded parameters:

$$RoC = 19.7654 + 1.49889A + 0.76778B - 0.22556C$$
$$+0.38889D - 0.13AB + 0.19875AD - 0.32088A^2 \quad (7)$$

Regression equation is in terms of actual parameters:

$$RoC = -94.5438 + 1.69975A + 0.454676B - 0.09212C$$
$$-1.74364D - 0.00186AB - 0.00519AD - 0.00655A^2 \quad (8)$$

These predicted models were verified by conducting confirmation tests within the selected range of the WEDM parameters.

3.2 Effect of WEDM parameters on performance characteristics

Response surface graphs (Figs. 6 to 8) have been plotted to analyse the influence of WEDM parameters on performance characteristics, namely CS, SR and RoC. Surface plots have been plotted for combined effect of two factors while keeping other two factors at their mid-values.

3.2.1 Effect of WEDM parameters on cutting speed

Figure 6a to c show that cutting speed increases with increasing T_{on} and WF, while it decreases with increasing T_{off} and

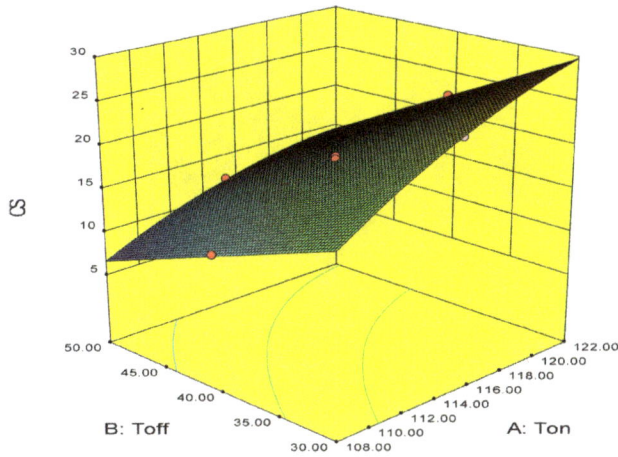

Figure 6a. Combined effect of T_{on} and T_{off} on CS.

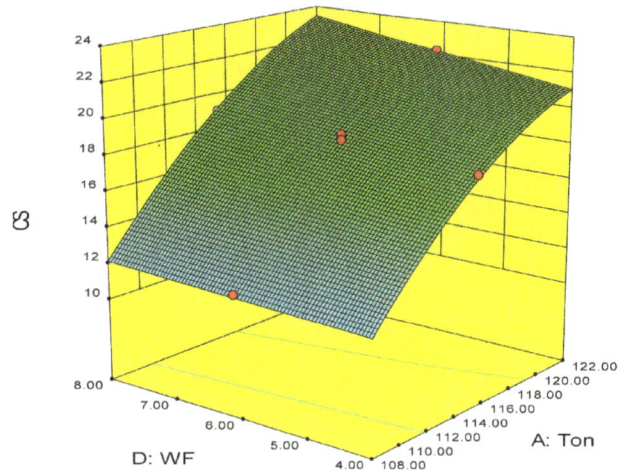

Figure 6c. Combined effects of T_{on} and WF on CS.

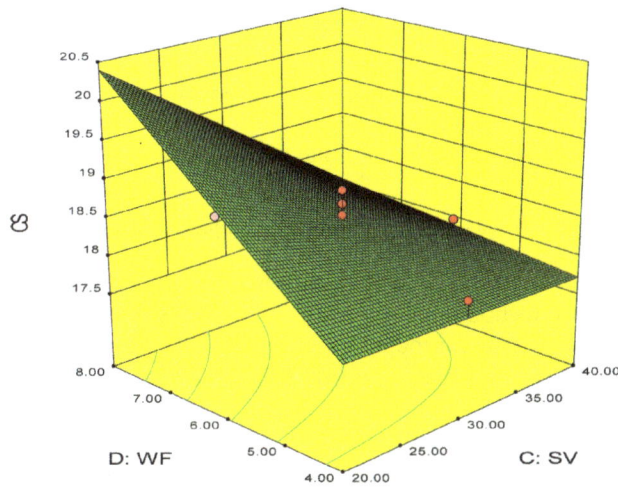

Figure 6b. Combined effect of SV and WF on CS.

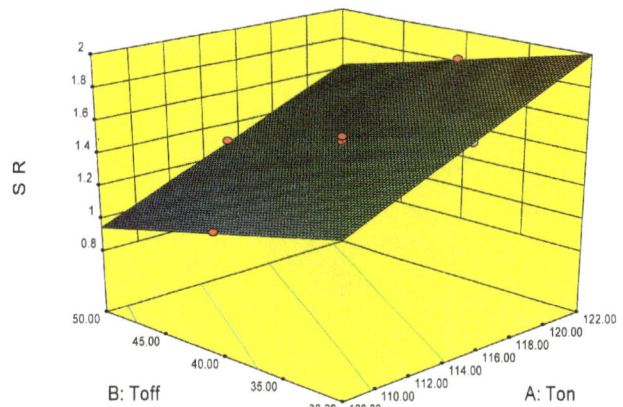

Figure 7a. Combined effects of T_{on} and T_{off} on SR.

SV. In WEDM, cutting speed depends on the melting of work surface and then flushing of the eroded material out of the spark zone. Melting of work surface depends on the thermal conductivity of work material and energy consumed per spark which is supposed to be the function of discharge current (I_p), pulse duration (T_{on}) and discharge voltage (SV) (Luo, 1995). Increasing pulse duration (T_{on}) increases the heat generation at the work surface which increases the cutting speed. Decreasing servo voltage closes the spark gap which results in rapid and large ionization of dielectric fluid which gives rise to more melting of work material and hence increases the cutting speed.

The rise in cutting speed can be highly noticed at high value of wire feed. High discharge energy results in more melting and evaporation of the work material causing liberation of large number of carbide debrises which coagulates in the spark gap and hence affects the machining process by producing arcs. Increasing wire feed rate leads to the easy and rapid escape of the eroded material out of the spark gap

and hence increases the cutting speed. This can be observed from the combined effect of WF with SV and T_{on} as shown in Fig. 6b–c.

Increasing pulse-off time (T_{off}) value decreases the discharge frequency and increases the overall machining time. Also long pulse-off time at high dielectric flow rate produces the cooling effect on work material and hence decreases the CS. The interaction among WEDM parameters (T_{on} and T_{off}; SV and WF) can be observed from curvature of the contour plots on response surface graphs. The low F-value (< 0.05) for these interactions effect shows a significant influence on performance characteristics as shown in Table 4.

3.2.2   Effect of WEDM parameters on surface roughness

Surface roughness increases with increasing pulse-on time (T_{on}) and decreasing servo voltage (SV) as shown in Fig. 7a–b. In WEDM, SR is mainly described by the shape and size of the surface craters which mainly depends on the discharge energy and re-deposition of melted material on work surface.

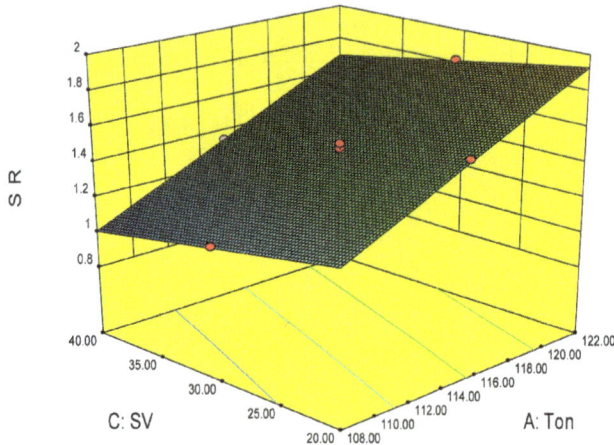

Figure 7b. Combined effects of T_{on} and SV on SR.

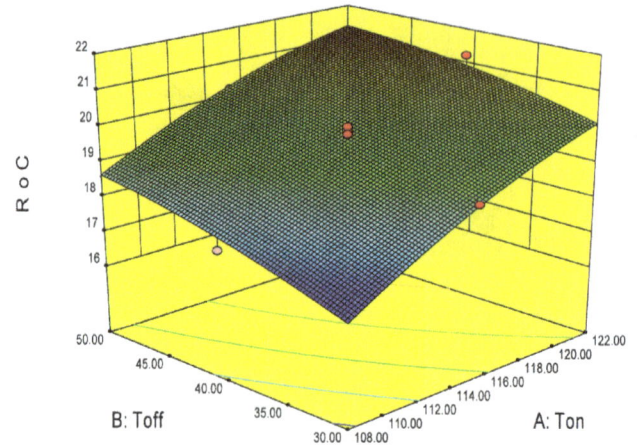

Figure 8a. Combined effects of T_{on} and T_{off} on RoC.

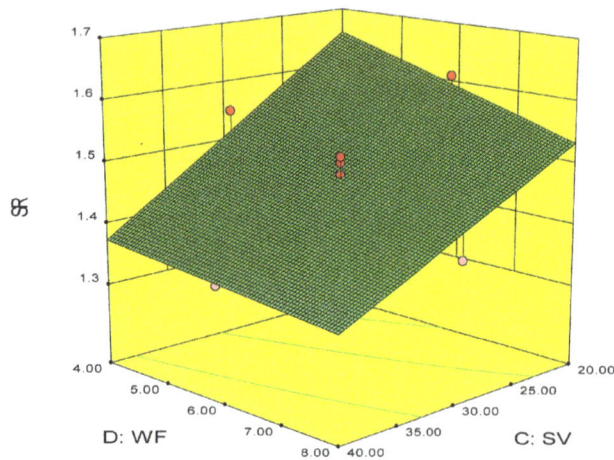

Figure 7c. Combined effects of SV and WF on SR.

residuals and hence low surface roughness. Curved contour on response graphs between WF and SV shows some interaction between these parameters. To get good surface finish, it is desirable to keep the electrical discharging energy at smaller level, by setting low T_{on} and high T_{off}-value.

3.2.3 Effect of WEDM parameters on radial overcut

RoC is the thickness of material removed perpendicular to the cutting direction as shown in Fig. 2. RoC helps to achieve accurate wire offset for rough cutting operation for obtaining precise dimensional tolerance. RoC depends on the amount of material melted and flushing it out of the spark gap. Increasing pulse-on time (T_{on}) results in high discharge energy which causes more melting and erosion of work surface which increases the RoC. At high value of pulse-off time (T_{off}), eroded carbide debris is easily flushed away from the spark gap. Therefore, RoC increases with increase in T_{on} and T_{off} as shown by surface plots in Fig. 8a, c and d. At low value of T_{off}, tendency of recasting of residue material on the work surface is high, which decreases the RoC.

Increasing wire feed rate helps to clear off the spark gap quickly which results in low re-deposition of eroded material and hence increases the RoC as shown in Fig. 8b. Decreasing servo voltage (SV) results in more ionization of the spark gap which results in more melting of work material and hence increases the RoC. Curved contours on response graph show the interaction between WEDM parameters. Using response surface graphs, values of WEDM parameters can be selected for the desired value of performance characteristics.

4 Conclusions

In present work, experimental investigation has been reported on machining performance of WC-5.3 % Co composite on WEDM. Response surface methodology (RSM), a statistical technique, has been utilised to investigate the

Increasing pulse-on time and decreasing servo voltage increases the discharge energy across the electrodes which results in deeper erosion crater on the work surface and hence increases the surface roughness (Lee and Li, 2003).

Due to the large difference between melting temperatures for WC and Co, discharge energy tends to melt, evaporate and remove cobalt even before the melting of WC. As a result WC grains may release without melting or in semi-solid state which coagulate in the spark gap (Saha et al., 2008). At high discharge energy, the probability of re-deposition of melted material of work surface is high. Increasing T_{off}-value increases the time between two consecutive sparks which results in complete flushing of the carbide debris out of the spark gap, and hence low re-deposition of eroded material results in low surface roughness as shown by the combined effect of T_{on} and T_{off} in Fig. 7a.

Surface roughness shows mild tendency to decrease with increasing wire feed rate (WF) as shown in Fig. 7c. Increasing wire feed results in easy escape of carbide debris out of the spark gap which results in low re-deposition of melted

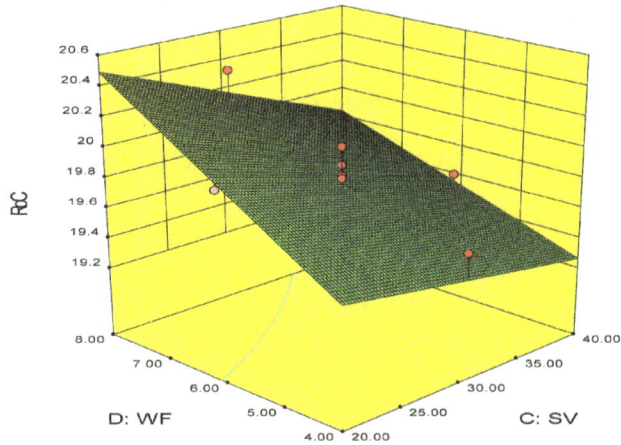

Figure 8b. Combined effects of SV and WF on RoC.

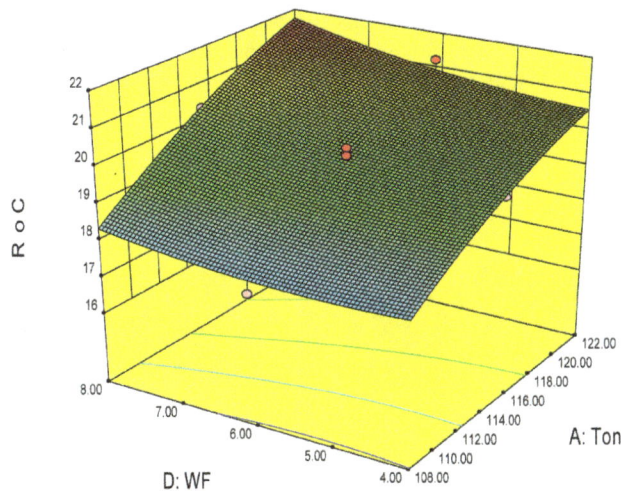

Figure 8d. Combined effects of WF and T_{off} on RoC.

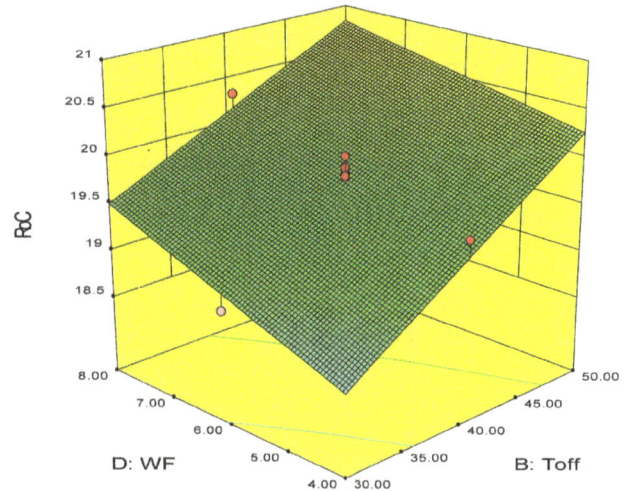

Figure 8c. Combined effects of T_{on} and WF on RoC.

T_{on} and WF produces significant effect. Using mathematical models, optimal parameters can be determined easily for desired performance characteristics. Hence a trade-off can be made among various performance characteristics.

Edited by: B. Azarhoushang

References

Çaydaş, U., Hasçalik, A., and Ekici, S.: An adaptive neuro-fuzzy inference system (ANFIS) model for wire-EDM, Expert Syst. Appl., 36, 6135–6139, 2009.

Chen, H.-C., Lin, J.-C., Yang, Y.-K., and Tsai, C.-H.: Optimization of wire electrical discharge machining for pure tungsten using a neural network integrated simulating annealing approach, Expert Syst. Appl., 37, 7147–7153, 2010.

Cochran, G. and Cox, G. M.: Experimental design, Asia Publishing House, New Delhi, 1962.

Engqvist, H., Ederyd, S., and Axen, N.: Grooving wear of single-crystal tungsten carbide, Wear, 230, 165–174, 1999.

Hewidy, M. S., El-Taweel, T. A., and El-Safty, M. F.: Modeling the machining parameters of wire electrical discharge machining of Inconel 601 using RSM, J. Mater. Process. Tech., 169, 328–336, 2005.

Jangra, K.: Study of unmachined area in intricate machining after rough cut in WEDM, Int. J. Indust. Eng. Comput., 3, 887–892, 2012.

Jangra, K., Jain, A., and Grover, S.: Optimization of multiple-machining characteristics in wire electrical discharge machining of punching die using grey relational analysis, J. Sci. Ind. Res., 69, 606–612, 2010.

Jangra, K., Grover, S., Chan, F. T. S., and Aggarwal, A.: Digraph and matrix method to evaluate the machinability of tungsten carbide composite with wire EDM, Int. J. Adv. Manuf. Tech., 56, 959–974, 2011.

Jangra, K., Grover, S., and Aggarwal, A.: Optimization of multi machining characteristics in WEDM of WC-5.3 % Co composite

influence of four important WEDM parameters – pulse-on time (T_{on}), pulse-off time (T_{off}), servo voltage (SV) and wire feed (WF) – on three performance characteristics: cutting speed (CS), surface roughness (SR) and radial overcut (RoC). Face centered central composite design was employed to conduct the experiments and to develop a correlation between the WEDM parameters and each performance characteristics. Analysis of variance (ANOVA) on experimental data shows that quadratic vs. two-factor interaction (2FI) model as the best fit for CS and RoC, while two-factor interaction (2FI) has been proposed for SR.

Response surface graphs were used to describe the influence of WEDM parameters on each performance characteristic. T_{on}, T_{off}, SV and WF produce significant influence on each performance characteristic. In case of cutting speed, interaction between T_{on} and T_{off}; T_{on} and WF; T_{off} and WF; SV and WF produces significant effect, while, in SR, interaction effect of SV and WF has been found significant. In case of RoC, interaction among T_{on} and T_{off};

using integrated approach of Taguchi, GRA and Entropy method, Frontiers of Mechanical Engineering, 7, 288–299, 2012.

Jia, K. and Fischer, T. E.: Sliding wear of conventional and nanostructured cemented carbides, Wear, 203–204, 310–318, 1997.

Kanagarajan, D., Karthikeyan, R., Palanikumar, K., Paulo, and Davim, J.: Optimization of electrical discharge machining characteristics of WC/Co composites uing non-dominated sorting genetic algorithm (NSGA-II), Int. J. Adv. Manuf. Tech., 36, 1124–1132, 2008.

Kansal, H. K., Singh, S., and Kumar, P.: Parametric optimization of powder mixed electrical discharge machining by response surface methodology, J. Mater. Process. Tech., 169, 427–436, 2005.

Kim, C. H. and Kruth, J. P.: Influence of electrical conductivity of dielectrical fluid on WEDM of sintered carbide, KSME Int. J., 15, 1276–1282, 2001.

Lauwers, B., Liu, W., and Eeraerts, W.: Influence of the composition of WC-based cermets on manufacturability by wire-EDM, J. Mater. Process. Tech., 8, 83–89, 2006.

Lee, S. H. and Li, X. P.: Study of the effect of machining parameters on the machining characteristics in electrical discharge machining of tungsten carbide, J. Mater. Process. Tech., 115, 344–358, 2001.

Lee, S. H. and Li, X. P.: Study of the surface integrity of the machined workpiece in the EDM of tungsten carbide, J. Mater. Process. Tech., 139, 315–321, 2003.

Liu, K. and Li, X. P.: Ductile cutting of tungsten carbide, J. Mater. Process. Tech., 113, 348–354, 2001.

Liu, K., Li, X. P., and Rahman, M.: Characteristics of high speed micro cutting of tungsten carbide, J. Mater. Process. Tech., 140, 352–357, 2003.

Luo, Y. F.: An energy-distribution strategy in fast cutting wire EDM, J. Mater. Process. Tech., 55, 380–390, 1995.

Mahdavinejad, R. A. and Mahdavinejad, A.: ED machining of WC-Co, J. Mater. Process. Tech., 162–163, 637–643, 2005.

Myers, R. H. and Montgomery, D. C.: Response Surface Methodology, New York, Wiley, 11, 535–561 and 12, 570–615, 1995.

Patil, N. G. and Brahmankar, P. K.: Determination of material removal rate in wire electro-discharge machining of metla matrix composites using dimensional analysis, Int. J. Adv. Manuf. Tech., 51, 599–610, 2010.

Saha, P., Singha, A., and Pal, S. K.: Soft computing models based prediction of cutting speed and surface roughness in wire electro-discharge machining of tungsten carbide cobalt composite, Int. J. Adv. Manuf. Tech., 39, 74–84, 2008.

Sarkar, S., Sekh, M., Mitra, S., and Bhattacharyya, B.: Modelling and optimization of wire electrical discharge machining of γ-TiAl in trim cutting operation, J. Mater. Process. Tech., 205, 376–387, 2008.

22

Initial prediction of dust production in pebble bed reactors

M. Rostamian[1]**, S. Arifeen**[2]**, G. P. Potirniche**[2]**, and A. Tokuhiro**[1]

[1]Department of Mechanical Engineering, University of Idaho, 1776 Science Center Dr, Idaho Falls, ID, USA
[2]Department of Mechanical Engineering, University of Idaho, 440902 Moscow, ID, 83844-0902, USA

Abstract. This paper describes the computational simulation of contact zones between pebbles in a pebble bed reactor. In this type of reactor, the potential for graphite dust generation from frictional contact of graphite pebbles and the subsequent transport of dust and fission products can cause significant safety issues at very high temperatures around 900 °C in HTRs. The present simulation is an initial attempt to quantify the amount of nuclear grade graphite dust produced within a very high temperature reactor.

1 Introduction

The gas-cooled graphite-moderated pebble bed reactor is a leading concept for the Next Generation Nuclear Plant, a Very High Temperature Reactor (VHTR) under consideration in the US. In the proposed reactor, spherical graphite pebbles (Fig. 1) are used as fuel elements. These graphite pebbles contain thousands of tristructural-isotropic (TRISO) fuel particles which are mainly made of enriched uranium. The graphite pebbles are inserted into the reactor from the top and they move downward due to gravity to reach the outlet chute where they go through burnup assay or fuel utilization. Here, the burnup limit of graphite pebbles is assessed to determine if they are to be recirculated or stored in the storage tank.

In VHTRs, the potential graphite dust generation is caused by several sources. These sources are pebble-pebble contact, pebble-wall contact, fuel handling which proceeds burnup assay and oxidation from impurities present in the helium coolant (Cogliati and Ougouag, 2008).

There are many disadvantages to dust production at very high temperatures in pebble bed reactors (PBR). In the presence of graphite dust, helium can become radioactive, reducing the efficiency of heat exchangers (Cogliati and Ougouag, 2010). In direct cycle HTGRs, the generated graphite particulates collide with turbine blades and can considerably decrease their service life (Cogliati and Ougouag, 2010). The amount of graphite dust generated is a key input for design safety review (US Nuclear Regulatory Commission, 2008; IAEA, 2003). The only previous prediction of dust production was made for the German **A**rbeitsgemeinschaft **V**ersuchs**r**eaktor (AVR), which was around 3 kg yr^{-1} (Xiaowei et al., 2005).

2 Computational model

In the present work, the simulations of pebble contact were conducted using the finite element software ABAQUS (ABAQUS FEA, 2010) to predict an initial estimate for PBR graphite dust production.

Two different configurations are considered: (1) two Quarter Spheres in contact (QS) model illustrated in Fig. 2a, and (2) a Body Centered Cubic (BCC) model 1, in one pebble is in contact with eight other ones, as shown in Fig. 2b. The QS model is a simple model to start performing simulations but the BCC model is a more realistic configuration which better represents the conditions of pebbles in a PBR (du Toit et al., 2009).

Two types of simulations are also considered: (1) static simulations where pebble-pebble forces and friction at contact points are considered without pebble movements with respect to each other, and (2) dynamic simulations, where pebble-pebble forces, friction at contact points and rotation of the center pebble with respect to the neighboring ones is considered.

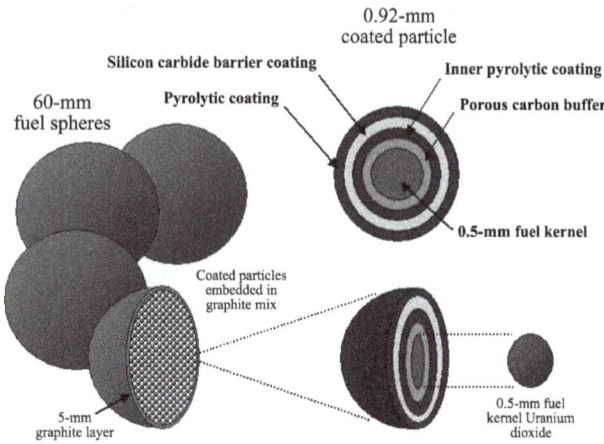

Figure 1. Graphite Fuel Elements in PBRs (Schaffer, 2007).

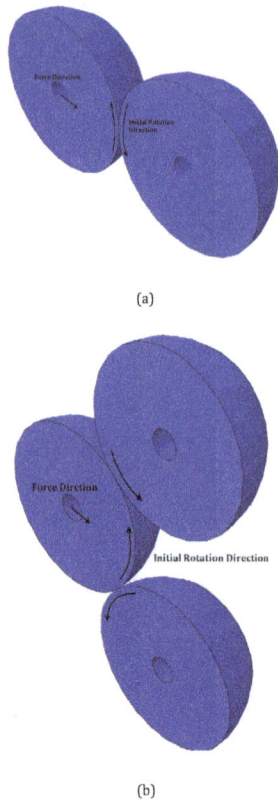

(a)

(b)

Figure 2. QS (**a**) and BCC (**b**) configurations. Symmery conditions are imposed.

The material properties were modeled by considering an elastoplastic stress-strain behavior coupled with ductile damage effects. A mesh refinement was considered to determine the number of elements with damage values exceeding a minimum predetermined value. The extent of damage was used to determine the amount of mass removed. Finally, the results are discussed and compared with previous works (Cogliati and Ougouag, 2010; Xiaowei et al., 2005).

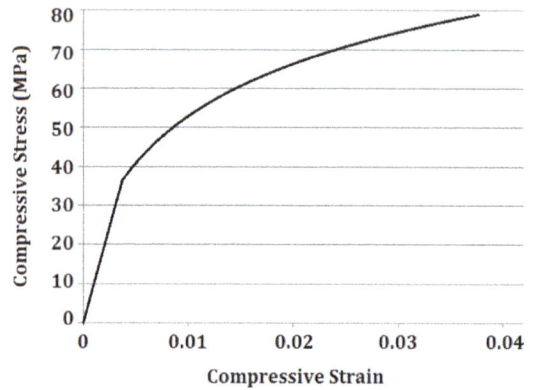

Figure 3. Stress-strain curve for IG-11.

Table 1. Isotrpic plastic model data.

Yield Stress (MPa)	Plastic Strain
45 005 500	0
48 779 700	0.0015747
53 811 900	0.0040155
58 765 500	0.0066925
63 247 300	0.0099206
67 414 600	0.0136999
71 739 100	0.0185028
74 962 900	0.0230694
77 557 600	0.0271636
79 287 400	0.0309429

2.1 Material properties

The nuclear grade graphite IG–11 with elastoplastic behavior at the strain rate of $0.0011s^{-1}$ is considered to determine the extent of pebble damage due to contact and wear forces (Fig. 3 and Table 1).

2.2 Isotropic elastic model

From the stress-strain diagram shown in Fig. 3, the elastic properties are derived. The Young's modulus and yield stress are found to be $E = 9.8$ GPa. The Poisson ratio is also known to be 0.126 for this nuclear grade graphite (Yokoyama et al., 2008).

2.2.1 Isotropic plastic model

From the plastic portion of this stress-strain curve, the yield stress and the corresponding plastic strain are derived. The plastic stress-strain data input for ABAQUS to replica is shown in Table 1.

Table 2. Damage model data.

Fracture Strain	Stress Triaxiality	Strain Rate (Yokoyama et al., 2008)
0.0448246	0.333	0.0011 s^{-1}

2.2.2 Ductile damage model

A ductile damage criterion (Johnson and Cook, 1985) is used in this numerical simulation. The fracture strain (at specific strain rate) and stress triaxiality are needed to set this criterion. Once this limit is exceeded, elements are capable of being removed from the surface of the contact point. In this study, the damage is investigated macroscopically and IG-11 is considered as a continuous homogenous material.

In Table 2 stress triaxiality is simply found to be 0.333 because of material homogeneity in three directions.

2.3 Model setup

The simulation setup consists of two sections with two models for pebble-pebble contact the QS model and the BCC model unit cell configuration. The latter represents a more realistic configuration of a pebble condition in contact with eight other ones.

2.3.1 QS Model

In this model, two quarter spheres are in contact and the boundary conditions and loads are set as illustrated in Fig. 4. In this figure, u represents translational velocity in m s^{-1} and ur represents rotational velocity in radians.

2.3.2 BCC Model

In this model, one quarter sphere is in contact with two eight-spheres representing a portion of a unit cell model, which is cut at its symmetrical planes. The boundary conditions and loads are set as in Fig. 5. In this figure, the model is reduced by considering the existing symmetry in the BCC configuration.

2.4 Simulation setup

The simulations are also carried out in two phases. A static simulation is performed where stationary pebble-pebble forces and friction at contact points are considered. In the static simulations, due to the linearity of the simulations, ABAQUS/Standard module has been implemented.

The boundary conditions, loads and constrains are illustrated in Figs. 4a and 5a.

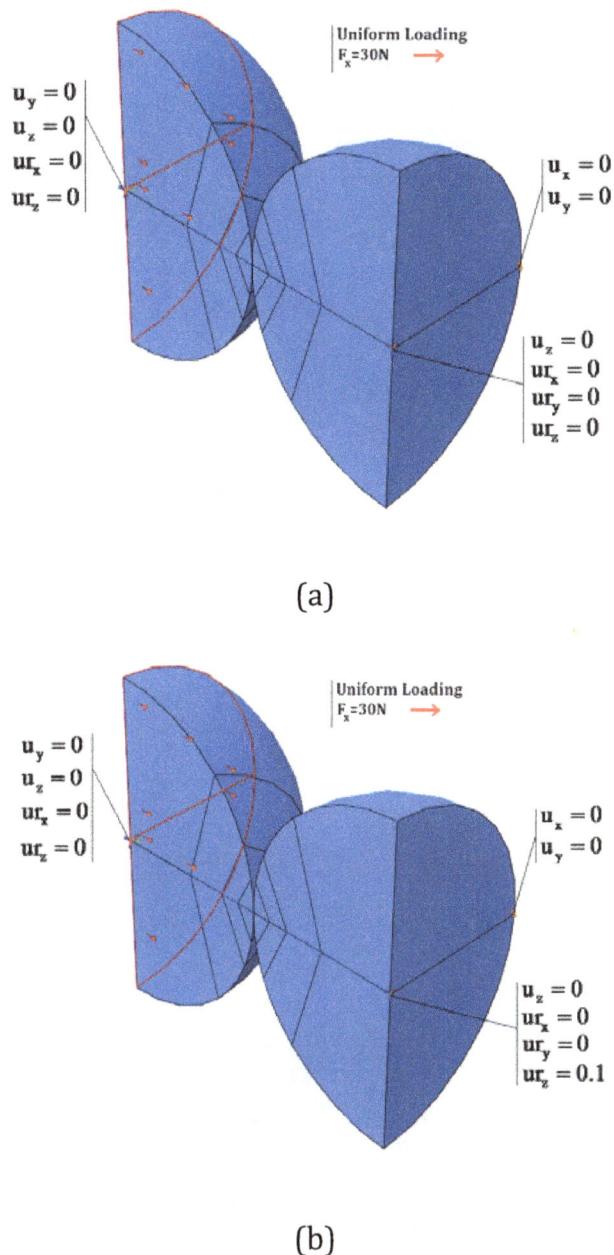

(a)

(b)

Figure 4. QS model boundary conditions and loads in static (**a**) and dynamic (**b**) phases (rotation in radians).

The dynamic simulation considers pebble-pebble forces, friction at contact points and rotation of the center pebble. In the dynamic simulations, ABAQUS/Explicit module has been implemented to handle the nonlinearities caused by rotational velocity and contact mechanics. The boundary conditions, loads and constrains are illustrated in Figs. 4b and 5b.

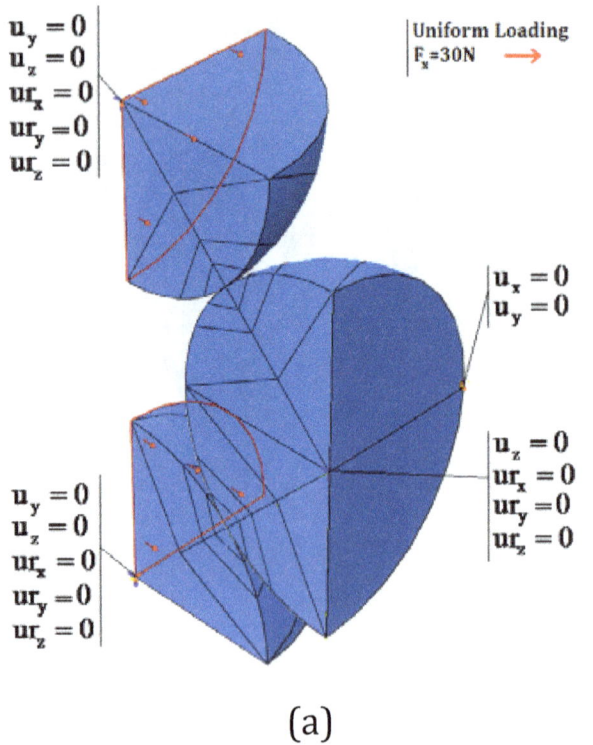

(a)

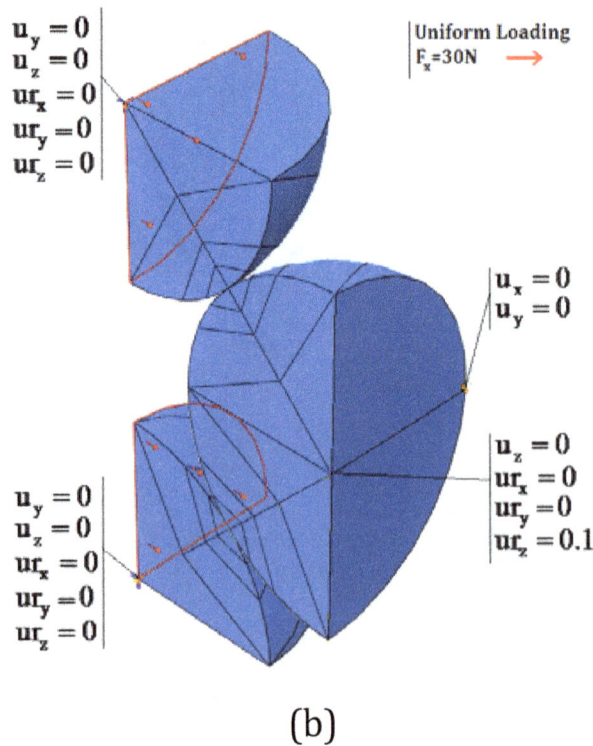

(b)

Figure 5. BCC model boundary conditions and loads in static (**a**) and dynamic (**b**) phases (rotation in radians).

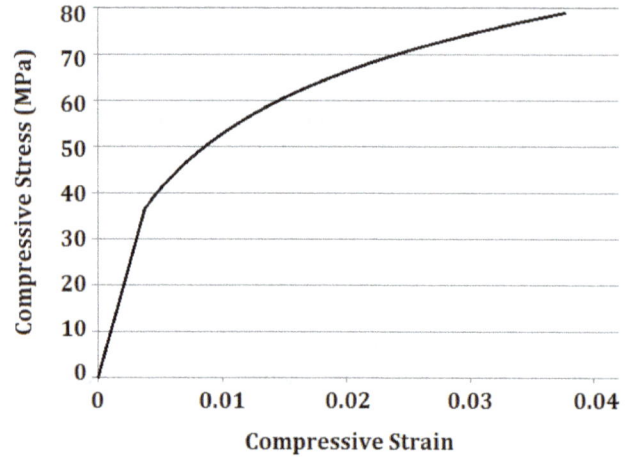

Figure 6. Mesh refinement study results: the optimum element size and stress.

2.5 Mesh refinement study

Considering the elastic model, simulations were run for different element sizes to perform a mesh refinement study (MRS); the MRS led to the optimal element size of 0.022 cm for the contact zone (Fig. 6).

This optimal element size helps reduce computational costs and time. The optimum mesh as identified from the MRS results in the optimal refinement of mesh in the contact zone, where results are sensitive to element size. Figure 6 presents the optimal mesh near the contact zone.

3 Results and discussions

In this section, first the von Mises stress contours and damage contours are illustrated and discussed briefly. Finally, an estimate of the mass removal from the surface of graphite pebbles is discussed.

3.1 Static simulation results

As seen in Fig. 7, in both configurations higher stresses are experiences at the tip of the contact area than those at the center-point. This is due to the deflection at the tip of the contact area which is itself a flat circle-shaped region perpendicular to the figure plain.

As seen in Fig. 8, higher damage is at the edges of the contact area due to higher local deformations.

3.2 Dynamic simulation results

As seen in Fig. 9, in both models higher stresses are experiences at the tip of the contact area than that at the center-point. This is due to the deformation at the tip of the contact area.

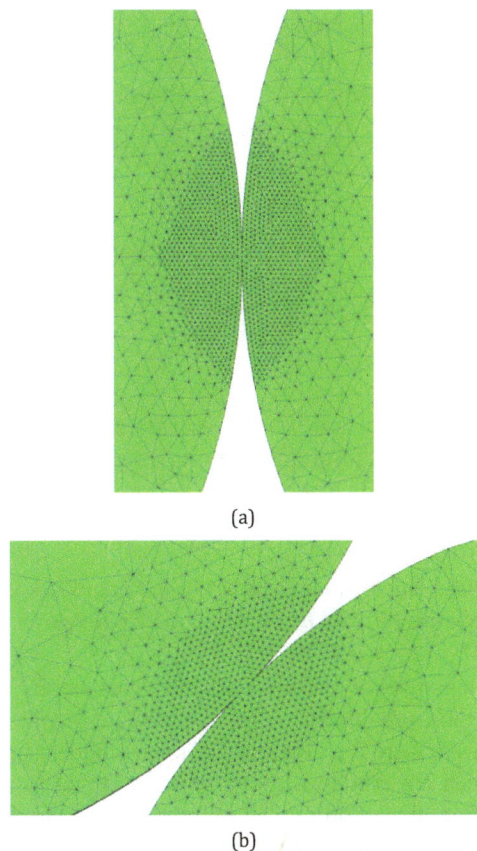

(a)

(b)

Figure 7. Contact zone refined mesh for (**a**) the QS and (**b**) BCC models.

Note that, in the dynamic simulation, the damaged zones are not completely facing each other due to the axial rotation of one pebble against the other.

As seen in Figs. 10 and 11, higher stress and damage are experienced at the edges of the contact area due to higher local deformations.

The contact zone is a circular area which is perpendicular to the figure plains whose upper and lower edges experience high stresses and therefore high damages.

4 Dust mass prediction

Based on the damage results, by probing the elements in the damaged region and tallying the damage criterion at each damaged element, the number of elements that were capable of removal due to excessive damage was counted and therefore an estimate of the mass removal was predicted.

In the Static simulation, no elements passed the damage criterion in the QS model and as a result the mass removal estimate was predicted to be zero; while in the static BCC model given the density of $1.77\,\mathrm{g\,cm^{-3}}$ (Yoda and Eto, 1983), an estimate of $5.9\,\mathrm{g\,yr^{-1}}$ was predicted.

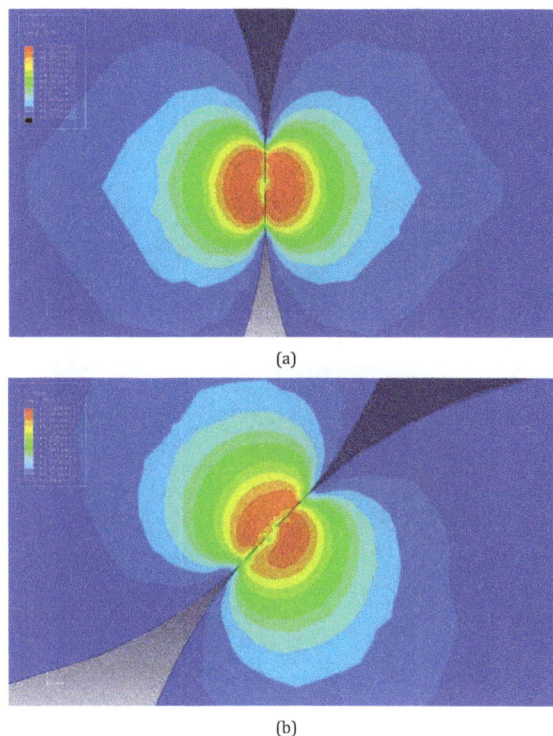

(a)

(b)

Figure 8. Von Mises stress contours at the contact points for (**a**) QS and (**b**) BCC models.

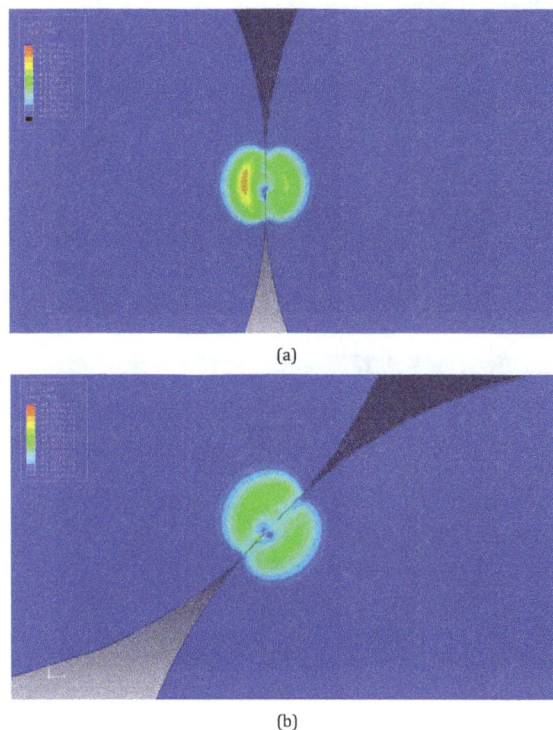

(a)

(b)

Figure 9. Damage contours at the contact points for (**a**) QS and (**b**) BCC models.

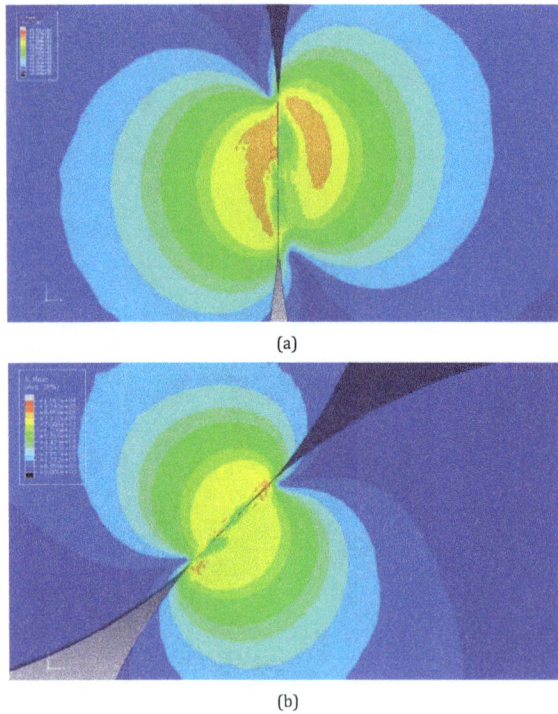

Figure 10. Von Mises stress contours at the contact points for (a) QS and (b) BCC models.

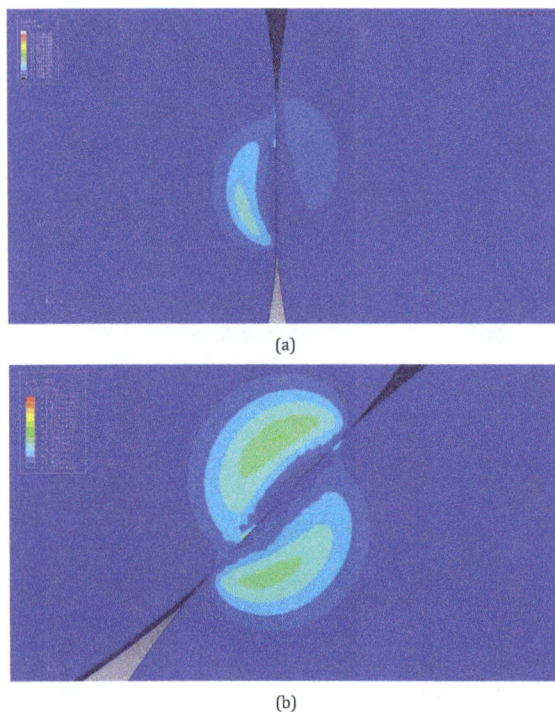

Figure 11. Damage contours at the contact points for (a) QS and (b) BCC models.

Table 3. Mass removal prediction.

Sinulation Model	Mass (g yr^{-1})
Static QS	0.0
Static BCC	5.9
Dynamic QS	6.5
Dynamic BCC	6.7
Cogliati's Results (Cogliati and Ougouag, 2010)	4.0

In the dynamic simulation which was a more realistic one, the QS model mass removal estimate was 6.5 g yr^{-1}, while it was 6.7 g yr^{-1} for the BCC model. Table 3 shows a comparison between the results of the current model and those by Cogliati (2010).

The dust production estimate of 6.5 to 6.7 grams per year is very low compared to an estimated of 3 kg per year in the German AVR (Moormann, 2009).

There are a number of reasons which make the present initial estimate more reliable as compared to AVR observations. The main reason there is a difference of three orders of magnitude between our results and that of AVR is that in this dust production prediction model, only pebble-pebble contacts are considered while, as mentioned earlier, there are a number of sources to graphite dust generation (Cogliati and Ougouag, 2010).

Another influential factor is the material properties; while temperatures of about 900 °C are experienced in a PBR, the only material parameters available at this point are at room temperature from (Yokoyama et al., 2008). Also, it is uncertain how much of the dust produced in AVR, was a result of mechanical wear, as opposed to other sources. The dust production in AVR was observed by annually removing the filter and measuring the amount of dust, which is generated from different sources such as oil ingress, air ingress, metallic components and pebble handling.

5 Conclusions

The advantage of performing numerical analyses is to achieve an estimate of the mass of graphite removed per working year of an NGNP during its service period. This order of magnitude of estimate can be an approximate input for safety design of PBRs. This estimate is a good agreement with those in the literature. However, more investigations are in progress to enhance the accuracy of the prediction of dust generation in PBRs. Considering a wear model to calculate the wear mass based on the contact normal forces leads to results of higher accuracy and dependability. Performing simulations for various pebble-pebble configurations can also lead to better evaluation the wear mass.

Acknowledgements. Thanks are due to different people who have been of various sorts of help. At University of Idaho the following people were of continuous assistance: Ahmad Abdelnabi, Bryan Riga and Karl Rink. The authors acknowledge the contributions of Abderrafi Ougouag and Joshua J. Cogliati at Idaho National Lab. This project, NEUP-90151, is supported by DOE.

Edited by: A. Barari

References

ABAQUS FEA: SIMULIA web site, Dassault Systèmes, www.simulia.com, last access: May 2010.

Cogliati, J. and Ougouag, A. M.: Pebble Bed Reactor Dust Production Model, Proceedings of ASME 4th International Topical Meeting on High Temperature Reactor Technology, HTR2008-58289, available at: http://www.inl.gov/technicalpublications/Documents/4074915.pdf, 2008.

Cogliati, J. J. and Ougouag, A. M.: Dust Production Model for HTR-10, INL Report, 2010.

IAEA: Evaluation of High Temperature Gas Cooled Reactor Performance: Benchmark Analysis Related to Initial Testing of the HTTR and HTR-10, IAEA-TECDOC-1382, 2003.

Johnson, G. R. and Cook, W. H.: Fracture Characteristics of Three Metals Subjected to Various Strains, Strain rates, Temperatures and Pressures, Eng. Fract. Mech., 21, 31–48, 1985.

Moormann, R.: A safety re-evaluation of the AVR pebble bed reactor operation and its consequences for future HTR concepts, JUEL-4275, retrieved 2 April 2009.

Schaffer, M. B.: Nuclear power for clean, safe and secure energy independence, Foresight, 9, 47–60, 2007.

du Toit, C. G., van Antwerpen, W., and Rousseau, P. G.: Analysis of the Porous Structure of an Annular Pebble Bed Reactor, Final Paper 9123, ICAPP, 2009.

US Nuclear Regulatory Commission: Next Generation Nuclear Plant Phenomena Identification and Ranking Tables (PIRTs), NUREG/CR-6944, Main Report, 2008.

Xiaowei, L., Suyaun, Y., Zhen-sheng, Z., and Shu-yan, H.: Estimation of Graphite Dust Quantity and Size Distribution of Graphite Particle in HTR-10, Nuclear Power Engineering, 26, 0258-0926, 2005.

Yoda, S. and Eto, M.: The tensile deformation behavior of nuclear-grade isotropic graphite posterior to hydrostatic loading, J. Nucl. Mater., 118, 214–219, 1983.

Yokoyama, T., Nakai, K., and Futakawa, M.: Compressive Stress-Strain Characteristics of Nuclear-Grade Graphite IG-11: Effect of specimen size and strain rate, Journal of Japan Society of Nuclear, 7, 66–73, 2008.

23

On mechanical properties of planar flexure hinges of compliant mechanisms

F. Dirksen and R. Lammering

Institute of Mechanics, Helmut-Schmidt-University/University of the Federal Armed Forces Hamburg,
Holstenhofweg 85, 22043 Hamburg, Germany

Abstract. The synthesis of compliant mechanisms yield optimized topologies that combine several stiff parts with highly elastic flexure hinges. The hinges are often represented in finite element analysis by a single node (one-node hinge) leaving doubts on the physical meaning as well as an uncertainty in the manufacturing process.

To overcome this one-node hinge problem of optimized compliant mechanisms' topologies, one-node hinges need to be replaced by real flexure hinges providing desired deflection range and the ability to bear internal loads without failure. Therefore, several common types of planar flexure hinges with different geometries are characterized and categorized in this work providing a comprehensive guide with explicit analytical expressions to replace one-node hinges effectively.

Analytical expressions on displacements, stresses, maximum elastic deformations, bending stiffness, center of rotation and first natural frequencies are derived in this work. Numerical simulations and experimental studies are performed validating the analytical results. More importance is given to practice-oriented flexure hinge types in terms of cost-saving manufacturability, i.e. circular notch type hinges and rectangular leaf type hinges.

1 Introduction

In order to create machine tools for small scale applications, compliant mechanisms (CM) have become more popular in the last years competing against rigid body systems connected by conventional pin joints. CM are flexible, monolithic structures that gain their motion from the (elastic) deformation of certain parts, so-called flexure hinges. CM are potentially more accurate, better scalable, cleaner, less noisy and most importantly more cost-saving in manufacturing and maintenance. However, designing CM is more difficult and non-intuitive due to its inherent complex overall deformation.

Several approaches have arisen to address this drawback by applying numerical topology design and optimization procedures. Relevant contributions have been made by various research teams, in particular, Ananthasuresh and Kota (1995), Frecker et al. (1997), Saxena and Ananthasuresh (2000), Howell (2001), Bruns and Tortorelli (2001), Ansola et al. (2002), Bendsøe and Sigmund (2003), Mattson et al. (2004), Bendsøe and Sigmund (2008). All these techniques lead in a systematic manner to final, optimized topologies, i.e. an optimal distribution of material over the design domain is obtained to meet the user-specified motion requirements. As a key result, one-node hinges (often called pseudo-hinges) with doubtful physical meaning arise. As an example, a gripping mechanism and a close up of a one-node hinge, obtained by a topology optimization procedure without any regularization, is shown in the upper box in Fig. 1. Although some techniques exist circumventing this critical issue, e.g. Poulsen (2002), Yoon et al. (2004) or Sigmund (2009), a more consequent way is to use the already known data from the finite element calculation used in the topology optimization process. Since nodal displacements for a given topology are known, the required deflection range and (internal) nodal forces are available without additional costs, as well. These information can be used to replace one-node hinges with real flexure hinges that meet the deflection and load bearing requirements as a result of their specific shape, dimension and material data.

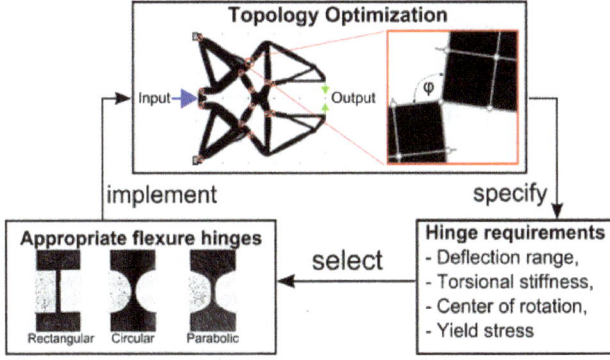

Figure 1. Beneficial procedure for non-intuitive synthesis of compliant mechanisms: Replacing artificial one-node hinges by appropriate flexure hinge types meeting specified, known hinge requirements.

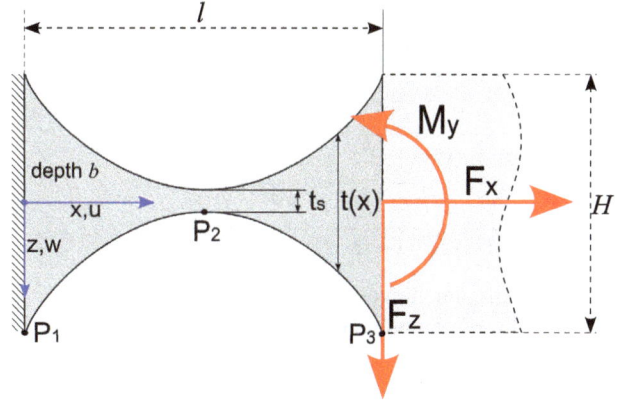

Figure 2. Planar, flexure hinge characterized by length l, depth b, height H, variable thickness $t(x) \geq t_s$ and common points P_1, P_2, P_3 to resist external (nodal) loads F_x, F_z, M_y.

Necessary mechanical properties of flexure hinges have been investigated by a few authors. Paros and Weisbord (1965) did pioneer work yielding approximate compliances of flexure hinges decades ago. Smith (2000) provided in his book a good background on flexure elements and some flexure systems. Lobontiu (2003) analytically investigated flexure hinges based on energy principles to calculate desired properties at individual single points of hinges. Recently, Raatz (2006) demonstrated in her dissertation the potential of flexure hinges in compliant parallel mechanisms using superelastic shape memory alloys.

In spite of the aforementioned research, the mechanical behavior of flexure hinges is not yet fully characterized in terms of the synthesis of compliant mechanisms and, thus, leaving a gap between final, optimized topologies and appropriate flexure hinges. In order to bridge this gap, mechanical properties of flexure hinges are derived and validated in this work to provide a comprehensive guide from a topology optimization standpoint. The overall scheme is shown in Fig. 1.

2 Objectives

For the synthesis of compliant mechanisms it is crucial to characterize and categorize individual flexure hinges in terms of their mechanical properties as a result of geometric shape and material data. Therefore, relevant mechanical properties are derived, such as:

– **Displacements** $u(x,z)$, $w(x)$, to gain a better understanding of the deformation of the whole flexure hinge.

– **Mechanical stresses** $\sigma_x(x,z)$, $\tau_{xz}(x,z)$, to identify critical regions that are not apparent.

– **Stiffness** k_z and **bending stiffness** c_ψ, to be able to model flexure hinges appropriately by spring joints.

– **Center of rotation** and its motion with deflection, to identify and compensate a change of kinematics under certain loading conditions.

– **Maximum (elastic) deformation**, to identify deflection limits and avoid material failure.

– **Natural frequencies** f, to understand the behavior under dynamic load conditions and to check the quality of numerical simulations against experimental data.

Analytical expressions are derived using a standard x-z-coordinate system, as shown in Fig. 2, by applying different established theories and models. If possible, numerical simulations and experimental data are used to validate the analytical calculations.

In this work, planar flexure hinges of different geometries are examined: rectangular, circular and parabolic flexure hinges, denoted by superscripts R, C, P respectively, are used due to an easy manufacturability (R, C) and convenient mathematical handling (R, P). In particular, circular shape is approximated by parabolic function using Taylor expansion to avoid complicated expressions. The geometric approximation error was checked and is negligible in all loaded regions.

The geometry of flexure hinges is described by length l, height H and variable thickness $t(x) \geq t_s$ as well as common points $P_1(0, \frac{H}{2})$, $P_2(\frac{l}{2}, \frac{t_s}{2})$ and $P_3(l, \frac{H}{2})$, as shown in Fig. 2. The depth is set to uniform $b = 10$ mm over the entire hinge, which is sufficient for the majority of planar applications.

Key aspect for the following calculations is the geometric shape given by the variable thickness $t(x)$ of each type of flexure hinge

$$t^R(x) = t_s, \tag{1}$$

$$t^P(x) = 2\left(c_1 + c_2 x + c_3 x^2\right)$$
$$= H - \frac{4x(H-t_s)}{l} + \frac{4x^2(H-t_s)}{l^2}, \tag{2}$$

$$t^C(x) = 2(z_M + \sqrt{r^2 - (x-x_M)^2})$$
$$= \frac{H^2 - t_s^2 + l^2}{4(H-t_s)} - \sqrt{\frac{(l^2 + (H-t_s)^2)^2}{(4(H-t_s))^2} - \frac{(l-2x)^2}{4}}. \tag{3}$$

Parabolic and circular hinges are first written in a general form denoted by polynomial coefficients c_1, c_2, c_3 and circle's center coordinates x_M, z_M and radius r, respectively. In the second lines of Eqs. (2) and (3), relevant geometric boundary conditions

$$c_1 = \frac{H}{2}, \qquad c_2 = \frac{-2(H-t_s)}{l}, \qquad c_3 = \frac{2(H-t_s)}{l^2},$$
$$x_M = \frac{l}{2}, \qquad z_M = \frac{t_s}{2} + r, \qquad r = \frac{l^2 + (H-t_s)^2}{4(H-t_s)}, \tag{4}$$

are applied. Throughout this paper, the formulations $t^{R,P,C} = t^{R,P,C}(x, H, l, t_s)$ are used to keep the solution adaptable to specific problems.

In order to compare analytical results with numerical and experimental data, a high strength aluminum wrought alloy AlCu4Mg1 (EN AW 2024) that is often used in applications of CM due to its high fatigue strength and high elastic strain, is considered throughout this work. The relevant material specifications are

$$E = 70\,\text{GPa}, \qquad \nu = 0.33, \qquad \rho = 2790\,\text{kg m}^{-3}.$$

Although it remains unchanged throughout this publication, the analytical formulas hold for other isotropic materials as well.

3 Mechanical properties of flexure hinges

Relevant mechanical properties of individual flexure hinges under quasi-static loading conditions are described and discussed in this section. The *total* behavior of CM consisting of *several* flexure hinges is not described here and is subject to further investigations. Since flexure hinges are mainly used in CM to allow rotational motion, the main focus is on axial bending caused by external nodal forces F_x, F_z and moment $M_y(x)$ as illustrated in Fig. 2.

3.1 Moments of area

The areas of the cross section $A(x) = bt_s(x)$, first moments of area $S_y(x,z)$ and second moments of area $I_y(x)$ were calculated and are listed in Table 1 for all considered flexure hinges using thicknesses $t(x)$ given in Eqs. (1)–(3). Note, that the first moment of area is calculated from z to $t(x)/2$.

The listed moments of area are used to calculate stresses and displacements in the following sections.

Table 1. First and second moments of areas of rectangular (R), circular (C) and parabolic (P) flexure hinges.

	$S_y(x,z) = \int_{A^*} z^* dA$	$I_y(x) = \int_A z^2 dA$
R	$\frac{b}{8}(t_s^2 - z^2)$	$\frac{bt_s^3}{12}$
C	$\frac{b\left(\left(z_M + \sqrt{r^2 - (x-x_M)^2}\right)^2 - z^2\right)}{2}$	$\frac{b((x-x_M)^2 + 2r(z_M - r))^3}{12r^3}$
P	$\frac{b\left((Hl^2 + 4(H-t_s)(x^2 - lx))^2 - 4l^4 z^2\right)}{8l^4}$	$\frac{b(H(l-2x)^2 + 4t_s(l-x)x)^3}{12l^6}$

3.2 Stresses

The normal stresses $\sigma_x(x,z) = \frac{F_x}{A(x)} + \frac{M_y(x)}{I_y(x)}z$, and shear stresses $\tau_{xz} = \frac{F_z(x)S_y(x,z)}{I_y(x)b}$ depend on the external loads, moments of area $S_y(x,z)$, $I_y(x,z)$ and depth b, where a linear-elastic, isotropic stress-strain relation is assumed. Furthermore, the normal stresses σ_y, σ_z and shear stresses τ_{yz}, τ_{xy} are assumed to be negligible. Thus, the relevant normal stresses are

$$\sigma_x^R(z) = \frac{1}{bt_s}F_x + \frac{12(x-l)z}{bt_s^3}F_z + \frac{12z}{bt_s^3}M_y,$$

$$\sigma_x^P(x,z) = \frac{l^2}{bh_*^3(x)}F_x + \frac{12l^6 z}{bh_*^9(x)}M_y + \frac{12l^6(x-l)z}{bh_*^9(x)}F_z, \tag{5}$$

$$\sigma_x^C(x,z) = \frac{1}{2bh_{**}(x)}F_x + \frac{3z(x-l)}{2bh_{**}(x)}M_y + \frac{3z}{2bh_{**}(x)}F_z,$$

and shear stresses are

$$\tau_{xz}^R(z) = \left(\frac{-6z^2}{bt_s^3} + \frac{3}{2bt_s}\right)F_z,$$

$$\tau_{xz}^P(x,z) = \left(\frac{3l^2\left(-4l^4 z^2 + h_*^6(x)\right)}{2bh_*^9(x)}\right)F_z, \tag{6}$$

$$\tau_{xz}^C(x,z) = \frac{3\left(r^2 - z^2 - (x-x_M)^2 + \left(z_M - 2\sqrt{r^2 - (x-x_M)^2}\right)z_M\right)}{4bh_{**}(x)}F_z,$$

where $h_*^3(x) = H(l-2x)^2 + 4(l-x)xt_s$ and $h_{**}(x) = z_M - \sqrt{r^2 - (x-x_M)^2}$ are introduced to keep the expressions short. Note, that any stress concentration effects are not yet taken into account as they will later, in Sect. 3.4.

3.3 Displacements

The displacements $u(x,z)$, $w(x)$ due to external loads F_x, F_z, M_y, as shown in Fig. 2, are calculated. Later, they are used to calculate stiffness and bending stiffness in Sect. 3.5.

In order to calculate displacements $u(x,z)$, $w(x)$ and bending slope $\psi(x)$, different beam theories are supposed to be applicable: *Euler-Bernoulli's* beam theory assumes that the (shear-indeformable) cross section remains perpendicular to the neutral axis and $\psi \approx \tan\psi = -w'(x)$, which is sufficient for slender beams (e.g. rectangular flexure hinges) undergoing small and moderate bending angles. *Elastica* beam theory lifts the latter limitation using the correct, non-linear expression $\frac{w_E''(x)}{(1+(w_E'(x))^2)^{3/2}} = -\frac{M_y(x)}{EI_y(x)}$ and, thus, it also holds for large bending angles. *Timoshenko's* beam theory holds for small and moderate bending angles, as Bernoulli's theory does, but it takes the shear deformation caused by arising shear stresses into account. Usually, this has a minor effect on the displacements considering "long" rectangular hinges ($t(x) \ll l$). However, it cannot be neglected in the case of "thick" hinges with an increased "effective" thickness ($t^{\text{eff}} = \frac{1}{l}\int t(x)dx \approx l$) compared to hinge length l, such as most circular and parabolic hinges. Further details can be found in standard literature; particularly the influence of large deformations and shear stresses are described in Love (1920) and Wang et al. (2000), respectively.

In this work, Timoshenko's beam theory is used to calculate the required displacements, since flexure hinges do not undergo large rotations and shear deformation cannot be neglected. The displacement expressions are

$$w'(x) = -\psi(x) + \frac{F_z}{\alpha_S GA(x)},$$

$$\psi'(x) = \frac{M_y(x)}{EI_y(x)}, \tag{7}$$

$$u(x,z) = z\,\psi(x) + \int_0^x \frac{F_x}{EA(x^*)}dx^*.$$

Here, the angles $w'(x)$ and $\psi(x)$ differ by an additional shear deformation term, where α_s is a shear correction factor compensating non-uniform shear stresses τ_{xs} in the cross section. Furthermore, the displacement $u(x,z)$ is expanded by an additional axial displacement term caused by axial forces F_x.

Based on Eq. (7), the displacement expressions can be calculated for different types of flexure hinges. As an example, the displacements for a rectangular flexure hinge based on Timoshenko's theory become

$$w^{\text{R}}(x) = \frac{12(1+\nu)t_s^2 x + 30lx^2 - 10x^3}{5Ebt_s^3}F_z - \frac{6x^2}{Ebt_s^3}M_y,$$

$$\psi^{\text{R}}(x) = -\frac{12lx - 6x^2}{Ebt_s^3}F_z + \frac{12x}{Ebt_s^3}M_y, \tag{8}$$

$$u^{\text{R}}(x,z) = \frac{x}{Ebt_s}F_x - \frac{(12lx - 6x^2)z}{Ebt_s^3}F_z + \frac{12xz}{Ebt_s^3}M_y.$$

The derived displacements expressions are used in Sect. 3.5 to calculate stiffness and bending stiffness of different flexure hinges. Note, that anti-clastic bending effects are neglected, as suggested by Conway and Nickola (1965).

3.4 Maximum elastic deformation

Flexure hinges can undergo smaller rotational deformation than conventional pin joints that have practically no limits. The maximum elastic deformation of flexure hinges can be estimated by combining the occurring stresses derived above to an equivalent stress σ_V which has to be lower than the yield stress $R_{p0.2}$: $\sigma_V \leq R_{p0.2}$. Among various established yield criteria, von-Mises yield criterion $\sigma_V = \sqrt{\sigma_x^2 + 3\tau_{xz}^2}$ is mainly used for ductile materials and, thus, applicable to the majority of materials in compliant mechanisms.

Static load cases and quasi-static motions are considered; fatigue effects and durability are not yet fully investigated and will be subject of future investigations and publications.

Maximum normal stress $\sigma_{x,\text{max}}$ can be found at the thinnest cross section $x = x(t = t_s)$ at the upper or lower edge $z = \pm t_s/2$. Whereas maximum shear stresses $\tau_{xz,\text{max}}$ occur at the *center* of the thinnest cross section $x = x(t = t_s)$ at $z = 0$ and is zero at the edges $\tau_{xz}(z = \pm t_s/2) = 0$. Typically in applications considered here, normal stresses are more dominant than shear stresses suggesting to neglect shear stress. However, maximum shear stresses are taken into account in the equivalent stress due to safety reasons in this work. Therefore, equivalent stresses become

$$\sigma_{V,\text{max}} = \sqrt{(\sigma_{\text{max},x}K_{tx} + \sigma_{\text{max},b}K_{tb})^2 + 3\tau_{\text{max}}^2}$$

$$= \sqrt{\left(\frac{F_x}{bt_s}K_{tx} + \frac{6M_y(x)}{bt_s^2}K_{tb}\right)^2 + 3\left(\frac{F_z}{2bt_s}\right)^2}, \tag{9}$$

where stress concentration factors K_{tx} and K_{tb} for axial and bending loads (second indices x, b) are introduced.

For rectangular leaf type hinges, with uniform thickness $t(x) = t_s = \text{const.}$, the critical section is solely determined by the maximum bending moment $M_{y,\text{max}}^{\text{R}}(x = 0) = M_y - lF_z$. In contrast to this, for parabolic and circular notch type hinges, the critical section is determined by the thinnest cross section t_s, as well, leading to a critical section very close to the thinnest cross section at $x \approx l/2$, where the bending moment becomes $M_{y,\text{max}}^{\text{C,P}}(x \approx \frac{l}{2}) = M_y - \frac{l}{2}F_z$. Thus, the maximum equivalent stresses are

$$\sigma_{V,\text{max}}^{\text{R}} = \sqrt{\left(\frac{F_x}{bt_s}K_{tx}^{\text{R}} + \frac{6(M_y - lF_z)}{bt_s^2}K_{tb}^{\text{R}}\right)^2 + 3\left(\frac{F_z}{2bt_s}\right)^2},$$

$$\sigma_{V,\text{max}}^{\text{C,P}} = \sqrt{\left(\frac{F_x}{bt_s}K_{tx}^{\text{C,P}} + \frac{6(M_y - \frac{l}{2}F_z)}{bt_s^2}K_{tb}^{\text{C,P}}\right)^2 + 3\left(\frac{F_z}{2bt_s}\right)^2}. \tag{10}$$

The stress concentration factors for rectangular leaf type hinges K_{tx}^{R}, K_{tb}^{R} strongly depend on the corner radius and can be found in Pilkey and Pilkey (2008). For circular and parabolic hinges, stress concentration factors can be approximated following Haibach (2006)

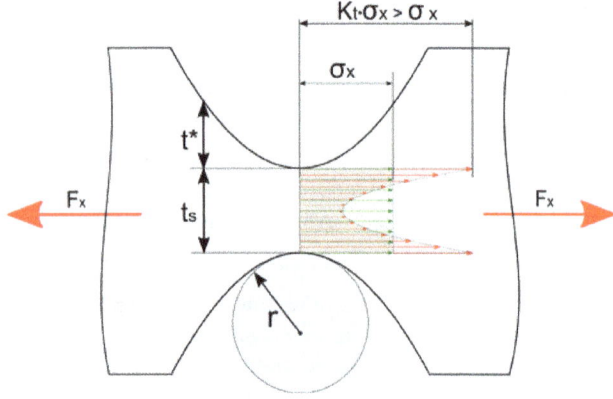

Figure 3. Stress concentration at flexure hinges due to notch effect.

Table 2. Maximum elastic deformation of circular (C) and parabolic (P) flexure hinges undergoing bending due to pure shear force F_z.

Shape [–]	l [mm]	t_s [mm]	K_{tx} [–]	K_{tb} [–]	F_z [N]	$w(l)$ [μm]	$w'(l)$ [rad]
C	8	2	1.119	1.051	47.6	52.0	−0.013
P	8	2	1.281	1.133	44.1	36.1	−0.009
C	9	1	1.040	1.015	12.2	93.2	−0.020
P	9	1	1.102	1.043	11.8	65.8	−0.015
C	9.5	0.5	1.014	1.004	3.1	145.9	−0.031
P	9.5	0.5	1.037	1.013	3.0	103.4	−0.022

Generally, the occurring stresses depend directly on the radii of curvature, i.e. smaller radii of curvature result in higher stress concentration factors leading to higher stresses. This is not a surprising result, however Eqs. (9)–(13) provide the reader with analytical expressions to calculate the range of elastic deformation of flexure hinges in compliant mechanisms prior to any modeling or manufacturing efforts.

$$K_{tx}^{C,P} = 1 + \left[0.1\left(\frac{r}{t^*}\right) + 0.7\left(1 + \frac{t_s}{2r}\right)^2 \left(\frac{t_s}{2r}\right)^{-3} \right.$$

$$\left. + 0.13\left(\frac{t_s}{2r}\right)\left(\frac{t_s}{2r} + \frac{t^*}{r}\right)^{-1}\left(\frac{t^*}{r}\right)^{-1.25} \right]^{-\frac{1}{2}}, \tag{11}$$

$$K_{tb}^{C,P} = 1 + \left[0.08\left(\frac{r}{t^*}\right)^{0.66} + 2.2\left(1 + \frac{t_s}{2r}\right)^{2.25}\left(\frac{t_s}{2r}\right)^{-3.375} \right.$$

$$\left. + 0.2\left(\frac{t_s}{2r}\right)\left(\frac{t_s}{2r} + \frac{t^*}{r}\right)^{-1}\left(\frac{t}{r}\right)^{-1.33} \right]^{-\frac{1}{2}},$$

where $t^* = \frac{H - t_s}{2}$ and the radii of curvature r are

$$r^C = \frac{l^2 + (H - t_s)^2}{4(H - t_s)} = \text{const.} \tag{12}$$

and

$$r^P\left(x = \frac{l}{2}\right) = \frac{l^2}{4(H - t_s)}, \tag{13}$$

for circular and parabolic hinges as shown in Fig. 3. Here, the geometric properties given in Eq. (2) and corresponding derivatives $t'(x)$, $t''(x)$ were applied to calculate the radius of a parabola $r^P(x) = \left| \frac{(1 + t'(x)^2)^{3/2}}{t''(x)} \right|$.

Finally, the equivalent stresses can be calculated using Eq. (9) for known (nodal) loads and all considered types of flexure hinges considered in this work.

As an example, Table 2 illustrates the maximum elastic deformation $w(l)$ of differently-sized parabolic and circular flexure hinges based on given geometric parameter: length l, height $H = 10$ mm, depth $b = 10$ mm and smallest thickness t_s. It can be noted, that circular flexure hinges provide a larger deflection range than parabolic counterparts maintaining the aforementioned common material points P_1, P_2, P_3.

3.5 Stiffness and bending stiffness

The stiffness k_x, k_z and, in particular, the bending stiffness c_ψ of a flexure hinge is important for modeling of compliant mechanisms using discrete spring joints or reduced finite element models. They are calculated analytically and compared to experimental data for all flexure hinge types considered in this work. The loads $\boldsymbol{F} = \left(F_z, M_y, F_x\right)^T$ and displacements $\boldsymbol{u} = (w(x), \psi(x), u(x,z))^T$ are coupled by the compliance expressions $\boldsymbol{u} = \mathbf{N}\,\boldsymbol{F}$ as given in Eq. (8). For modeling and topology optimization purposes, it is beneficial to convert this relation to

$$\mathbf{K}\boldsymbol{u} = \boldsymbol{F}. \tag{14}$$

The stiffness matrix $\mathbf{K}$ represents all mechanical properties (for quasi-static problems) that are crucial for modeling purposes, topology optimization problems and (embedded) finite element calculations using efficient, reduced models.

Generally, these expressions are quite large, especially, for parabolic and circular flexure hinges. Due to conciseness, the stiffness matrix of a rectangular flexure hinge is presented solely. However, the calculation of stiffness matrices for parabolic and circular hinges is similar and straightforward. Using the derived relations between loads and corresponding displacement from Eq. (8) yield

$$\mathbf{K} = Ebt_s^3 \begin{pmatrix} \frac{5}{l(12(1+\nu)t_s^2 + 5l^2)} & \frac{5}{24(1+\nu)t_s^2 + 10l^2} & 0 \\ \frac{5l}{2l(12(1+\nu)t_s^2 + 5l^2)} & \frac{10l^2 + 6(1+\nu)t_s^2}{l(12(1+\nu)t_s^2 + 5l^2)} & 0 \\ 0 & 0 & \frac{1}{lt_s^2} \end{pmatrix}. \tag{15}$$

To compare these analytical calculations with experimental data with superposed external loads, scalar values

Table 3. Bending stiffness c_ψ for rectangular (R), circular (C) and parabolic (P) flexure hinges: analytical calculations and experimental results.

Shape	l	t_s	$c_{\psi,\text{ana}}$	$c_{\psi,\text{exp}}$	error
[−]	[mm]	[mm]	[$\frac{\text{Nm}}{\text{rad}}$]	[$\frac{\text{Nm}}{\text{rad}}$]	[%]
P	8	2	199.37	180.06	10.72
C	8	2	143.64	133.71	7.43
R	8	2	58.33	61.18	−4.66

for bending stiffness c_ψ are desirable. Therefore, the stiffness matrix is decomposed (diagonalized) into $\mathbf{K_D} = \text{diag}(\lambda_1, \lambda_2, \lambda_3)$, where $\lambda_1, \lambda_2, \lambda_3$ are the eigenvalues of $\mathbf{K}$. The resulting eigenvectors $\mathbf{b_1}, \mathbf{b_2}, \mathbf{b_3}$ are used to form the orthogonal transformation matrix $\mathbf{T} = \mathbf{T}(\mathbf{b_1}, \mathbf{b_2}, \mathbf{b_3})$ which is negligibly close to identity $\mathbf{I}$ ($\text{norm}(\mathbf{T} - \mathbf{I}) \leq 10^{-4}$) for every case considered in this work. Thus, Eq. (14) can be rewritten as

$$\underbrace{\begin{pmatrix} \lambda_1 = k_z & 0 & 0 \\ 0 & \lambda_2 = c_\psi & 0 \\ 0 & 0 & \lambda_3 = k_x \end{pmatrix}}_{\mathbf{K_D}} \underbrace{\begin{pmatrix} w \\ \psi \\ u \end{pmatrix}}_{u} \approx \underbrace{\begin{pmatrix} F_z \\ M_y \\ F_x \end{pmatrix}}_{F} \quad (16)$$

yielding a desired, decoupled relation among loads, stiffness and deflection.

The bending stiffness c_ψ for rectangular (R), circular (C) and parabolic (P) flexure hinges are listed in Table 3. Here, the analytical calculations differ from the experimental results by a maximum relative error $\leq 11\,\%$, which is acceptable, considering manufacturing imperfections in z-direction and its enormous effect on the stiffness as described in detail in Ryu and Gweon (1997). Therefore, the aforementioned analytical expressions represent a good prediction for superposed, application-oriented loading conditions.

3.6 Center of rotation

The center of rotation and its motion with deflection of bodies connected by flexure hinges are crucial for a correct modeling of compliant mechanisms. Ignoring the particular center of rotation of flexure hinges can lead to parasitic motion or failure of the entire mechanism due to unwanted behavior, e.g. snap through effects.

The center of rotation is usually considered for rigid-body-motions. However, many parts of a compliant mechanism that are connected by flexure hinges are very stiff and can be treated in a similar way. In this work, overall center of rotation $\mathbf{P}_{\text{eff}}^{01}$ refers to a fixed point considering undeformed (0) and maximum elastically deformed state (1) as illustrated in Fig. 4, whereas its motion refers to the herpolhode, i.e. motion of instantaneous center of rotation with deflection.

In order to calculate the center of rotation of a rigid body attached to the flexure hinge, the position of two single points

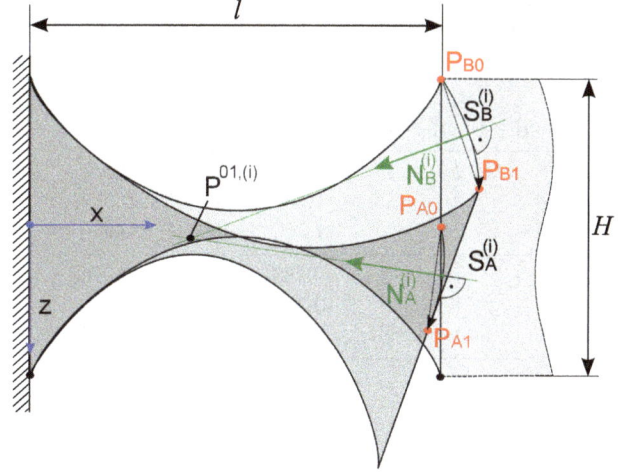

Figure 4. Determination of instantaneous center of rotation $\mathrm{P}^{01,(i)}$ of a flexure hinge.

P_A, P_B and the corresponding displacements $u_A = u(l,0)$, $w_A = w(l,0)$ and $u_B = u(l,-H/2)$ are considered. This yields an overall center of rotation

$$\mathbf{P}_{\text{eff}}^{01} = \frac{1}{2} \begin{pmatrix} \frac{w_A}{u_A - u_B} H + u_A + u_B + 2l \\ \frac{u_A}{u_B - u_A} H - \frac{u_A u_B}{w_A} + w_A \end{pmatrix}, \quad (17)$$

which holds for all types of flexure hinges.

The motion of the (instantaneous) center of rotation can be calculated considering an infinite number of intermediate steps between undeformed and maximum elastically deformed step. Therefore, the motion paths:

$$S_A^{(i)} = P_A + \begin{pmatrix} u_A^{(i)} \\ w_A^{(i)} \end{pmatrix}, \qquad S_B^{(i)} = P_B + \begin{pmatrix} u_B^{(i)} \\ w_B^{(i)} \end{pmatrix}, \quad (18)$$

the tangent vectors:

$$T_A^{(i)} = \dot{K}_A = \begin{pmatrix} \dot{u}_A^{(i)} \\ \dot{w}_A^{(i)} \end{pmatrix}, \qquad T_B^{(i)} = \dot{K}_B = \begin{pmatrix} \dot{u}_B^{(i)} \\ \dot{w}_B^{(i)} \end{pmatrix}, \quad (19)$$

and the corresponding normal vectors:

$$N_A^{(i)} = \begin{pmatrix} \dot{w}_A^{(i)} \\ -\dot{u}_A^{(i)} \end{pmatrix}, \qquad N_B^{(i)} = \begin{pmatrix} \dot{w}_B^{(i)} \\ -\dot{u}_B^{(i)} \end{pmatrix}, \quad (20)$$

need to be calculated first. Determining the point of intersection of $N_A^{(i)}$ and $N_B^{(i)}$ leads to the center of rotation for all intermediate steps, i.e. the desired motion with deflection:

$$\mathbf{P}^{01,(i)} = \begin{pmatrix} \frac{2(u_A + l)\dot{u}_A - 2(u_B + l)\dot{u}_B + H\dot{w}_A}{2(\dot{u}_A - \dot{u}_B)} \\ w_A - \frac{2(u_A - u_B)\dot{u}_B + H\dot{w}_A}{2(\dot{u}_A - \dot{u}_B)\dot{w}_A} \dot{u}_A \end{pmatrix}, \quad (21)$$

Table 4. Overall center of rotation for rectangular (R), circular (C) and parabolic (P) flexure hinges.

Shape [−]	l [mm]	t_s [mm]	F_z [N]	M_y [Nm]	$P^{01}_{\text{eff},\Delta x}$ [mm]	$P^{01}_{\text{eff},z}$ [mm]
R	8	2	35	−1.260	−0.1479	0.0499
C	8	2	35	−1.260	−0.0657	0.0191
P	8	2	35	−1.260	−0.0615	0.0143
R	9	1	8	−0.288	−0.1607	0.1168
C	9	1	8	−0.288	−0.0142	0.0307
P	9	1	8	−0.288	−0.0070	0.0222

Table 5. Natural frequencies for rectangular (R), circular (C) and parabolic (P) flexure hinges connected to a rigid body ($H = 10$ mm, $L = 50$ mm, $b = 10$ mm).

l [mm]	t_s [mm]	f_{ana} [Hz]	f_{num} [Hz]	f_{exp} [Hz]	$f^{\text{dm}}_{\text{exp}}$ [Hz]
R	9.47	1.18	146.2 (7.3 %)	144.0 (5.7 %)	136.2	141.0 (3.9 %)
C	9.09	0.99	218.6 (8.3 %)	219.7 (7.8 %)	238.4	242.5 (1.7 %)
P	9.09	1.07	284.8 (0.2 %)	274.0 (4.0 %)	285.3	288.5 (1.1 %)
R	8.42	2.28	408.4 (10.9 %)	400.9 (8.9 %)	368.1	370.0 (0.9 %)
C	8.08	2.00	534.5 (2.9 %)	524.4 (4.8 %)	550.9	541.6 (1.7 %)
P	8.03	2.13	676.9 (6.0 %)	643.3 (0.8 %)	638.4	636.4 (0.3 %)

The effective centers of rotation for rectangular (R), circular (C) and parabolic (P) flexure hinges are illustrated in Table 4 for two different loading conditions. Here, the x-coordinates refer to the center of the flexure hinge denoted by Δx. It can be noted, that the effective center of rotation is shifted to $(-x, +z)$ direction for all flexure hinges for the given load case. The motion of the center of rotation for rectangular flexure hinges is clearly larger due to the deflection of the entire hinge length l, whereas it is very small for circular and parabolic flexure hinges. Thus, it is crucial to consider the center of rotation and its motion with deflection for rectangular flexure hinges in order to ensure an appropriate modeling.

3.7 Natural frequency

The natural frequency of a system consisting of a rigid body connected to a flexure hinge as shown in Fig. 5 (left) is relevant for compliant mechanisms under dynamic loading conditions. In addition, it is a good quality measure comparing analytical, numerical and experimental studies.

The natural frequencies are first calculated analytically using the aforementioned Timoshenko's beam theory. Extending Eq. (7) to a dynamic state yield the differential equations:

$$\rho A(x)\ddot{w}(x) - (GA_*(x)(w'(x) + \psi))' = q,$$
$$\rho I_y(x)\ddot{\psi} - (EI_y(x)\psi')' + GA_*(w' + \psi) = 0. \tag{22}$$

Applying standard boundary conditions at fixed end $x = 0$

$$w = 0,$$
$$\psi = 0, \tag{23}$$

and free end $x = l$

$$EI_y\psi' = 0,$$
$$GA_*(w' + \psi) = 0, \tag{24}$$

the differential Eq. (22) can be solved. However, solving these equations analytically for a variable, unspecified thick-

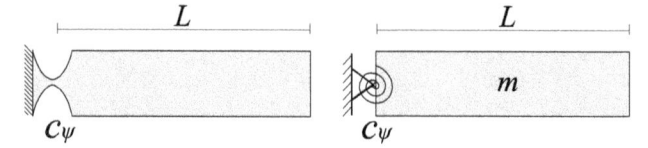

Figure 5. Continuous flexure hinge (left) and equivalent discrete torsion spring model (right).

ness $t(x) \neq$ const. is not always possible. Therefore, a numerical approach, namely the Rayleigh quotient

$$\omega_1^2 = \frac{\max E_p}{\max E_k} \tag{25}$$

is chosen, where the first natural circular frequency ω_1 is approximated by the ratio of maximum values of potential and kinetic energies E_p and E_k. Following Tabarrok and Karnopp (1967) yield

$$\omega_1^2 = \frac{\int_0^l EI_y(x)\Psi'(x)^2 + GA_*(x)(W'(x) + \Psi(x))^2 dx}{\rho \int_0^l A(x)W(x)^2 + I_y(x)\Psi(x)^2 dx}, \tag{26}$$

where displacement and bending angle are described by appropriate test functions $\Psi(x)$ and $W(x)$. In order to determine $\Psi(x)$ and $W(x)$, the displacements and bending angle occuring from a uniform transverse load q_0 with a resulting bending moment $M_y(x) = -\frac{1}{2}q_0(l - x)^2$, as suggested in Rao (2007), are used.

Table 5 lists the analytically, numerically and experimentally determined first natural frequencies $f_{\text{ana}}, f_{\text{num}}, f_{\text{exp}}, f^{\text{dm}}_{\text{exp}}$ for different flexure hinges connected to a rigid body, as shown in Fig. 5 (left). The analytical calculations f_{ana} are based on Eq. (26) using Timoshenko's beam theory. The numerical calculations f_{num} are obtained by a numerical modal analysis using the commercial software package Abaqus 6.9. The experimental data f_{exp} is gathered by a experimental modal analysis using non-contact laser scanning vibrometer system. The frequencies $f^{\text{dm}}_{\text{exp}}$ are calculated using the standard relation

$$f^{\text{dm}}_{\text{exp}} = \frac{1}{2\pi} \sqrt{\frac{c_{\psi,\text{exp}}}{I_m}}, \tag{27}$$

with mass inertia I_m, that holds for discrete models of a torsional (bending vibration, as illustrated in Fig. 5 (right).

It can be noted, that the analytical calculation agrees well with the numerical results on all types of flexure hinges. Compared to the experimental data, a relative error less than 11 % can be noted. Due to imperfections in the manufacturing process of the specimen, this error seems acceptable to the authors; cf. Ryu and Gweon (1997). Comparing the experimental data f_{exp} and f_{exp}^{dm} with each other, a very small relative error of less than 4 % can be noted. This implies, that flexure hinges can be modeled by discrete torsional springs as illustrated in Fig. 5, using the bending stiffness c_ψ calculated in Sect. 3.5.

4   Conclusion: benefits for the synthesis of compliant mechanisms

In this work, planar flexure hinges are investigated in terms of their application in the synthesis of compliant mechanisms, where one-node hinges occur as an artificial artefact of many topology optimization methods. In order to replace these pseudo-hinges by flexure hinges efficiently, a characterization of different types of flexure hinges was done.

Relevant mechanical properties, such as displacement and bending angle, mechanical stresses, bending stiffness, center of rotation, maximum elastic deformation and first natural frequencies were derived analytically and agree well with numerical and experimental data.

The analytical expressions were derived based on Timoshenko's beam theory taking into account shear deformation of flexure hinges. In order to calculate an elastic deflection range, von-Mises yield criterion was chosen. Numerical simulations were performed using commercial software package Abaqus 6.9. Experimental results of bending stiffness and natural frequencies were gathered using a tension test machine and a non-contact scanning laser vibrometer system.

More importance was given to practice-oriented flexure hinge types in terms of cost-saving manufacturability, i.e. circular notch type hinges and rectangular leaf type hinges, as well as well-customizable parabolic hinges. Comparing different types of flexure hinges of similar dimensions, the following conclusion can be drawn:

Rectangular geometry of flexure hinges yield low bending stiffness and very high rotational deflection, while the location of the center of rotation and its motion with deflection needs to be taken into account.

Circular geometry of flexure hinges yields moderate bending stiffness and high rotational deflection, while the center of rotation remains close to the center point of the hinge.

Parabolic geometry of flexure hinges yield high bending stiffness and low rotational deflection, while the center of rotation remains very close to the center point of the hinge.

Some of these conclusions are not surprising, however the key results of this work are the analytical expressions that enable the reader:

- to calculate the relevant mechanical properties of flexure hinges explicitly and

- to select the appropriate type of flexure hinge based on the (known) nodal loads and displacements resulting from the synthesis of compliant mechanism,

prior to any modeling or manufacturing efforts. Thus, the synthesis and manufacturing process of compliant mechanisms can be accelerated.

Acknowledgements. This research was supported by priority programme grant SPP1476 from the German Research Foundation (DFG) to both authors, and by a doctoral research stipend from the German Academic Exchange Service (DAAD-D/10/46834) to the first author. In addition, the authors would like to thank several students, in particular M. Anselmann, for their studies on flexure hinges.

Edited by: N. Tolou

References

Ananthasuresh, G. and Kota, S.: Designing compliant mechanisms, Mech. Eng., 117, 93–96, 1995.

Ansola, R., Canales, J., Tárrago, J., and Rasmussen, J.: An integrated approach for shape and topology optimization of shell structures, Comput. Struct., 80, 449–458, 2002.

Bendsøe, M. and Sigmund, O.: Topology optimization: theory, methods, and applications, Springer Verlag, 2003.

Bendsøe, M. and Sigmund, O.: Optimization of structural topology, shape, and materials, ORBIT, 2008.

Bruns, T. and Tortorelli, D.: Topology optimization of non-linear elastic structures and compliant mechanisms, Comput. Method Appl. M., 190, 3443–3459, 2001.

Conway, H. and Nickola, W.: Anticlastic action of flat sheets in bending, Exp. Mech., 5, 115–119, 1965.

Frecker, M. I., Ananthasuresh, G. K., Nishiwaki, S., Kikuchi, N., and Kota, S.: Topological Synthesis of Compliant Mechanisms Using Multi-Criteria Optimization, J. Mech. Design, 119, 238–245, 1997.

Haibach, E.: Betriebsfestigkeit: Verfahren und Daten zur Bauteilberechnung, Springer New York, 2006.

Howell, L.: Compliant mechanisms, Wiley-Interscience, 2001.

Lobontiu, N.: Compliant mechanisms: design of flexure hinges, CRC, 2003.

Love, A.: A treatise on the mathematical theory of elasticity, at the University Press, 1920.

Mattson, C., Howell, L., and Magleby, S.: Development of commercially viable compliant mechanisms using the pseudo-rigidbody model: Case studies of parallel mechanisms, J. Intel. Mat. Syst. Str., 15, 195–202, 2004.

Paros, J. and Weisbord, L.: How to design flexural hinges, Mach. Des., 25, 151–156, 1965.

Pilkey, W. D. and Pilkey, D. F.: Peterson's stress concentration factors, John Wiley & Sons, 2008.

Poulsen, T.: A simple scheme to prevent checkerboard patterns and one-node connected hinges in topology optimization, Struct. Multidiscip. O., 24, 396–399, 2002.

Raatz, A.: Stoffschlüssige Gelenke aus pseudo-elastischen Formgedächtnislegierungen in Pararellrobotern, Vulkan-Verlag GmbH, 2006.

Rao, S.: Vibration of continuous systems, John Wiley & Sons Inc, 2007.

Ryu, J. and Gweon, D.: Error analysis of a flexure hinge mechanism induced by machining imperfection, Precis. Eng., 21, 83–89, 1997.

Saxena, A. and Ananthasuresh, G.: On an optimal property of compliant topologies, Struct. Multidiscip. O., 19, 36–49, 2000.

Sigmund, O.: Manufacturing tolerant topology optimization, Acta Mech. Sinica, 25, 227–239, 2009.

Smith, S.: Flexures: elements of elastic mechanisms, CRC, 2000.

Tabarrok, B. and Karnopp, B.: Analysis of the oscillations of the Timoshenko beam, Z. Angew. Math. Phys., 18, 580–587, 1967.

Wang, C., Reddy, J., and Lee, K.: Shear deformable beams and plates, Elsevier, Amsterdam, 2000.

Yoon, G., Kim, Y., Bendsøe, M., and Sigmund, O.: Hinge-free topology optimization with embedded translation-invariant differentiable wavelet shrinkage, Struct. Multidiscip. O., 27, 139–150, 2004.

Permissions

All chapters in this book were first published in Mechanical Sciences, by Copernicus Publications; hereby published with permission under the Creative Commons Attribution License or equivalent. Every chapter published in this book has been scrutinized by our experts. Their significance has been extensively debated. The topics covered herein carry significant findings which will fuel the growth of the discipline. They may even be implemented as practical applications or may be referred to as a beginning point for another development.

The contributors of this book come from diverse backgrounds, making this book a truly international effort. This book will bring forth new frontiers with its revolutionizing research information and detailed analysis of the nascent developments around the world.

We would like to thank all the contributing authors for lending their expertise to make the book truly unique. They have played a crucial role in the development of this book. Without their invaluable contributions this book wouldn't have been possible. They have made vital efforts to compile up to date information on the varied aspects of this subject to make this book a valuable addition to the collection of many professionals and students.

This book was conceptualized with the vision of imparting up-to-date information and advanced data in this field. To ensure the same, a matchless editorial board was set up. Every individual on the board went through rigorous rounds of assessment to prove their worth. After which they invested a large part of their time researching and compiling the most relevant data for our readers.

The editorial board has been involved in producing this book since its inception. They have spent rigorous hours researching and exploring the diverse topics which have resulted in the successful publishing of this book. They have passed on their knowledge of decades through this book. To expedite this challenging task, the publisher supported the team at every step. A small team of assistant editors was also appointed to further simplify the editing procedure and attain best results for the readers.

Apart from the editorial board, the designing team has also invested a significant amount of their time in understanding the subject and creating the most relevant covers. They scrutinized every image to scout for the most suitable representation of the subject and create an appropriate cover for the book.

The publishing team has been an ardent support to the editorial, designing and production team. Their endless efforts to recruit the best for this project, has resulted in the accomplishment of this book. They are a veteran in the field of academics and their pool of knowledge is as vast as their experience in printing. Their expertise and guidance has proved useful at every step. Their uncompromising quality standards have made this book an exceptional effort. Their encouragement from time to time has been an inspiration for everyone.

The publisher and the editorial board hope that this book will prove to be a valuable piece of knowledge for researchers, students, practitioners and scholars across the globe.

List of Contributors

S. Henein
Centre Suisse d'Electronique et de Microtechnique (CSEM SA), Neuchâtel, Switzerland

G. Krishnan
Mechanical Engineering, University of Michigan, Ann Arbor, MI 48105, USA

C. Kim
Mechanical Engineering, Bucknell University, Lewisburg, PA 17837, USA

S. Kota
Department of Mechanical Engineering, University of Michigan, Ann Arbor, MI 48105, USA

M. J. Telleria
Massachusetts Institute of Technology, Cambridge, USA

M. L. Culpepper
Massachusetts Institute of Technology, Cambridge, USA

L. Birglen
Ecole Polytechnique of Montreal, Department of Mechanical Engineering, Ecole Polytechnique of Montreal, Montreal, QC, H3T 1J4, Canada

G. Chen
School of Mechatronics, Xidian University, Xi'an, Shaanxi 710071, China

S. Zhang
School of Mechatronics, Xidian University, Xi'an, Shaanxi 710071, China

W. O'Connor
UCD School of Mechanical and Materials Engineering, UCD Belfield, Dublin 4, Ireland

H. Habibi
UCD School of Mechanical and Materials Engineering, UCD Belfield, Dublin 4, Ireland

L. Rubbert
LSIIT, Université de Strasbourg-CNRS, Strasbourg, France

P. Renaud
LSIIT, Université de Strasbourg-CNRS, Strasbourg, France

W. Bachta
ISIR, Université Pierre et Marie Curie-CNRS, Paris, France

J. Gangloff
LSIIT, Université de Strasbourg-CNRS, Strasbourg, France

B. P. Mann
Mechanical Engineering and Material Science Department, Duke University, Durham, NC 27708, USA

M. M. Gibbs
Mechanical Engineering and Material Science Department, Duke University, Durham, NC 27708, USA

S. M. Sah
Mechanical Engineering and Material Science Department, Duke University, Durham, NC 27708, USA

C. Wang
Advanced Manufacturing and Enterprise Engineering, School of Engineering and Design, Brunel Univ., UK

Y. H. J. Au
Advanced Manufacturing and Enterprise Engineering, School of Engineering and Design, Brunel Univ., UK

Y. F. Liu
Complex and Intelligent System Laboratory, School of Mechanical and Power Engineering, East China University of Science and Technology, Shanghai, China

J. Li
Complex and Intelligent System Laboratory, School of Mechanical and Power Engineering, East China University of Science and Technology, Shanghai, China

Z. M. Zhang
Department of Mechanical Engineering, University of Saskatchewan, Saskatoon, Canada

X. H. Hu
Department of Mechanical Engineering, University of Saskatchewan, Saskatoon, Canada

W. J. Zhang
Complex and Intelligent System Laboratory, School of Mechanical and Power Engineering, East China University of Science and Technology, Shanghai, China
Department of Mechanical Engineering, University of Saskatchewan, Saskatoon, Canada

A. Jomartov
Institute Mechanics and Mechanical Engineering, Almaty, Kazakhstan

X. Pei
School of Mechanical Engineering and Automation, Beihang University, Beijing 100083, China

J. Yu
School of Mechanical Engineering and Automation, Beihang University, Beijing 100083, China

A. G. Dunning
Faculty of Mechanical, Maritime and Materials Engineering, Department of Biomechanical Engineering, Delft University of Technology, Delft, The Netherlands

N. Tolou
Faculty of Mechanical, Maritime and Materials Engineering, Department of Biomechanical Engineering, Delft University of Technology, Delft, The Netherlands

J. L. Herder
Faculty of Mechanical, Maritime and Materials Engineering, Department of Biomechanical Engineering, Delft University of Technology, Delft, The Netherlands

G. R. Hayes
The Pennsylvania State University, University Park, PA, 16802, USA

M. I. Frecker
The Pennsylvania State University, University Park, PA, 16802, USA

J. H. Adair
The Pennsylvania State University, University Park, PA, 16802, USA

S. A. Lopez
Mechanical Engineering Dept., Massachusetts Institute of Technology, Cambridge, MA, USA

L. R. Hernley
Mechanical Engineering Dept., Massachusetts Institute of Technology, Cambridge, MA, USA

E. N. Bearrick
Mechanical Engineering Dept., Massachusetts Institute of Technology, Cambridge, MA, USA

L. M. Tanenbaum
Health Sciences and Technology Dept., Massachusetts Institute of Technology, Cambridge, MA, USA

M. A. C. Thomas
Mechanical Engineering Dept., Massachusetts Institute of Technology, Cambridge, MA, USA

T. A. Toussaint
Mechanical Engineering Dept., Massachusetts Institute of Technology, Cambridge, MA, USA

J. J. Romano
Hospital Medicine, Massachusetts General Hospital, Boston, MA, USA

N. C. Hanumara
Mechanical Engineering Dept., Massachusetts Institute of Technology, Cambridge, MA, USA

A. H. Slocum
Mechanical Engineering Dept., Massachusetts Institute of Technology, Cambridge, MA, USA

H. Bararnia
Department of Mechanical Engineering, Babol University of Technology, Babol, Iran

D. D. Ganji
Department of Mechanical Engineering, Babol University of Technology, Babol, Iran

M. Otomori
Kyoto University, Kyoto, Japan

T. Yamada
Nagoya University, Nagoya, Japan

K. Izui
Kyoto University, Kyoto, Japan

S. Nishiwaki
Kyoto University, Kyoto, Japan

A. Eskandari
Department of Aerospace Engineering, Ryerson University, Toronto, Canada

P. R. Ouyang
Department of Aerospace Engineering, Ryerson University, Toronto, Canada

M. F. Wang
LARM: Laboratory of Robotics and Mechatronics, DICeM-University of Cassino and South Latium, Cassino (Fr), Italy

M. Ceccarelli
LARM: Laboratory of Robotics and Mechatronics, DICeM-University of Cassino and South Latium, Cassino (Fr), Italy

G. Carbone
LARM: Laboratory of Robotics and Mechatronics, DICeM-University of Cassino and South Latium, Cassino (Fr), Italy

K. Jangra
Department of Mechanical Engineering, YMCA University of Science and Technology, Faridabad 121006, India

S. Grover
Department of Mechanical Engineering, YMCA University of Science and Technology, Faridabad 121006, India

M. Rostamian
Department of Mechanical Engineering, University of Idaho, 1776 Science Center Dr, Idaho Falls, ID, USA

S. Arifeen
Department of Mechanical Engineering, University of Idaho, 440902 Moscow, ID, 83844-0902, USA

G. P. Potirniche
Department of Mechanical Engineering, University of Idaho, 440902 Moscow, ID, 83844-0902, USA

A. Tokuhiro
Department of Mechanical Engineering, University of Idaho, 1776 Science Center Dr, Idaho Falls, ID, USA

F. Dirksen
Institute of Mechanics, Helmut-Schmidt-University/University of the Federal Armed Forces Hamburg, Holstenhofweg 85, 22043 Hamburg, Germany

R. Lammering
Institute of Mechanics, Helmut-Schmidt-University/University of the Federal Armed Forces Hamburg, Holstenhofweg 85, 22043 Hamburg, Germany